Chemicals for Crop Protection

and

Pest Control

Other titles of interest

Books

KOREN:	Environmental Health and Safety
MOHTADI:	Man and His Environment
SCORER:	Air Pollution
STEEDMAN et al:	Chemistry for the Applied Sciences
STRAUSS:	Industrial Gas Cleaning, 2nd edition

*Journals**

Chemosphere

Progress in Water Technology

*Free specimen copy available on request

Chemicals for Crop Protection
and
Pest Control

BY

M. B. GREEN

G. S. HARTLEY

AND

T. F. WEST

PERGAMON PRESS

OXFORD · NEW YORK · TORONTO · SYDNEY · PARIS · FRANKFURT

U.K.	Pergamon Press Ltd., Headington Hill Hall, Oxford OX3 0BW, England
U.S.A.	Pergamon Press Inc., Maxwell House, Fairview Park, Elmsford, New York 10523, U.S.A.
CANADA	Pergamon of Canada Ltd., 75 The East Mall, Toronto, Ontario, Canada
AUSTRALIA	Pergamon Press (Aust.) Pty. Ltd., 19a Boundary Street, Rushcutters Bay, N.S.W. 2011, Australia
FRANCE	Pergamon Press SARL, 24 Rue des Ecoles, 75240 Paris, Cedex 05, France
WEST GERMANY	Pergamon Press GmbH, 6242 Kronberg-Taunus, Pferdstrasse 1, West Germany

First edition 1977

Library of Congress Cataloging in Publication Data

Green, Maurice Berkeley.
Chemicals for crop protection and pest control.

First ed. by G. S. Hartley and T. F. West published in 1969 under title: Chemicals for pest control.
Includes bibliographies and index.
1. Pesticides. I. Hartley, Gilbert Spencer, joint author. II. West, Trustham Frederick, joint author. III. Hartley, Gilbert Spencer. Chemicals for pest control. IV. Title.
TP248.P47H3 1977 668'.65 77-4881

ISBN 0-08-021174-7 (Hardcover)
ISBN 0-08-019013-8 (Flexicover)

In order to make this volume available as economically and rapidly as possible the authors' typescripts have been reproduced in their original form. This method unfortunately has its typographical limitations but it is hoped that they in no way distract the reader.

Printed in Great Britain by William Clowes & Sons, Limited
London, Beccles and Colchester

CONTENTS

PREFACE TO SECOND EDITION

"Chemicals for Pest Control" was originally published in 1969 as one of a series of monographs commissioned by Sir Robert Robinson as manuals for teaching senior students about various aspects of chemical industry, and it formed part of the Commonwealth and International Library of Science, Technology and Engineering. It was very well received, not only by students, but by all those concerned with the manufacture and use of pesticides, because it was the first book which was based on a consideration of pesticides as products of manufacture and trade by the chemical industry. It is still unique in that respect. Many other books on pesticides have been published since this one, but they have all approached the subject from the biological viewpoints of activity, selectivity, modes of action and fate in the environment.

It was, therefore, with great pleasure that I agreed to prepare a revised second edition. In doing so I have not made changes just for the sake of doing so. I have retained much of the material in the first edition because I considered that it could hardly be improved upon. However, in the past ten years many new types of pesticides have been introduced and views on crop protection and pest control have altered considerably, so I have attempted to include all important new products and to reflect the changes which have taken place in the attitudes of the public, of registration authorities, of the growers and of the chemical industry.

My grateful thanks are due to Miss Yvonne Testal who prepared camera-ready typescript from sometimes almost illegible handwriting.

M.B. Green

INTRODUCTION

Pests have many unexpected and far-reaching consequences. In his book Islands of the South Pacific, Sir Harry Luke relates that there is one bishopric in Christendom, the islands of Wallis and Futuna, whose creation is entirely due to their influence. Rhinoceros beetle, a serious pest of coconuts, was introduced by sea from Ceylon to Samoa in 1910 and its ravages led to the imposition by the "clean" Pacific ports of strict controls on vessels coming from infected territories. This delayed the bishop's journey between "clean" Wallis Island and partially infected Tonga and made excessive inroads into his time. Monseigneur Blanc saw no solution (he said this to Sir Harry Luke in 1958 - Blanc was then retired and living in Tonga) but to recommend to the Vatican the detachment of Wallis and Futuna Islands from his jurisdiction and their erection into a separate diocese.

Many other strange examples of unexpected intrusions of pests into human affairs could be quoted, yet it is sometimes suggested that the pesticides industry has overstated the significance of pests in order to create a market which it can then supply. This error is unlikely to be made in the developing tropical countries where pest activity is all too often disastrous. It should not be made in Ireland where the great famine of 1842-7, which sent a quarter of the population to the grave and another quarter to the New World, is still strong in the national memory. It should not be made by a reader of the Old Testament in which are many colourful and horrific references to plagues of locusts, to vines "eaten by the worms" and the olive "that has cast his fruit".

Earliest man must have attempted to scratch away his lice and fleas, but had probably no idea of agricultural pests as long as he lived as a hunter or nomad, feeding off the land as he found it. When he started to cultivate land, however, he must have become aware of the problem of pests, even if at first only the larger herbivores were recognized as such. His domestic animals were potential pests of his crops, as they are today, and the fence and the tether must have been early inventions. Zinc-coated steel wire could perhaps be regarded as a very important contribution of the chemical industry to modern pest control, but in this book a pesticide is considered, conventionally, to be a compound acting in some way chemically upon the pest. Its particular function is to extend the protection of a fence to keep off pest animals or fungi too small to be restrained mechanically.

Protection of the crop from premature consumption has a much greater significance than is often appreciated for the biological economy of the cultivated world. Not only has man become the dominant species in influence, numbers and sheer weight of living flesh (the "biomass"), but he has arranged that the world supports a total biomass, including that of all his domesticated animals, greater than was supported in the wild state. To make this possible, mechanical cultivation, plant breeding and the use of plant foods have all been necessary, but not least the organization of time of consumption. Two blades of grass must be conserved for fodder where only half a blade was allowed to survive before. It is particularly to this process of agricultural conservation that the chemical industry has made such important contributions with synthetic pesticides.

So great has been the increase of agricultural productivity under the influence of applied science that the nineteenth-century warnings by Malthus about the limited feeding capacity of the world seemed for a time unimportant. The control of pests has, however, been matched by control of disease. The increase of crop productivity has been matched by that of human population. There can be no relaxation in the development of pest control methods while mankind is solving the enormous problem created by his own increase.

When a successful method of control of some agricultural pest is known, comparisons can be made of the yields from similar treated and untreated plots and a reasonable estimate of economic return can be made. Often the value of the extra yield is many times the expenditure on pesticide. The economics are often difficult to assess. The gain from treatment of one crop in one season may be reaped mainly in a succeeding crop in the next season. The true loss may be hidden because an uncontrollable pest, or, more usually, a disease, may necessitate replacement of a preferred crop, in a region where soil and climate seem suitable, by a less valuable alternative.

Some soils in the U.K. will, at their best, grow a far heavier crop of potatoes than others. Potatoes are, on these soils, at their best, the most profitable crop. Continued cropping, however, builds up a population of eelworm which can eventually reduce the cropping capacity almost to nil. Chemical control is very expensive. Natural recovery takes several years. The farmer is best served by restricting his potentially most valuable crop to one year in four. The regular alternation of different crops through a standard cycle of years, is called rotation. It may bring other benefits, associated with maintenance of soil fertility and timing of labour and machinery loads, but "pest evasion" is the most important factor deciding many old-established rotational practices.

Man must eat throughout the year. Most staple foods are gathered in a short harvest period and must therefore be held in store for several months. A great deal of stored foodstuff can be lost to pests. A fair estimate, the world over, is that pests take one third of the crop before harvest and another third in storage. A proportion of pesticides goes on to stored food or its containers. Much less, however, is used in this way than on growing crops, for two reasons.

Firstly, the toxic residue problem is more critical on stored grain because the compound is protected from weathering and, being applied to a dry and dormant plant organ, is less subject to biological destruction. Moreover it is applied directly to the edible portion of the crop, while most field-applied pesticides are applied before the harvested organs are accessible. These considerations impose severe limits on the number and quantity of pesticides which can be used on stored products, although the fact of their storage in closed spaces permits the use of fumigants which are of little use in open agriculture.

Secondly, non-chemical alternative means of protection are much more effective in the case of stored products. High value foods such as meat and fruit and vegetables can be stored in sealed cans or deep-frozen. Grain can be stored in rat-proof buildings and pests in it destroyed by low temperature or anaerobic conditions. Most of the heavy losses of stored grain in under-developed countries could be prevented by impermeable (but more expensive) containers and better organization.

Not only foods, but also structural materials, are vulnerable to damage by insects and fungi. It might be thought that the most important contribution of the chemical industry to this problem is the provision of synthetic structural materials. It would be unwise, however, to underestimate the ability of insects, fungi and micro-organisms to evolve strains adapted to wholly new nutrients. Cases are already known of "biodeterioration" of synthetic materials. It must also be appreciated that replacement of natural structural materials has still not gone very far. Annual world production of timber is about 300 million tons. This is about half the tonnage of basic food grains. Of this, over 80% is used as structural timber, the rest being processed in various ways into composition boards, paper and fibres. Cotton is still by far the most used fibre. The total of truly man-made fibres is still less than that of sheep-made wool.

This book, like others in this series, is intended mainly for the student of chemistry. Its authors are chemists by training. It contains, however, a lot which is not chemistry. This is for the very good reason that the subject is not unified by its chemistry but rather by the practical objectives for which the chemicals are used. This is more fully explained in Chapter 1. Moreover these books should sketch the surroundings in which the chemist must work, the impact that the chemist has made on the market which uses his products and the unsolved problems which still require a chemical solution. The authors hope that they have put the chemical work in perspective and that the book will not be without value to students of agriculture and other technologies where what is now being called "biodeterioration" is important.

The enforced miscellany of subject matter makes it impossible to find an entirely satisfactory chapter sequence. For the most part, the conventional classification into insecticides, herbicides and fungicides has been followed and some more specialized chemicals are also classified according to their biological function, but in two of the main technical chapters the classification is according to the physicochemical aspects of application. These are the chapters on Formulation and on Fumigation, dealing respectively with the preparation of compounds in a form convenient for application and with compounds which distribute themselves by gas-phase diffusion. One chapter is concerned with application itself.

Chemicals for crop protection pest control have had a dramatic rise in number and quantity over the last 30 years. Problems of their use have created new legislation and a great deal of public interest. Not only will new chemicals appear in the next few years but changes also in their use. Even the type of agriculture in which they are used may undergo drastic changes. A final chapter therefore departs from a roughly balanced picture of the present and attempts a glimpse into the future.

The book does not aim to provide a complete catalogue of the structures and properties of all chemicals in current use as pesticides. This could not have been done in the space available without drastic elimination of other matter. Moreover such a catalogue is of limited value unless arrangements are made to keep it up to date. References to works designed to do this are given at the end of Chapter 1. The chemicals of which brief description has been given in the present book have been selected to illustrate various aspects of the whole subject. It is hoped that they provide a reasonable representation of types of chemical structure and mode of action and also of problems created, such as difficulties of application or persistence of undesirable residues. The student going into this field should know what is set out here but the particular chemicals on which he will find himself working are probably not listed here. Very likely, they could not be listed here for the good reason that they are not disclosed at the time of writing.

Some words very familiar in agricultural and pesticide technology may be unknown in their context to a student of chemistry. A short glossary has therefore been provided. Detailed references to original literature are not given except in a few cases where the information has not got into text books or review articles. Suggestions for further reading are given at the end of each chapter and most of the works referred to include extensive references to original literature.

Chapter 1
THE SHAPE OF THE INDUSTRY

A Convergent Industry

This book is about chemicals for crop protection and pest control, considered as products of manufacture and trade by a section of the chemical industry.

Within this trade, these products are often referred to as agrochemicals. The trade association, which began as the Association of British Insecticide Manufacturers, is now known as the British Agrochemicals Association because its members realized that a major part of their trade was in herbicides and that fungicides should also logically be included. It is accepted by the association that the term "agrochemicals" does not include the large-tonnage fertilizers - inorganic salts supplying ammonium, potassium, nitrate and phosphate ions for major crop nutrition, although it is sometimes held to include more specialized nutrient products, such as organic chelates of iron and manganese, aimed at curing deficiency "diseases" of plants. It also includes compounds having some other ancillary use in crop production, such as the encouragement of rooting of cuttings or of setting of fruit. Agrochemicals could not logically be held to include compounds used for control of non-agricultural pests, but it would, of course, be absurd either in the organization of the industry or of this book, to include DDT if used for killing pea-moth and not DDT if used for killing bed-bugs. It would be equally absurd to include simazine if used as a selective herbicide in the maize crop but not if used at a higher rate to keep all vegetation out of a railway siding.

The title Chemicals for Crop Protection and Pest Control has therefore been chosen as most comprehensive and least likely to convey a wrong impression outside the trade. Although fertilizers are excluded, chemicals having functions in the maintenance of crop health other than the control of visible pests are included. Competing plant species - weeds - are classed as pests and also the often unseen micro-organisms responsible for plant diseases.

Chemicals for crop protection and pest control - known collectively as pesticides - include a very wide range of structures and their manufacture requires a wide range of processes. The industry is not unified at the stage of chemical production. Technically, it is unified, not by common starting materials or processes, but by the manner and purpose of use of the end products. Commercially, it is unified more by its sales organization than by its production organization. This structure is dictated by the need for a highly technical sales force which must deal with a very large number of technical customers. On both sides, however, the need is for expertise in pest or agricultural technology rather than chemical technology. If single descriptive words are wanted, one can most suitably call the pesticide industry a convergent one, which brings products from a variety of sources on to a group of closely related targets. These targets are related biologically and economically, not chemically.

The heavy chemical industry provides a marked contrast. It must be organized on lines dictated by its major processes. A chlorine plant which discarded its simultaneously produced alkali could not be economic. The integration of the related processes of the early "LeBlanc complex" is described in A History of the Modern British Chemical Industry. Such industry might fairly be called divergent. A good example was the form of the Albright and Wilson Company up to the time of recent acquisitions and diversification. This company was based on the processing of a single difficult and important element - phosphorus. The only practicable way, for a long time, to obtain most phosphorus compounds in adequate purity was via elementary phosphorus. This was a difficult substance to produce and demanded special methods of handling. It was a natural step, economically, to harness as many orthodox secondary processes as possible to this highly specialized primary process.

The pesticide industry is often described as "research intensive". There are still many pest problems not solved by any existing chemicals. Requirements in agricultural pest control are constantly changing owing to the impact of changes in cultural methods (including pesticide use) on pest problems. The development of strains of insects resistant to insecticides is a serious problem in both agriculture and public health. The requirements of Governmental registration authorities for safety to the consumer and to the environment have become much more stringent. The research chemist is therefore always seeking new structures and the research biologist new tactics and strategy. It is rather unusual for a compound to command a large share of the market for a long time. The compound may therefore cease to be commercially viable before means for its most economic production have been worked out.

Fertilizer production provides an informative contrast. Crops will always need available nitrogen, phosphate and potash (NPK). Compounds which are worth considering to supply this need are very limited. Research on the economics of packaging and transport and the problems of serviceability at the user end are of course common to all industries producing consumer goods, but, whereas a large proportion of fertilizer research effort must go into minimizing cost of production, the need in the pesticide industry is to get down to an acceptable cost quickly rather than to the lowest possible cost eventually.

Another contrast with the fertilizer industry illustrates another important point. NPK will not only always be needed by plants of any kind known to man throughout history, but will be needed by crop-producers every season in not very variable amount. The farmer, however, always hopes that he will not need pesticides, sales of which are, in consequence, subject to wide variation from one season to the next. In the phenomenally wet summer of 1958 in the U.K., the industry completely exhausted its supplies of copper fungicides for use against potato blight, but, in the following, phenomenally dry, summer, such products could hardly be given away. Both the farmer and the manufacturer are reluctant to tie up capital in stock-piling. It is therefore a very desirable feature of pesticide production processes that they be elastic with regard to rate of output.

The products of the pesticides industry are toxic to some living species and are valued for just this reason. Few are so highly selective that their ingestion in accidental amounts would be quite without significance in man or farmstock. Indeed a comprehensive term for these products, often used in the U.S.A., is "economic poisons", which conveniently emphasizes both their value and the hazards latent in their use. During the last 20 years an increasing effort has been put into the study of these hazards and of means to reduce them and many countries have introduced elaborate legislation to control the use of pesticides. This legislation, or the voluntary schemes which operate in some countries, including the U.K., now imposes on the industry an expensive burden of closely specified and expert investigation which will be described later. This public responsibility of the industry is another powerful reason why it is unified at the using, rather than the manufacturing end.

Production of a multiplicity of complex chemical products, mostly on a comparatively small scale, poses considerable economic problems, and, for this reason, few chemical companies are in the pesticide manufacturing business alone. The general tendency of large financial groups to ensure their future stability by diversification of interests is therefore strengthened. Association with other manufacturing facilities enables the pesticide section of a large company to concentrate on the peculiar problems of its multi-customer, highly technical market, while leaving some of the basic production to be integrated with other production for quite different markets. Association with groups having other interests in research on new chemicals enables the expert "screening" facilities of the pesticide group to have a larger number of speculative new chemicals to test for activity in its own field.

Although the processes of manufacture of active compounds are diverse, the types of "formulation" in which they are offered to the user are more limited and characteristic of the industry and the market it supplies. The properties of the active compound may restrict the choice of formulation, but what is desirable at the user end is determined by considerations of packaging, transport, application and biological efficiency and has little in common with the formulation of cosmetics and pharmaceuticals. Technical aspects of formulation of pesticides are dealt with in a later chapter, but it should be pointed out here that one result of the diversity of active compounds and the close association of formulation with market requirements is a tendency for formulation to be carried out in local small units, taking their active and ancillary compounds from different sources. This tendency is much greater in countries such as the U.S.A., Africa and South America, where long transport hauls must be made from the manufacturing bases, than in small and industrialized agricultural countries like the U.K. There are two reasons for this. Firstly, many formulations require dilution of the active compound with water, solvents or powdered minerals, and economics dictate that these diluents be added locally. Secondly, various factors in local conditions, including customer preferences and compatibility with other locally used products, have a big influence on choice of formulation. As the larger land units are more diverse in climate and agricultural practice, the local formulator, with his knowledge of local factors, is better able to meet the requirements than a remote manufacturer.

Many manufacturers have their own formulation branch factories in important regions. Others may supply their active chemical products to a local formulator. The parties will, of course, find it desirable to protect their interests by some specification agreement, the formulator to protect himself against difficulties due to variation in the technical product, the manufacturer to avoid his product coming into disrepute through bad formulation. Where the active compound has ceased to be protected by patents, manufacture may pass into the hands of general chemical companies who may have little knowledge of agriculture or pest control but have facilities for cheap production of the compound. Some pesticides have in this way become commodity products.

The companies developing and manufacturing new pesticides carry very heavy costs for discovery and development and for research into safety and use problems, about which more is said below. There is no possibility of recouping these costs if the product can be made and sold freely by a purely manufacturing company incurring only process research costs. Pesticide companies are therefore very concerned to exploit patented products. They may formulate commodity items in order to complete their sales range, but if adequate control of quality can be exercised, will buy in these products from a commodity manufacturer. They may therefore be strongly competitive in their exclusive products but buy other products from one another.

Relation with Pharmaceutical Industry

The problems of production raised by the great multiplicity of chemical structures are in large measure shared by the pharmaceutical industry, except that the latter generally confines its interests to more complex and expensive compounds, used medicinally. The technology of chemicals for pest control is often called, especially in France, "phytopharmacy" but it covers a wider field than attack on diseases produced by internal micro-organisms. The mechanical advance of civilization has reduced the significance for human health of many predators and vectors of disease. In remote communities where war must be waged on the leopard or the wolf, the part played by the chemical industry is to provide propellents for bullets rather than poisons. If attack must be made on the malaria-carrying mosquito or the typhus-carrying body louse, the necessary chemicals are obtained from the pesticide rather than the pharmaceutical industry.

Human pharmacy and plant pharmacy are not therefore, in practice, fully comparable. Plant pharmacy is much more concerned with gross parasitism and predation than with diseases in the generally accepted sense, while in human pharmacy the relative importance is reversed. One should not conclude that plant diseases produced by microscopic or submicroscopic parasites are either unimportant or unrecognized. Historically, the first diseases to be clearly attributed to the growth of microorganisms were fungal diseases of plants - blight of potatoes, rust of wheat, clubroot of cabbage. The discovery of bacteria took place only a few years later, the first disease to be shown to be caused by bacteria being anthrax in sheep, followed rapidly by a "vegetable anthrax" - fire blight in pears. That filter-passing, proliferating substances, later called viruses, were responsible for some diseases was discovered almost simultaneously in plants (tobacco mosaic) and cattle (foot and mouth disease). One can say broadly that, in the 50 years from 1860 to 1910, man's knowledge of the causes of disease of plants and animals developed from vague ideas about miasmas and murrains to the sound foundations on which an extensive detailed structure has since been built.

Prevention, amelioration and cure of the true diseases have made much more rapid progress in animal, and especially human, pharmacy than in plant pharmacy. The reasons, biological, sociological and economic, are outside the scope of this book, but it is well to be reminded that, as agriculture increases control over competition by weeds and consumption and damage by insects, then, in the treatment of true diseases, delay of invasion and control of insect vectors may be replaced by curative treatments of which much could be learnt from human medicine. The development of systemic fungicides, described in a later chapter, is an example of this progression.

The science and industry of veterinary chemicals is rather naturally intermediate between that of agricultural pest control and human medicine. There is more concern with cure of animal sickness than of plant sickness, but more concern than in human medicine with control of the larger parasites. By the larger parasites are meant insect and acarid "ecto"-parasites, such as blowflies, lice and ticks, and helminth and insect "endo"-parasites, such as lung worms, intestinal worms and warble-fly larvae.

There are obvious advantages in the integration of the pest control and pharmaceutical branches of the chemical industry and the majority of companies with an interest in the one have association of some sort with a company interested in the other. Experimentation on veterinary medicines has obviously a great deal in common with that on human medicines, but a sales channel for a successful compound may be opened more easily in the agriculturally oriented sister company. Preparation of speculative new chemicals may be intended to provide a cure for the common cold but may show, instead, a useful action against an agricultural pest if tested for this purpose. Occasionally there may be an exploitable activity in both fields, a classic example being warfarin which is used both curatively in humans suffering from thrombosis and as an effective poison for rats.

In the field of human medicine, more is known about the biochemistry of pathological conditions and the mode of action of curative drugs than in the much wider and less intensively studied field of pest control. There is more possibility for chemical logic in the design of new molecules, but even here the logic has in most cases to be confined to minor changes on an established structure. In the field of pest control, almost every "breakthrough" into an entirely new type of active structure has come from wide-range hit-and-miss research by teams of organic chemists and biological "screeners". One could perhaps claim as a partial exception the discovery of 2,4-D and MCPA as suggested by natural plant hormones and consider that Wain's finding of more selective herbicidal action in the 3-phenoxybutyric than in the phenoxyacetic compounds was neither minor nor accidental. While it is at present true that successful wholly new structures have been found by guesswork and that, if anything is known of the biochemistry of the toxic action, the knowledge has come after the event, predictive biochemistry is likely to become increasingly important. Increasing effort is therefore going into the study of biochemical mechanisms in both academic and industrial research.

Research Collaboration

Until predictive biochemistry takes over, the present strategy of synthesis of new compounds in wide variety is likely to be maintained for some years to come. It is therefore an obvious advantage for the pesticide manufacturer to have close association, at the compound-testing stage, not only with a company working in the related pharmaceutical field, but also with any company producing a wide variety of chemical compounds which may be of interest directly or as intermediates. The dyestuffs industry is a useful ally since it must produce compounds in great variety, (a) to meet the whims of fashion and (b) to solve the problems set by advances in synthetic fibres. There is little doubt that the early development of dinitrophenol compounds, and much later, of the substituted diamino-s-triazines, was initiated by availability of compounds from the associated dyestuffs industry. Aminotriazole was an intermediate in colour-photographic dye production before it was known as a herbicide.

Other technologies have contributed. The dithiocarbamate fungicides were first produced as ancillary compounds in rubber processing. Organotin compounds had their origin in purely chemical exploration of the combining properties of tin. Their first exploitation was for stabilization of polyvinylchloride against photochemical attack.

At the research stage in the industry, therefore, diversity of chemical synthesis is an important characteristic, often augmented through compound-exchange and collaboration with other laboratories. An efficient biological screening department is an equal necessity. It must be able to test compounds on a small scale, quickly and all the year round, for activity on a representative range of plant and pest species. If, perhaps, two or three compounds in 1000 show some preliminary promise, they must be tested under more varied and realistic conditions and a start made on the eventually extensive investigations on all aspects of safety, including means of quantitative estimation in trace quantities on crops. For these extended investigations the chemist must meet the problem of preparation on a scale exceeding that convenient in ordinary laboratory practice. Most of these first-selection compounds will fail to meet one or another requirement for commercial success. One commercially successful compound out of 10 000 must be regarded as satisfactory. To avoid expensive waste of effort it is therefore necessary that all the steps in the follow-up research on a candidate compound be closely coordinated and further work stopped as soon as failure is certain.

It is in the important exploration of detailed chemical variations on a promising new theme - a "lead" as it is always called in the trade - that physiochemical and biophysical logic can play a part. Observation of the effects of changes in parameters, such as acidity of a molecule and its oil-water partition coefficients, on biological activity within a particular chemical group can facilitate predication of the structure of the most active compound within that group. The advent of the computer has made it possible to handle the considerable amount of data needed, to calculate the required parameters for thousands of possible compounds, and to home in on the optimum combination of those parameters. Nevertheless, physical parameter/biological activity relationships are not yet so certain that the best compound can be identified unequivocally; the field of search can be greatly narrowed but rapid biological testing, without limitations of season, in an efficient screening department working closely in association with the synthetic chemists, is still an important requisite to identify the compound which has the best chance of commercial success.

The industry, no doubt in common with many others, is facing changing problems. Some, such as arise from diversity of chemical processes, it must sort out for itself. Others may require increasing collaboration with research organized by governments, with agricultural trade organizations or with other bodies outside the chemical industry.

Further Reading

Bradbury, F.R. & Dutton, B.G., Chemical Industry: Social and Ecnomic Aspects, (Butterworth, London, 1972)

Gregory, J.G., (Ed), Modern Chemistry in Industry, (Society of Chemical Industry, London, 1968)

Jones, D.G., (Ed), Chemistry and Industry, (Clarendon Press, Oxford, 1967)

National Agrochemicals Chemicals Association, Pesticide Industry Profile Study, (Washington, D.C., 1971)

Shreve, R.N., Chemical Process Industries, (McGraw-Hill, London & New York, 1967)

Williams, T.I., The Chemical Industry, (Pergamon Press, London, 1953)

Chapter 2

TECHNOLOGICAL ECONOMICS OF PESTICIDES

Economics of Pesticides for the Manufacturer

The modern development of crop protection chemicals came almost entirely from the chemical industry. In this respect it differed from many technologies which evolved from initial discoveries made during the scientific pursuit of knowledge in universities and other academic institutions. The object of research in industrial companies is not, as in universities, to increase scientific knowledge for its own sake, but to find new ways in which the company can increase its business and earn profits. The chemical industry has no obligation to discover, develop and manufacture new pesticides. It has done so because it has proved to be an acceptably profitable investment for new capital. It will continue to do so only so long as it is more profitable to do so than to invest available money in discovery, development and manufacture in other commercial areas such as plastics, pharmaceuticals, cosmetics, synthetic fibres and similar technologies.

The pesticides industry in the modern fine-chemical sense dates from World War 2. Before that pesticides were mainly inorganic materials such as sulphur or lead arsenate together with a few naturally-occurring organic materials such as nicotine and pyrethrum. The advances made in the science of organic chemistry during the latter half of the nineteenth century led, in the first half of the twentieth century, to commercial exploitation of organic chemicals, first in dyestuffs, and then in pharmaceuticals. Before World War 2 a number of chemical companies were considering the possibility of using organic chemicals to control the pests and diseases of plants. However, in the 1930's, farming was in a depressed state in the U.S.A. and Europe and it did not seem likely that farmers would be willing to pay the high prices which seemed likely to be needed for such chemicals. Nor was it clear, at that time, that any organic chemicals could, in fact, be discovered which would be useful in treating horticultural and agricultural pests and diseases. So it appeared, at that time, to be a highly risky and uncertain undertaking for the chemical industry to invest money in research and development in this area.

After World War 2, the picture had changed. Food prices had increased considerably, standards of living in developed countries were rising rapidly and farming was becoming a much more profitable occupation. The discoveries of DDT in Switzerland, of the organophorphorus insecticides in Germany and of the phenoxyacetic herbicides in the U.K. had all demonstrated that useful crop protection products could, in fact, be discovered and that these need not necessarily be costly to produce. The organic chemical industry now saw pesticides as a profitable diversification of their manufactures, and so the pesticides industry had a phenomenal growth rate during the period 1945 to 1975, as illustrated by the figures in Table 1.

TABLE 1

Growth of World Pesticides Industry
(thousands of tonnes)

	1945	1955	1965	1970	1975
World output	100	400	1000	1500	1800

This remarkable growth was augmented by the fact that the new post-war petro-chemicals industry was producing all kinds of novel organic starting materials from oil, and many of the larger manufacturers of primary petro-chemicals therefore came into crop protection and pest control as an outlet for their products. However, it cannot be too strongly stressed that the fact that the first effective organic pesticides which were discovered were cheap and easy to manufacture paved the way to acceptance by farmers of much more expensive products and was a major factor in the rapid growth of the industry. The post-war development of the pesticides industry is a good example of the recipe for commercial success - "the right product in the right place at the right time at the right price" - and illustrates how successful innovation depends on technological, economic and social factors all being favourable. If the phenoxyacetic herbicides had been discovered in 1930 it is doubtful whether their commercial impact on agriculture at that time would have been very substantial.

Pesticides have always been a high risk investment. Development of a new crop protection chemical is a long and expensive business. Many compounds have to be synthesized and screened before a "lead" is discovered, extensive synthetic work has then to be undertaken to locate the most suitable compound in the "lead" area for commercial development and then this selected compound has to be evaluated world-wide on a range of crops, soils, climates and environments and, concurrently, far-reaching toxicological, residue and environmental studies have to be carried out. Economical processes of manufacture have to be discovered, pilot quantities produced and full-scale manufacturing plant designed, constructed and commissioned. The way in which development costs build up is shown in Table 2.

The type of cash flow which this produces is shown in Figure 1. This is the cash flow for a typical successful pesticide which shows the cumulative difference between money laid out and money coming in, adjusted to take account of the fact that the value of money several years hence is considerably less than the value of money now. This is what is called a "discounted cash flow" and is a good guide to the profitability and risk of any proposed investment.

The cash flow in Figure 1 illustrates both the risk and uncertainty of development of new pesticides as a commercial investment. It goes $ 6M into the "red" and breaks even only after 10 to 11 years. It is only after this "break-even" point that profit is made but, by then, there remain only a few years out of the total life of the patent (16 years in the U.K.) before competitors who have had to bear none of the heavy development costs can come in and force down prices. This is because the patent will almost certainly have been taken out when the original discovery was made. The cash flow shown in Figure 1 is that of a commercial "winner", the development of which has gone smoothly and without major setbacks. Performance in the field

TABLE 2

Estimate of development costs of a pesticide in 1970

Year		No of compounds	Cost per compound	Total cost
			$	$'000's
1	Synthesis	8000	140	1640
	Screening		65	
	Survival rate 1:100			
2	Glasshouse trials	80	5000	400
	Initial field trials			
	Survival rate 1:5			
3	Field trials	16	5000	112
	Initial toxicology		2000	
	Survival rate 1:4			
4	Field evaluation		50 000	
	Toxicology	4	20 000	480
	Formulation and process		50 000	
	Survival rate 1:2			
5	World wide evaluation		500 000	
6	Toxicology, environment, ecology		350 000	
	Formulation and process	2	200 000	2850
	Production		200 000	
	Registration and patent		175 000	
	Survival rate 1:2			
			Total cost	5482

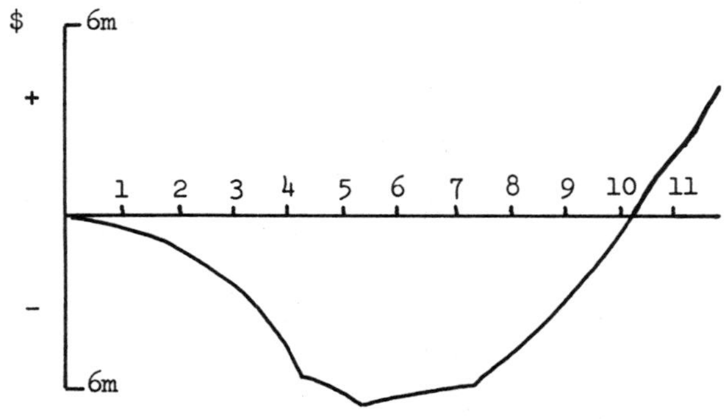

FIGURE 1 Discounted cash flow of successful pesticide

which is not consistent over a wide range of soils and climates, unexpected toxicological or environmental effects, discovery by competitors of cheaper or more effective alternatives, development of resistant strains of the pest or disease, could all "kill" a promising new product, often after a considerable amount of money has been spent abortively on its development.

The total cost of development of a new pesticide in 1970 was \$ 5.5M, but the chance that sales would ever exceed \$ 5M per year was estimated in that year to be about 10 to 1 against. Of all the pesticides introduced to Western Europe between 1960 and 1970 only 12 achieved annual sales in Western Europe greater than \$ 2.5M. To put large sums of money at high risk for long periods in the hope of eventual profit is acceptable only to the largest chemical manufacturers who have a great liquidity of assets and a diversity of interests to buffer them against misfortune. This is why the number of firms engaged in discovery and development of new pesticides is comparatively small and why these are the large diversified international chemical organisations.

Such companies have a responsibility to their shareholders to invest their money wisely and to their employees to maintain viable businesses so, if pesticides cease to be an acceptable financial risk, discovery and manufacture of them will cease. It has already been pointed out that, before World War 2, the risk was not an acceptable one. From 1940 to 1970 the risk was still there but it was now acceptable because the profits to be made from a successful product were very large. Nevertheless, it is interesting to note that such highly profitable major products are few and far-between and that each major producer of pesticides in the world has one, or, at the most, two such products to his name which supply most of the profit for his pesticide business, augmented by a number of minor products which help to maintain an acceptable cash flow. The interesting question whether discovery and development of new pesticides is once again becoming an unacceptably risky investment is currently being hotly debated in the industry.

We have seen that after World War 2, everything was right for the developing pesticides industry. Farming was becoming a prosperous occupation, agricultural land prices were rising rapidly and agricultural labour was becoming scarce and expensive so there was great incentive to improve yields per hectare. Food prices were rising and the populations of developed countries were increasing in number and becoming more affluent and able, therefore, to demand and pay for a wider variety of more expensive foods. Marketing methods and food processing both called for fruit and vegetables with no visible signs of pest or disease damage. Why, then, should things now be going wrong for the pesticide industry?

Firstly, the costs of discovery and development of new pesticides are rising rapidly as shown in Table 3. The percentage of sales income spent on research, and the time needed for development of a new product are both increasing, as shown in Table 4. At the same time the number of candidate compounds which have to be screened to obtain one commercial product is increasing, as shown in Table 3, and the success rate is falling.

TABLE 3

Increase in development costs of a pesticide

Year of estimate	1956	1964	1967	1969	1970	1972
Cost of development ($ millions)	1.2	2.9	3.4	4.1	5.5	10.0
Number of compounds screened per marketed product	1800	3600	5500	5040	8000	10 000

TABLE 4

Research profile of US pesticide industry (Ernst & Ernst Trade Association Department, May 1971)

	1967	1970
Total sales	$ 639 m	$ 722 m
R & D on new products	$ 52.3 m	$ 69.9 m
Percentage	8.2	9.7
No of compounds screened per marketed product	5481	7430
Cost of development of each marketed product	$ 3.4 m	$ 5.5 m
Time from discovery to marketing	60 months	77 months

TABLE 5

Minimum requirements for world-wide registration

	1950	1960	1970
Toxicology			
	Acute toxicity 30-90 day rat-feeding	Acute toxicity 90 day rat-feeding 90 day dog-feeding 2 year rat-feeding 1 year dog feeding	Acute toxicity 90 day rat-feeding 90 day dog-feeding 2 year rat-feeding 2 year dog-feeding Reproduction 3 rat generations Teratogenesis in rodents Toxicity to fish Toxicity to shellfish Toxicity to birds
Metabolism			
	None	Rat	Rat and dog Plant

TABLE 5 (Continued)

	1950	1960	1970
Residues	Food crops 1 ppm	Food crops 0.1 ppm Meat 0.1 ppm Milk 0.1 ppm	Food crops 0.01 ppm Meat 0.1 ppm Milk 0.005 ppm
Ecology	None	None	Environmental stability Environmental movement Environmental accumulation Total effects on all non-target species

Secondly, the requirements for toxicological and environmental testing have become much more stringent in recent years, as shown in Table 5, and are likely to become even more demanding - and, therefore, more expensive - in the future. The cost of providing the necessary evidence of safety falls entirely on the company which wishes to market the new product, and may amount to $ 2M. This cost has to be recouped from the profits of future sales if the investment is to be one which is commercially justified. By and large, the costs of searching for new biologically active compounds and of toxicological, residue and environmental studies on candidate products, are both largely independent of the size of the eventual market for the product. The consequence is that discovery and development of new pesticides is becoming more and more an acceptable undertaking for the chemical industry only if there is a reasonable chance of finding a product which has a wide spectrum of activity andwhich can command a large market in a major crop. As there are a limited number of major crops in the world the markets for new pesticides are becoming increasingly fragmented as more and more products are introduced to the same markets, and the chances of achieving a substantial share of total market for a new product are decreasing steadily. The incidental consequence that minor crop or minor pest problems are not getting sufficient attention and that, often, compounds which are discovered in the laboratory which could be efficacious in dealing with them are not developed commercially because it is not profitable to do so, may be regrettable for agriculture but is not one which the manufacturers can avoid so long as the economics of pesticide development are as they are.

We have already seen that the chance of a new pesticide making large profits is not high, and is getting steadily less. The reason is that, during the past thirty years, a large number of new pesticides have been developed, and many of these have achieved large tonnages and have become patent-free so that they can be made comparatively cheaply because of scale of manufacture and competetive production. More and more pesticides are coming into this category every year. There are, therefore, reasonably satisfactory answers available to the farmer for most crop protection and pest control problems at a reasonable price. The choice for the farmer is not as it was twenty years ago between using a pesticide or suffering substantial crop loss but between using a cheap established pesticide which gives reasonable control or a new and more expensive pesticide which performs marginally better. The available markets for new pesticides are, as we have said, becoming increasingly fragmented and there will be smaller market opportunities and lower tonnage requirements for new products. The chances of achieving sales sufficient to recover development costs and to make an acceptable profit are, therefore, diminishing. The future of discovery and development of new pesticides is doubtful unless the financial prospects for the manufacturer can be made more attractive.

It is an ironic consequence of the increased costs of safety and environmental studies that it has become unprofitable to discover and develop compounds of highly specific activity for restricted and specialized markets. Such compounds are the least likely to give rise to side-effect problems. If all the pesticide needs of the farmer could be covered by a multiplicity and diversity of specialized products the possibility of cumulative risk to the consumer or the environment would be far less, than that for a few very widely used compounds of unspecific activity. Yet, a proper concern by the regulating authorities to try to ensure maximum safety to the consumer and the environment is having the effect of making only the least intrinsically safe types of pesticides, namely, those with a wide spectrum of activity intended for extensive use on major crops, economically justifiable as targets for discovery and development.

The Size of the Market for Pesticides

Although there may be doubts about the financial viability of development of new pesticides there is no doubt that world demand for established pesticides will increase. Although alternative methods of crop protection and pest control may be introduced it is certain that chemicals will be the main weapons in the farmer's armoury against pests and diseases for many years to come. The number of companies which are engaged in discovery and development of new pesticides is very small, but there are thousands of small companies who are engaged in formulation and marketing of established pesticides which are no longer protected by patents. Thus, in the U.S.A. there are about 800 registered active ingredients for pesticides but at least 80,000 different commercial labels.

TABLE 6

Herbicides

	Amount manufactured in U.S.A. in 1971	Amount used by U.S.A. farmers in 1971
	tonnes	tonnes
Inorganic herbicides	13,600	820
Organic arsenicals	15,900	3,600
Dalapon	2,300	450
Chloramben	9,100	4,400
Other benzoic acids	500	230
2,4-D	20,400	15,700

TABLE 6 (Continued)

	Amount manufactured in U.S.A. in 1971 tonnes	Amount used by U.S.A. farmers in 1971 tonnes
Other phenoxyacetics	3,000	2,400
Propachlor	10,700	10,700
Alachlor	9,100	6,700
Propanil	3,000	3,000
Naptalam	2,000	1,500
Other anilides	500	360
Dinitrophenols	4,000	3,300
Fluorodifen	600	590
Trifluralin	11,400	5,200
Nitralin	3,600	1,200
Carbamates	10,000	8,300
Linuron	900	820
Diuron	2,700	550
Fluometuron	1,800	1,500
Noruron	1,000	590
Other ureas	200	90
Atrazine	40,900	26,000
Other triazines	3,000	2,400
Other organics	2,200	4,000
Total	172,400	104,400
Petroleum oils	-	66,100

TABLE 7

Fungicides

	Amount manufactured in U.S.A. in 1971 tonnes	Amount used by U.S.A. farmers in 1971 tonnes
Copper sulphate	5,000	3,500
Other copper compounds	1,500	1,000
Other inorganics	4,000	2,800
Maneb)		1,800
Zineb)	18,000	900
Ferbam)		600
Other dithiocarbamates)		2,600
Captan	8,100	3,000
Other phthalimides	1,000	450
Other organics	4,000	2,600
Total	41,600	19,250
Sulphur	68,000	51,000

TABLE 8

Insecticides

	Amount manufactured in U.S.A. in 1971 tonnes	Amount used by U.S.A. farmers in 1971 tonnes
Inorganic insecticides	500	300
TDE	200	110
DDT	20,400	6,500
Methoxychlor	4,500	1,400
Toxaphene	22,700	17,000
Aldrin and endrin	4,500	4,200
Other organochlorines	3,000	2,200
Disulfoton	3,600	1,800
Methyl parathion	20,400	12,500
Parathion	6,800	4,300
Malathion	15,900	1,600
Dichlorvos	4,100	1,100
Diazinon	4,500	1,400
Phorate	3,600	1,900
Azinophos methyl	1,800	1,200
Ethion	1,200	1,100
Other organophosphorus	6,000	5,100
Carbaryl	20,400	8,100
Carbofuran	3,600	1,300
Metalkamate	2,000	1,600
Methomyl	1,800	490
Other organics	200	150
Total	151,700	75,350
Petroleum oils	-	34,000

The most reliable statistics for production and use of pesticides come from the U.S.A. In 1974 U.S.A. manufacturers produced 643,000 tonnes of synthetic organic pesticides having a total value of \$ 1732M. In Tables 6, 7 and 8 details are given of the amounts of the most important herbicides, fungicides and insecticides which were manufactured in the U.S.A. in 1971 and the amounts used by U.S.A. farmers in that year. It will be seen that production and sales are dominated by a comparatively few products. In herbicides, 25% of the total use of synthetic organics was accounted for by atrazine, 15% by 2,4-D and a further 26% by chloramben, propachlor, alachlor and trifluralin. The total usage of petroleum oils for weed control was 64% of the total usage of synthetic organics. In fungicides nearly all the organics were accounted for by the dithiocarbamates and captan but the major fungicide was sulphur, the total usage of which was 2.6 times that of all other fungicides put together. In insecticides, toxaphene, methyl parathion and carbaryl together accounted for 50% of the synthetic organics while the total usage of petroleum oils for insect control was 45% of the total usage of synthetic organics.

The fact that the market for crop protection and pest control chemicals is dominated by a few major products which can be produced comparatively cheaply because of scale of manufacture, should be borne in mind when considering the multiplicity of chemicals described in this book. Many of these have achieved only modest commercial success.

TABLE 9

U.K. Pesticide Production 1975

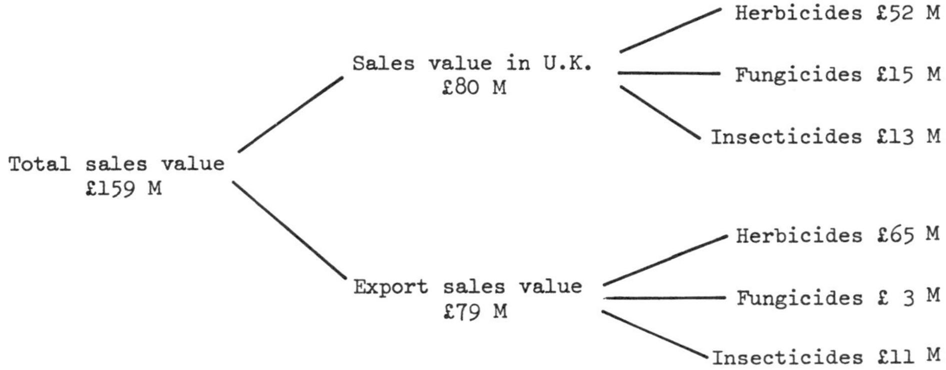

TABLE 10

World Pesticide Production 1974
Millions of dollars

	U.S.A.	Japan	U.K.	World
Herbicides	1,058	194	187	2,190
Fungicides	116	164	29	961
Insecticides	491	259	38	1,822
Fumigants	20	-	-	69
Growth Regulators	47	-	-	96
	1,732	617	254	5,138

TABLE 11

U.S.A. Pesticide Production

	1945 tonnes	1955 tonnes	1964 tonnes	1969 tonnes	1974 tonnes
Herbicides	2,000	20,000	103,000	178,000	274,000
Fungicides	2,000	5,000	51,000	64,000	74,000
Insecticides	24,000	106,000	202,000	259,000	295,000
Total	28,000	131,000	356,000	501,000	643,000

The value of pesticides produced in the U.K. in 1975 is shown in Table 9 and of pesticides produced in the U.S.A., Japan, U.K. and in the world as a whole in Table 10. During the ten years 1964 to 1974 total world production of organic pesticides doubled and production specifically of herbicides trebled. This trend is illustrated by the figures for the U.S.A. in Table 11. The overall annual growth rate for sales of organic pesticides was about 8% from 1964 to 1969 and about 5% from 1969 to 1974. It is thought by many that, in the next decade, the average annual growth rate will slow to about 3% for all organic pesticides and about 6% for herbicides specifically. However, some much more optimistic market forecasts have been given, and these are shown in Table 12.

TABLE 12

Projected World Demand for Pesticides
Millions of Dollars (1975 Price Levels)

	1975	1980	1985	1990
Herbicides				
Triazines	650	900	1,240	1,710
Carbamates	280	360	460	600
Phenyl ureas	190	315	520	860
Phenoxys	150	260	450	780
Benzoics	140	215	330	500
Arsenicals	30	25	20	20
Others	860	1,275	2,120	3,230
Total	2,300	3,450	5,140	7,700
Fungicides				
Dithiocarbamates	450	580	735	900
Captan	350	500	580	665
Coppers	150	186	215	260
Mercurials	50	40	20	0
Others	35	45	50	55
Total	1,035	1,345	1,600	1,880
Insecticides				
Organophosphates	1,100	1,440	1,870	2,450
Carbamates	470	610	800	1,050
Organochlorines	320	330	400	200
Arsenicals	20	10	0	0
Total	1,910	2,390	3,070	3,700
All Pesticides Total	5,245	7,185	9,810	13,280

For historical interest and as a comparison to the data in Tables 6, 7 and 8, with which they should be compared, the main pesticides manufactured in the U.S.A. in 1945 and 1955 are shown in Table 13.

TABLE 13

U.S.A. Pesticide Production
(excluding sulphur and petroleum oils)

	1945 tonnes	1955 tonnes
White arsenic	22,100	11,900
Calcium arsenate	11,600	1,300
Lead arsenate	32,000	6,700
Copper sulphate	113,900	70,900
BHC	1,000	4,600
DDT	15,000	56,800
parathion	1,200	2,300
TEPP	820	90
2,4-D	400	15,400
2,4,5-T	360	1,400
Other organics*	9,000	50,000
Total inorganic	179,600	90,800
Total organic	27,780	130,590

* mainly allethrin, aramite, aldrin, captan, chlordane, dieldrin, dithiocarbamates, endrin, methoxychlor, malathion and toxaphene.

Economics of Pesticides for Farmers

In 1971 U.S.A. farmers spent \$ 1,000M on crop protection and pest control chemicals. Expenditure by farmers in other developed countries has been comparable, and the total value of world sales of pesticides to farmers was probably around \$ 5,000M in 1976. The amounts spent by farmers in developed countries has been increasing steadily. In the U.S.A. they were \$ 427M in 1964, \$ 1,002M in 1971 and \$ 1,346M in 1973, and they are likely to continue to rise.

The reason why farmers in developed countries have been willing to spend money on pesticides is that such expenditure has proved to be financially rewarding. It has been estimated that each dollar spent on pesticides in the U.S.A. produces an average of about four dollars additional income for the farmer. In the U.K. the estimate is that each £1 spent on pesticides produces £5 additional income.

It is not easy to calculate precise figures for the financial benefits of pesticides for individual crops grown by individual farmers, as these vary widely from crop to crop, from farm to farm, and from season to season, and depend very much on the skill and economy with which the farmer uses them. Also attacks by pests and diseases are unpredictable. At one extreme pesticide application may prevent total loss of a crop on which a farmer is totally dependent and thus avoid total loss of income to him for that year. In developed countries, the small farmers who grew an assortment of crops and kept various types of animals is tending to disappear and many farmers grow just one type of crop, e.g. wheat, corn, soya, cotton, beet. At the other extreme, money spent on safeguarding against pests and diseases a crop which was eventually ruined for some other reason, such as storm damage, may be money spent needlessly. Money on pesticides is likewise wasted if the farmer cannot sell his crop at a satisfactory price. In the 1968-1969 season in the U.K. about 30% of the brussels sprouts grown were sold at less than the break-even price for the farmer so any money he had spent on crop protection would only have increased his financial loss.

Many farmers use pesticides as an insurance policy, that is, they adopt a regular programme of spraying as a preventative against possible losses. The economics of this procedure are becoming more comprehensible as the tendency towards contract growing and guaranteed prices increases. In such a situation a farmer can calculate accurately how much he can afford to spend on crop protection. From the farmer's point of view a crop which cannot be sold because it is blemished by pests and diseases is as financially disastrous as one which is totally destroyed. The consumer and food processor nowadays demand such high standards of freedom from insect damage or disease blemish that use of pesticides is mandatory and, in the case of contract growing, the programme of crop protection is often precisely specified for the grower in his contract. Nevertheless it is true that money spent on pesticides to safeguard against an attack which does not occur may be regarded as money wasted.

The economics of herbicide usage are clearer than those of insecticides or fungicides, as herbicides are essentially used as a substitute to mechanical or hand weeding and can be assessed on this basis. In 1975 it took 50 man-hours to grow and harvest a hectare of cotton using chemical weeding and defoliation and mechanical picking, compared with 200 man-hours in 1933 using manual hoeing and picking. Similarly, in 1975, it took 40 man-hours to grow and harvest one hectare of peanuts compared with 190 man-hours in 1935, and 22 man-hours to grow and harvest 5000kg of corn compared with 135 man-hours in 1945.

Nevertheless although the financial benefits of pesticide use vary widely from crop to crop, from location to location and from farmer to farmer, it is possible to make some generalisations. The most convincing evidence of benefit of pesticides to U.S. farmers is that they did spend $ 1346M in 1973 on pesticides, since farmers do not part with their money easily.

One or two specific studies have been made. A series of experiments with potatoes over a period of ten years indicated $ 6.71 average return for each $ 1 spent on pesticides, and a similar study on apples showed $ 5.17 average return for each $ 1 expended. In Canada, $ 1 spent in apple orchards on pesticides returned $ 13 in Nova Scotia, $ 5 in Quebec and $ 2.34 in Ontario. A study on tomatoes in the U.S.A. showed that use of pesticides increased farmers' gross incomes by 75%. The increase in value of crops harvested on ten million hectares in Canada treated with herbicides at a cost of $ 8M was estimated to be $ 58.8M equivalent to a 7:1 return. Studies on German farms growing cereals and root crops over a period of four years showed a gain in crop value of $ 47.2 per hectare for an outlay on crop protection chemicals of $ 20.3 per hectare.

Economics of Pesticides for the Nation

Farmers have used pesticides because it has proved profitable for them to do so, not from any altruistic motives. The consequence, for the nation, has been that pesticides have contributed substantially to the increases in yields per hectare and yields per man-hour for all major crops during the past thirty years. To quote one example, up to 1958 the average yield of wheat in the U.K. had exceeded 2500 kg/ha on only two occasions in 60 years. Now, in 1976, the average yield is about 4000 kg/ha. Yields of all other major crops have increased similarly. In the U.K., wheat is produced at 210 kg per man-hour compared with 2 kg per man-hour in subsistence agriculture in the third world. In the U.S.A. rice is produced at 240 kg per man-hour compared with 4 kg per man-hour by peasants in the tropics. Such yield increases are not entirely due to pesticides but also to higher-yielding varieties, irrigation and fertilizers. Nevertheless, it is sometimes possible to unravel the separate contribution of pesticides. Thus, between 1955 and 1970 in Illinois, when the same hybrid corn was grown and the same amounts of fertilizer used, average yields rose from about 3500 kg/ha to about 5000 kg/ha consequent on an increase in the proportion of the total crop treated with pesticides from about 3% to about 80% during that period. Pesticides have helped to make possible the fact that developed countries have been able to supply steadily increasing populations with adequate food at reasonable prices from a fixed amount of arable land using a decreasing labour force. The result is that, in the U.S.A. and U.K., one agricultural worker produces enough food for himself and 60 other people and the average person in these countries has to spend only about 20% of his disposable income on food. In India, by contrast, one agricultural worker feeds himself and only 4 other people, and the average person has to spend about 66% of his income on food.

It has been estimated that cessation of use of all pesticides in the U.S.A. would reduce total production of all crops and livestock by 30% and would increase the price of farm products to the consumer by 50% to 70%. This would mean that the average person would have to spend a substantially greater proportion of his income on food. As there are 25 million people in the U.S.A. who are on poverty level incomes, this section of the population would suffer privation.

As a specific example of the value of pesticides the United States Department of Agriculture calculated in 1970 that discontinuance of use of the herbicide 2,4-D would add $ 290M to the production costs of bread and other wheat products and would necessitate 20 million extra man-hours of work by farmers and their families without any increase in income to maintain present levels of wheat production. The use of 2,4-D in wheat cultivation has resulted in enough extra wheat in 1975 for 130 billion more loaves than in 1940.

That the adverse effects on the community of lack of crop protection and pest control are not pure conjecture is drawn by the experience in the U.S.A. in 1970 when Southern corn leaf blight got out of control and 20 billion kg of corn were lost. Corn prices rose from 5.3 c/kg to 6.6 c/kg and stayed at that level for a year with a total extra cost to the consumers of $ 2 billion.

Further Reading

British Agrochemicals Association, The Industry's Statistics, (London, 1975)

Food and Agricultural Organisation, Crop Loss Assessment Methods, (United Nations, Rome, 1970)

Green, M.B., Are Herbicides too Expensive?, Weed Science, 1973, 21, 374, (Washington, D.C.)

Green, M.B., Pesticides: Boon or Bane? (Elek Books, London & Westview Press, Boulder, Colorado, 1976)

Gregory, J.G., (Ed), Technological Economics of Crop Protection and Pest Control, S.C.I. Monograph No. 36, (Society of Chemical Industry, London, 1970)

National Academy of Sciences, Contemporary Pest Control Practices and Prospects, (Washington, D.C., 1975)

United States Department of Agriculture, Agricultural Research Service Economic Report, No. 194, (Washington, D.C., 1970)

United States Department of Agriculture, Losses in Agriculture, Agr. Handbook, No. 291, (Washington, D.C., 1965)

United States Department of Agriculture, The Pesticide Review, (Washington, D.C., 1975)

Chapter 3
PESTICIDES AND ENERGY

Agricultural productivity in developed countries is maintained at a level sufficient to feed their populations adequately only by injecting fossil fuel energy into agriculture either directly in mechanical devices such as tractors or, indirectly, as fertilisers and pesticides. Energy is used whenever material is processed or transported so that, at each stage of manufacture, there is an energy input which is carried on to the next stage. The final product has indirect energy inputs from the intrinsic energy of all the hydrocarbon materials used in its manufacture and direct energy inputs from heat and electricity used in processing.

Computer programs have been developed which sum up all these energy inputs for a given chemical compound. These programs try to take all energy inputs into consideration, including energy which goes into building and maintaining the manufacturing plant, as well as other ancillary energy contributions.

TABLE 14

Energy inputs for various crop protection chemicals (GJ/t)

	Naphtha	Fuel Oil	Natural Gas	Coke	Electricity	Steam	Total
MCPA	53.3	12.6	12.0		27.5	22.3	130
Diuron	92.3	5.2	63.1		85.6	28.3	270
Atrazine	43.2	14.4	68.8		37.2	24.7	190
Trifluralin	56.4	7.9	12.8		57.7	16.1	150
Paraquat	76.1	4.0	68.4		141.6	169.3	460
2,4-D	39	9			23	16	85
2,4,5-T	43	2	23		42	25	135
Chloramben	92	5	29		44	0	170
Dinoseb	49	11	9		3	8	80
Propanil	62	3	40		64	51	220
Propachlor	107	14	29		84	56	290
Dicamba	69	4	73		96	53	295
Glyphosate	33	1	93		227	100	454
Diquat	70	1	65		100	164	400
Ferbam			42	3	13	23	61
Maneb	27	9	23	8	25	7	99
Captan	38		14		52	11	115
Methyl parathion	37	2	24	6	73	18	160
Toxaphene	3	1	19		32	3	58
Carbofuran	137	44	63	1	127	82	454
Carbaryl	11	1	48	26	54	13	153

Table 14 shows the energy inputs for some widely-used herbicides, fungicides and insecticides. These figures are for 100 per cent active ingredient. In fact, formulation adds only about 2 to 5GJ/t active ingredient unless large quantities of some particular complex formulating agents are used.

TABLE 15

Energy Inputs to Crop Production (MGJ)

	U.S. total crops (1970)		U.K. total crops (1972)	
Direct use of fuel	686	44%	83.7	35%
Fertilizers	370	24%	91.2	38%
Pesticides	24	1.6%	2.1	0.9%
Machinery	303	20%	30.2	13%
Irrigation, transport, etc.	160	10%	29.8	13%

The total amounts of energy used in U.S. crop production are 2.2% of the nation's total energy usage for all purposes, and in the U.K. are 2.6%, so pesticides use about 0.04% of the nation's total energy.

The figures in Table 15 show that crop protection, including both the chemicals and their application, accounts for only a very small proportion - less than 2 per cent - of the total fossil fuel energy put into primary agricultural production. What do we get out of crop protection in energy terms? The increases in yields and the total or partial losses prevented are very variable quantities, so Price-Jones has suggested a different approach.

The figures in Table 16 show that prevention of 10 per cent loss of maize by use of pesticides saves on balance about 19 gallons of oil per hectare.

TABLE 16

Energy savings in U.S. maize production from pesticide usage (1970)

Average fossil fuel energy input for pesticide-treated maize	29.98GJ/ha
Average fossil fuel energy input for non-treated maize	29.57GJ/ha
Number of hectares of non-treated maize to give yield equal to that of 1ha of treated maize, assuming non-treatment reduces yield by 10 per cent	1.11
Fossil fuel energy input for 1.11ha non-treated maize	32.85GJ
Fossil fuel energy saved by pesticide treatment	2.87GJ/ha (19 gal oil)

TABLE 17

Metabolically utilisable energy of various foodstuffs

Crop	Condition	Metabolically utilisable energy (MJ/kg)	Weight of crop yielding 263 MJ (kg)
Wheat	Unprocessed	14.95	17.6
Barley	Unprocessed	13.10	18.0
Potatoes	Raw	3.18	83.7
Sugar beet	Unprocessed	2.64	99.6
Peas	Cooked	2.05	128.2
Broad beans	Cooked	2.89	91.0
Apples	Raw	1.92	137.0
Cabbage	Raw	1.17	224.8
Carrots	Raw	0.96	274.0
Celery	Raw	0.33	797.0
Mushrooms	Cooked	0.30	876.7

Table 17 shows the additional weights of a number of crops which would have to be obtained to provide metabolically utilisable energy equivalent to the fossil fuel energy which is used up in the production and application of a typical pesticide at 1 kg/ha. To compensate in energy terms for a pesticide applied at 1 kg/ha to wheat, for example, it is necessary to obtain 17.6 kg/ha extra crop. A typical wheat yield in the U.K. is 4300 kg/ha, so 17.6 kg/ha represents a yield increase of only 0.4 per cent. Even with a comparatively low energy-yielding crop like peas, which typically yield about 3000 kg/ha, to compensate for the use of the pesticide the yield increase needed is still only 4.3 per cent.

The crop losses prevented or the yield increases obtained by crop protection are generally very much larger than those that would be needed to compensate for the energy inputs shown in Table 17. Crop protection is consequently a good way to use limited resources of fossil fuel energy because we are using a small amount of such energy to increase greatly the total amount of solar energy which is made available to us via photosynthesis as metabolically utilisable food energy.

Herbicides are essentially used as substitutes for human and mechanical labour, that is, to carry out by chemical means weeding and other cultivation operations. It has been demonstrated by investment appraisal studies that this can result in significant financial savings as a result of the reduced labour requirements. For example, in the U.S.A. it has been estimated that, in 1974, it required 50 man-hours to grow a hectare of cotton compared with 300 man-hours in 1954. The financial gains from increased yields consequent on use of herbicides are well documented, for example, Hurtig estimated $ 58 million increase in the value of crops harvested on 10 million hectares in Canada treated with herbicides in 1960 at a cost of $ 8 million.

A typical value for the amount of energy used in carrying out a mechanical weeding operation is 0.56 GJ/ha (1.5 gal/acre of diesel fuel), and for a typical spraying operation is 0.056 GJ/ha (0.15 gal/acre of diesel fuel). A comparison of the total energy input in carrying out a mechanical weeding with the total energy inputs for carrying it out with herbicides, assuming typical rates of application, is shown in Table 18. Of course, in practice, for weed control throughout a whole season, the number of herbicide applications needed may differ from the number of mechanical weeding operations, the rates of application of herbicide may differ from crop to crop, and the amount of mechanical energy needed may depend on the nature of the terrain and weather conditions. Also if the herbicide can be applied pre-emergently in conjunction with a sowing operation then the mechanical energy of spraying will be saved. Nevertheless, it should be realised that, in general, at least two mechanical weeding operations are required to achieve the effect of one chemical treatment. However, the data in this table will enable the reader to make a fair approximation for any particular case in which he is interested.

Table 19 shows a comparison of energy inputs into mechanical and chemical weeding in forestry. This is based on the use of a Massey Ferguson 165 tractor travelling at 0.56 m/s ($1\frac{1}{4}$ m.p.h.) in a new forestry plantation with a row spacing of 2.1 m. The mechanical weed control achieved 2.6 ha/day and the chemical weed control 13.0 ha/day. The average fuel consumption of the MF165 tractor was 1.27×10^5 J/s (3.0 gal/h). The herbicide used was 2,4,5-T at 3.5 kg/ha active ingredient. The reason why chemical treatment is so much more economical on tractor usage is that the mechanical operation can cope with only one row at a time, i.e. 2.1 m working width, whereas a mist blower on the tractor gives weed control with the herbicide of a 10 m strip for each pass of the tractor. Furthermore, it is usual to have to carry out two mechanical weed control operations per year, but only one chemical treatment.

An interesting case for comparison is the use of paraquat in minimum tillage and direct drilling operations as an alternative to conventional ploughing and cultivation. An APAS/NIAE Farm Mechanisation Study quotes diesel fuel consumption figures for a range of large farm tractors operating on both light and heavy soils. Taking an average figure for a medium soil the amounts of diesel fuel consumed for conventional ploughing and cultivation and direct drilling respectively are shown in Table 20. To the direct drilling figure has been added the energy content of the paraquat used to give a total energy input.

TABLE 18

Energy used in mechanical and chemical weeding

Method	Fuel (litres/ha)	Energy (GJ/ha)	Total
Mechanical	16.9	0.56	0.56
Chemical Herbicide (MCPA 0.75 kg/ha=0.67 lb/acre)	1.7	0.06 0.10	0.16
Chemical Herbicide (Diuron 2.3 kg/ha=2.0 lb/acre)	1.7	0.06 0.62	0.68

TABLE 18 (Continued)

Method	Fuel (litres/ha)	Energy (GJ/ha)	Total
Chemical	1.7	0.06	0.27
Herbicide (Atrazine 1.13 kg/ha=1.0 lb/acre)		0.21	
Chemical	1.7	0.06	0.23
Herbicide (Trifluralin 1.13 kg/ha=1.0 lb/acre)		0.17	

TABLE 19

Comparison of mechanical and chemical weeding in forestry

	Mechanical	Chemical
Area treated per day	2.6 ha	13.0 ha
Energy used per 8 h day by tractor	3.48 GJ	3.48 GJ
Energy used per ha	1.34 GJ	0.27 GJ
Energy in 2,4,5-T used (3.5 kg/ha=3.1 lb/acre)		0.47 GJ
Total energy for two mechanical weedings	2.68 GJ/ha	-
Total energy for one chemical weeding	-	0.74 GJ/ha
Annual energy saving for chemical weeding	-	1.94 GJ/ha (=13 gal/ha of diesel fuel)

Energy content of 2,4,5-T=135 GJ/t.
Other data supplied by U.K. Forestry Commission.

TABLE 20

Energy used in conventional cultivation and direct drilling

Ploughing and cultivating		
Operation	Fuel (litres/ha)	Energy (GJ/ha)
Ploughing	22.5	0.75
Heavy cultivating (1 x 2)	22.5	0.75
Light harrowing	5.6	0.19
Drilling	11.2	0.38
Light harrowing	5.6	0.19
Total		2.26

TABLE 20 (Continued)

Direct drilling		
Operation	Fuel (litres/ha)	Energy (GJ/ha)
Spraying	1.7	0.06
Drilling	11.2	0.38
Harrowing	5.6	0.19
Paraquat (0.84 kg/ha =0.75 lb/acre)		0.39
Total		1.02

The results suggest that direct drilling and minimum tillage techniques can save about 1.0 GJ/ha of total energy (equivalent to 2.7 gals/acre of diesel fuel.

Further Reading

Green, M.B., Eating Oil, (Westview Press, Boulder, Colorado, 1977)

Green, M.B., Energy in Agriculture, Chem. & Ind., 1976, 641, (London)

Green, M.B. & McCulloch, A., Energy Considerations in the Use of Herbicides, J. Sci. Fd. Agric. 1976, 27, 95, (London)

Chapter 4
TYPES OF PESTICIDES

Classification

"Pesticide" is an ugly word but is with us to stay as an ombibus term to describe the various chemicals used in crop protection and pest control. The US Federal Environmental Pesticide Control Act defines a pesticide as (1) any substance or mixture of substances intended for preventing, destroying, repelling or mitigating any insect, rodent, nematode, fungus, weed or any other form of terrestrial or aquatic plant or animal life or virus, bacteria or other microorganism which the Administrator declares to be a pest, except viruses, bacteria or other micro-organisms on or in living man or other animals (2) any substance or mixture of substances intended for use as a plant regulator, defoliant or desiccant.

Broadly, the term includes all chemicals used in horticulture and agriculture, except fertilizers and veterinary products for internal illnesses or parasites, and all chemicals used to control pests of any kind, except internal pests or parasites of men and animals, in any non-agricultural situations.

The term "pest" is not descriptive of any intrinsic characteristics of a particular organism but merely of the way in which it behaves in certain circumstances. In broadest terms, any living organism which is somewhere that you do not wish it to be doing something that you do not wish it to do, is a "pest". "Weed" is not a description of a particular type of plant, but of behaviour, and a weed has been defined as "any plant growing in a place where you do not wish it to grow".

The main pests of economic importance are weeds, fungi and insects but there are also mammals, birds, molluscs, mites, nematodes, bacteria and viruses. The main groups of pesticides are therefore herbicides (weeds), fungicides and insecticides with minor groups rodenticides, avicides (birds), molluscicides, acaricides (mites), nematicides, bactericides and antivirals.

Pesticides are often sub-classified according to their "mode of action". However, this phrase means different things in different scientific disciplines, and misunderstanding often results. The field biologist may distinguish some herbicides as "pre-sowing" or "pre-emergent", being those best suited to apply to the soil to kill seedlings in the absence of a crop, and some as "contact", being those applied, with good coverage, to foliage to kill all leaves (of susceptible species) contacted but which probably allow regrowth from perennial rootstock. There are several equivalent classes of this kind and subdivisions dependent on, e.g. persistence. If a pre-sowing herbicide is a persistant one no crop could be sown shortly afterwards unless of a biochemically resistant species. Thus atrazine can be applied to soil before or after sowing maize because the crop is resistant to it, but paraquat can be used to clear ground before any sowing because it is inactivated by soil.

The biologist may describe an insecticide as "contact", usually adding the adjective "residual", if it stays on the surface of walls and kills insects which walk thereon, as a "stomach poison" if it kills caterpillars which eat treated leaves or as "systemic" if it enters the tissue fluids of the host plant and kills sucking insects on parts of the plant not directly treated.

The biochemist, on the other hand, will, under "mode of action", distinguish different disturbances of essential biochemistry which lead to the final observed symptoms. Some compounds arrest photosynthesis, others inhibit oxidative phosphorylation, others the enzymic hydrolysis of acetylcholine etc.

In this predominantly chemical book it might seem natural to adopt the biochemical classification, but this is not a book about biochemistry but more a book for chemists about pesticide practice and for the pesticide practitioner about the chemistry he is called upon to use. Our interest is therefore mainly in the field biologists' classification but we must try to sort out some inconsistencies which have arisen during the development of the subject - of the kind which tries to distinguish between green things and square things. For example "systemic" and "contact" insecticides are often contrasted, but "systemic" refers to behaviour of the compound in the tissues of the plant or animal host while "contact" refers to the mode of entry into the pest insect. We can quote no example of an insecticide arriving in a leaf only via the systemic route (from application to root or to another leaf) and killing an insect by contact, but such behaviour might be found. Certainly some insecticides have a useful direct contact effect (e.g. dimethoate on flies) but also have a useful systemic behaviour, killing sucking insects after systemic transfer.

Confusion arises mainly because the differences are not in the action of the compounds within the final target but in the means of transfer from the site of application to the target. To describe an insecticide as having "vapour action" is misleading. No "action" goes on in the gas phase, only transfer. What is meant is that the compound can reach the target from the site of application via the vapour. We know of no insecticide which has, exclusively, "stomach" or "contact action". The insecticide can get into the vital tissues from the body integument, from the tarsi, by ingestion or (as vapour) through the trachea. Which route is most important depends on the situation - i.e. the habitat and behaviour of the pest, the habit of the host and the means of application. It is not a property descriptive of the insecticide as such. One insecticide may best be used, and be the best to use, in a situation where direct contact is the main means of transfer, another where systemic transfer is called for, but change the situation and the choice may change.

"Volatile", "Superficial", "Systemic" might be the best main classifications of insecticide and fungicide, "Contact" and "Systemic" parallel classification of herbicides but the distinctions are not clear. We will consider the first of these at some length to illustrate the dependence on situation.

Volatile Pesticides

Some pesticides, in some applications, redistribute themselves in the environment by purely physical processes. These are compounds used as fumigants. They do not form a biological class since they are used against many widely different organisms. They do not form a chemical class. More logically they can be regarded as a physical class but this defines the compound and method of use together, rather than the compounds themselves. Obviously a fumigant must be sufficiently volatile to be effective in any particular situation but there is no absolute limit of volatility. The words "volatile", "involatile", "of negligible vapour pressure", are used very carelessly in a good deal of pesticide literature, and the last phrase particularly can be very misleading. The laboratory chemist naturally regards a substance as involatile if it can be handled by ordinary laboratory procedures and in at least milligram quantities without special precautions against loss. It may nevertheless be quite sufficiently volatile for a lethal dose to be transmitted through the vapour phase under favourable conditions. Thus, if flies are caged and the cage suspended in a closed vessel containing lindane they will be dead in a few hours although no direct contact has been made. When triallate is incorporated in the surface layer of soil to control wild oat, there is good reason to believe that the main route of access of the compound to the seedling is through the air spaces in the soil and the cuticle of the shoot rather than through the water phase and the root. Neither compound would be classed as volatile on ordinary laboratory standards.

Calculation, as well as example, may serve to emphasize this point. The vapour pressure of lindane at 20°C is 9×10^{-6} mm Hg, corresponding to a saturation concentration in air of 0.15 mg/m^3. If a man, breathing at an average rate of 15 l./min, were to absorb into his lungs all the lindane from a saturated atmosphere, he would take in only 3 mg in 24 hr. Several years would be necessary to take in an amount which would be lethal if ingested as a single dose, and the rapid excretion of most of the compound would make this prediction unrealistic. If a pellet of this substance 2 mm in diameter were suspended freely in air it would lose by evaporation at ordinary temperature only about 0.8 μg in 24 hr.

In both of these examples, lindane could apparently justifiably be said to have "negligible volatility". Suppose, however, that the 2 mm object in air is the body of a small insect and that it can take up the substance from saturated air just as efficiently as a pellet of the substance can lose weight into pure air. It gains 0.8 μg in 24 hr, but its body weight is only 4 mg. Its content of toxicant at the end of 24 hr is 200 mg/kg body weight, well into the lethal range for insects. In this context the volatility is far from negligible.

It may seem unrealistic to consider insects having free access to the saturated vapour in simple laboratory vessels, because, in nature, there will be many other substances present - soil particles, crop leaves, etc., all having adsorbing surfaces. The significance of these depends on whether the environment permits only reversible adsorption or whether chemical decomposition of the adsorbed substance occurs. It depends also on the distance through which the toxicant must diffuse and on the effect of speed of intake on toxicity. Lindane would be quite ineffective in killing insects distributed in a store of grain if it were itself only applied to an exposed surface. An intrinsically less toxic but much less adsorbed toxicant, such as methyl bromide, is far more effective. On the other hand, applied to the curling leaves in the centre of a young Brassica plant, lindane is effective even against aphids and the necessary transfer of toxicant occurs mainly in the vapour phase. An equal dose of methyl bromide would evaporate during spraying before the target was reached.

An insecticide having more rapid vapour action than lindane but not normally considered a fumigant is dichlorvos. It has a vapour pressure of 0.01 mm mercury at 20°C. The saturated vapour is rapidly lethal to insects but eventually lethal also to mammals. By using a slow-release source, the vapour concentration in ordinary rooms can be held in a range lethal to insects but harmless to mammals. A special formulation of this compound consisting of thick plastic pads containing the insecticide in slowly diffusible form can be hung up on walls, in number depending on room size, and effectively keep the rooms free of flying insects.

Volatility not only provides in some situations an effective means of transfer, it also provides a means of loss from the gross target. This is not wholly disadvantageous since toxic residues can in this way be reduced more quickly. Compromise must be made between conflicting requirements. Some compounds are so volatile that they are useful only in closed environments, from rabbit burrows to warehouses.

Transport Systems in Plants

There are two long-distance transporting systems in higher plants. The xylem system consists of continuous tubes formed of dead cells which carry water and mineral nutrients from root to leaf in response to some pumping mechanism in the root and suction due to evaporation from the leaf. The phloem system is more complex, less well understood, has a flow rate of at most a few centimetres per hour and is concerned in the transport of products of photosynthesis from mature leaves to growing tissues. The content of the phloem vessels is a fairly concentrated (about 16%) solution of sucrose - the universal fuel of plant metabolism - together with much smaller concentrations of amino acids and proteins. It is only in these systems that long-distance transport can take place and, as the moving fluid is essentially aqueous, only water-soluble materials can be transported.

The xylem vessels can transport any water-soluble substance. This is particularly easily demonstrated by the schoolboy device of putting the cut stalk of a white flower in a bottle of red ink. Within a half hour or so the main veins of the white petals have turned bright pink. This demonstration cannot be repeated if an intact root system is immersed, or if a damaged root system is placed in soil moistened with red ink. In these more realistic experiments water enters the xylem vessels by diffusion through cellular outer tissues which form an effective barrier to large molecules. Entry into the xylem rather than transport by it is the limiting factor. If compounds are to be effective systemically after application to foliage, they must enter the more elaborate phloem vessels after first diffusing through the cuticle or through fine protoplasmic strands which penetrate it and then through or around the epidermal and mesophyll cells.

The blood stream of mammals is a very much more rapid means of distribution than the xylem and phloem streams of plants. It has to be in order to fulfil its primary function of oxygen transport, whereas plants rely on a thin and wide extension of their actively growing tissues to secure adequate oxygen (and carbon dioxide) exchange with the air by diffusion. The blood is, moreover, adapted to the transport of liquid fats in emulsified form. Compounds can therefore be effective systemically in mammals even when they partition favourably to oil from water.

Insecticides Systemic in Plants

The first example of systemic behaviour in plants might be considered to be demonstrated by the natural toxicity (to mammals as well as insects) of crops growing on certain soils with a high selenium content. The first commercial exploitation of systemic behaviour of a deliberately applied compound was made with schradan. It was outstandingly successful in the protracted control of the cabbage aphis, *Brevicoryne brassicae*, without damage to the natural enemies of this pest.

All compounds effectively systemic in plants are much more soluble in water than oils, or are converted in the plant to water-favourable substances. Thus dimefox and schradan partition very strongly in favour of water from mineral and glyceride oils, dimethoate strongly so. Demeton and its relatives are rapidly oxidized at the thioether position to form the strongly water-favourable sulphoxides. The oxidation products are the systemic compounds. The applied demeton itself is effectively a transient superficial insecticide and much less selective than schradan. Amiton, in which the C_2H_5S of demeton is replaced by $(C_2H_5)_2N$ is water-favourable by virtue of the basic character of the tertiary amino group which is cationic at plant sap pH. This compound is a very effective and persistent systemic aphicide but too toxic to mammals for its use to be permitted.

For translocation within the plant, it is water:oil partition ratio rather than absolute solubility which must be high. The majority of good systemic insecticides are effective at a concentration of at most a few parts per million in plant sap. If, however, they were strongly oil-favourable, they would be held up in the lipid membrances, organelles and deposits in the plant tissues. The slow-moving water in the conducting vessels would not be able to translocate them effectively. The behaviour is similar to that in a chromatographic column where only compounds partitioning favourably to the moving phase can progress nearly as rapidly as the moving phase itself.

There is no really sharp distinction between "systemic" and "superficial" insecticides. Only exceptionally water-favourable and stable compounds like dimefox can be effective at remote sites, for example killing aphids on mature trees after application to the soil. Oil solubility, however, does not arrest transport, but only retards it, and unstable compounds can diffuse through short distances before decomposition. Thus even the oil-favourable lindane and the very rapidly hydrolysed TEPP can exert useful "translaminar" action, as it is called, i.e. they can kill insects feeding on the surface of a leaf after application only to the reverse surface. They cannot, however, exert useful control on leaves other than those actually sprayed.

Systemic insecticides, to be of value for control of insects in agriculture, must not damage the crop. Ideally, therefore, they should be completely without physiological effect on plants, although this ideal is never fully attained. Being almost without effect on the plant, they diffuse passively within the cells or through the apoplast between them. Their penetration and translocation is governed mainly by solubility, partition properties and molecular size and shape. Since diffusion through cellular tissue into the conducting tissue must occur, it is not surprising to find that diffusion from one set of conducting vessels to the other also occurs. Distribution becomes rather general. In all rapidly growing crop plants, which must have adequate water supply, the net water movement is upward and outward. This produces a tendency for the systemic insecticide to accumulate in rapidly transpiring young, but fully developed, leaves. There is no net movement down into the root. Even a soil-applied systemic insecticide is not accumulated in the root, but in the leaves a much higher concentration can build up than that present in the soil water.

Systemic insecticides are therefore effective against leaf- and stem-sucking insects but have not achieved any success against root-feeding insects. Even the aphids, which are in general most vulnerable to systemic insecticides, are safe from this method of attack when they inhabit roots only. The systemic insecticide has as yet made no contribution to the control of the root-sucking aphids responsible for the phylloxera disease of vines.

The water-soluble insecticides which behave systemically are more effective against the sucking insects, aphids and mealybugs, than against grossly phytophagous insects. It seems that the more generally effective insecticides are dominantly oil- rather than water-soluble, even when judged by direct spraying tests under laboratory conditions. When used by the systemic route only - for example by application to the soil to kill leaf-sucking aphids - an additional factor comes into play.

The sucking insects take in a very large supply of sap from the phloem vessels. Indeed they do not, strictly speaking, suck. They probe with their hollow stylets selectively into the phloem vessels where the strong sugar solution is under high osmotic pressure. The host then in fact pumps food solution into the passive insect. This device of the aphis is used by the plant physiologist to obtain samples of pure phloem contents. He lets the aphis establish its feeding posture and then cuts the head and body off the inserted stylets, from the cut ends of which the solution can be collected.

The phloem solution is over-rich in sugar for a complete animal diet and deficient in protein. The aphid metabolism rejects most of the sugar from a very massive intake, filtering out the more valuable nitrogen compounds. The rejected sugar, which is squirted out as fine droplets, forms the "honeydew" associated with a heavy aphid infestation. This is responsible for the dirty appearance of aphid-infested trees, since it collects mineral dust and cultures dark-coloured superficial fungi. A successful systemic insecticide is retained along with the food compounds and the aphis thus accumulates a disproportionate dose as compared with a grossly phytophagous insect taking a more balanced diet.

If attack on root-feeding insects or the very important nematodes is ever to result from spraying the aerial parts of a plant, it seems necessary either that the compound applied must be in some way fixed in the root tissue or that it should exert its effect indirectly via the chemical processes of the host itself. The phloem system can transport sucrose very effectively from leaves to roots and other necessary chemicals in much lower concentration, but none of these can accumulate in the roots unless they are fixed there or converted chemically to fixed substances - as occurs, of course, when new root tissue is formed from substances mainly supplied from the leaves. One should not dismiss the possibility of the discovery of a compound that will be so fixed and also be insecticidal, but the solution to the problem will depend on biochemical, not plant-physical, processes. One evident danger is that a compound so fixed, being involved in the chemical processes of the host, may well have phytotoxic properties.

Systemic Fungicides

Until recent years all commercially used fungicides were protective and unless sprayed in advance to stop the initial infection were of very little curative value except against very superficial fungi like the powdery mildews. During the past decade a large number of compounds have been discovered which move in the plant and act systemically on the fungal disease. These are described in detail in the chapter on systemic fungicides.

Contact Herbicides

It is doubtful if any herbicide has a truly local action. If it did, it could not be lethal to plants except following a coverage so uniform that it could not be realised in field practice. The incidental damage sometimes arising from insecticide or fungicide application, often referred to as "phytotoxicity" which should, of course, include all herbicide action, is more strictly local, often resulting in necrotic spots under the spray residues.

The dinitrophenols, cyanophenols and pentachlorphenol are usually called contact herbicides because they usually kill off top growth only and need reasonably good spray coverage to do so, but some translocation obviously occurs. Even paraquat has been called a contact herbicide despite clear evidence of upward translocation with lethal effect. The reason for this misnomer is that spray drops on many species do produce local scorch spots before the general systemic kill.

Systemic Properties of Herbicides

The translocation of herbicides is considerably more complex than that of the present plant-indifferent insecticides. A herbicide necessarily interferes with the plant's physiology and many, particularly both "hormone" and "scorching" herbicides, interfere drastically with their own translocation.

Despite the low solubility, due to high melting point, of the substituted phenylurea and aminotriazine compounds, they are water-, rather than oil-, favourable and are truly systemic. The low solubility restricts their effectiveness by foliage application, but they are taken up from dilute solution in the very extensive volume of the soil water and translocated to the photosynthetic cells of the shoot. There, non-interference with function, except that of chloroplasts, enables them to move in the xylem stream as inert compounds.

MCPA and 2,4-D are usually looked upon as translocated herbicides but, although they can enter leaves and produce bending of stems both above and below the leaves treated, they have a strong affinity for root tissue and are held there. They are not easily translocated from root to shoot and hardly at all from one part of the root system to another. They can effectively kill plants after foliage application, but they do so by concentrating in the upper part of the root system, to which they do irreparable damage, rather than by migrating throughout the whole plant. Death of a plant is very unsatisfactory evidence for systemic distribution of the herbicide. Death of aphids feeding on every part of a plant is very clear evidence of systemic behaviour of an insecticide. Not only is systemic movement of herbicides more complicated than that of insecticides but it produces results less easy to interpret without ambiguity.

Not all "hormone" herbicides are strongly fixed in particular tissues. CMPP and 2,4,5-T appear to be generally more mobile than MCPA and 2,4-D. 2,3,6-TBA is extremely mobile and always moves into, and deforms, newly developed shoots while having very little direct effect on root development.

The systemic effect of herbicide is often markedly dependent on season and on stage of growth of the weed. The effectiveness of MCPA, for example, on perennial weeds such as nettle is greatest when the herbicide is applied to young leaves early in the season or to mature, but not senescent, leaves late in the season. This herbicide is much more effective on bracken (*Pteridium aquilinum*) in early autumn than at any other time. Amitrole is more effective against perennial weeds when sprayed in autumn. This compound is, under favourable conditions, very effective in killing underground tissue after being applied to the leaves. It is probably fixed in roots by conjugation with sugars. Glyphosate also shows outstanding power to kill underground tissues when applied to leaves.

Nomenclature of Pesticides

Manufacturers market pesticides under trade names, which refer to the particular formulation, not specifically to the active ingredient. Thus, one active ingredient may be marketed in several different formulations under several different trade names, and often, with different trade names in different countries. When a pesticide comes out of patent protection and can be manufactured and sold by anybody who wishes to do so, this situation is multiplied. The result is that, in the U.S.A. for example, there are 800 registered active ingredients and 80,000 different trade names. This causes confusion to the farmer who wants to know exactly what he is buying so that he can compare costs accurately. Rapid and precise identification in case of accidental spillage or ingestion is also necessary. It is desirable, therefore, that there shall be some internationally agreed common name for every active ingredient and that the common name shall uniquely specify a definite chemical compound. This aim has not been entirely achieved.

The International Standards Organisation Technical Committee 81 is supported by most countries except the U.S.A. and U.S.S.R. It tries to get agreement on common names for pesticides between the National Standards Organisations of the various participating countries. In the U.K., the British Standards Institute is responsible through its committee PCC/1. It is often a long and tedious procedure to get agreement between member countries.

The U.S.A., through the American National Standards Institute Committee K62, decides its own common names for use in the U.S.A. In the past there have been considerable differences between A.N.S.I. and I.S.O. but they now work much more closely together and most common names are now the same in the U.S.A. and the rest of the world. The U.S.S.R. follow their own course without reference to anybody and there are considerable variations from I.S.O., although many I.S.O. names are used in the U.S.S.R.

The problem of coining a short common name becomes progressively more difficult as it must (a) be easily pronounceable in any language (b) not resemble any word in any language too closely (c) not conflict too closely with any trade-mark in any country (d) desirably have some relationship to the chemical name.

Nearly all registration authorities now require that labels shall bear, in addition to the trade-name, the percentage composition of active ingredient in terms of its common name.

Throughout this book, all pesticides are referred to by their common names.

Further Reading

BSI Standard 1831 and Supplements, Recommended Common Names for Pesticides, (British Standards Institution, London, 1969 onwards)

Frear, J., Pesticide Index, (Entomological Society of America, Washington, D.C., 1977)

Kilgore, W.W. & Doutt, R.L., Pest Control, (Academic Press, London & New York, 1967)

Martin, H. & Worthing, C.R., Pesticide Manual, 4th Edition, (British Crop Protection Council, 1974)

Martin, H., Scientific Principles of Crop Protection, (Arnold, London, 1928 and later editions)

National Academy of Sciences, Scientific Aspects of Pest Control, (Washington, D.C., 1961)

Pyenson, L.L., Elements of Plant Protection, (Chapman & Hall, London & Wiley, New York, 1951)

Rose, G.J., Crop Protection, (Hill, London, 1963)

Thomson, W.T., Agricultural Chemical Books I to IV, (Thomson Publications, Fresno, California, 1976)

Woods, A., Pest Control, (McGraw Hill, London & New York, 1974)

Chapter 5
OILS AS PESTICIDES

Except within the narrow context of some particular technology, where, for example, the motor mechanic has a particular conception of "gear oil" or the perfume manufacturer of "oil of lemon", the word "oil" has a wide and not very precise meaning. When one "pours oil on troubled waters" or speaks of the "oil phase" of an emulsion one is thinking quite generally of any liquid of very low water solubility and usually of low volatility. The chemist restricts the word to liquids which are predominantly hydrocarbon in composition but which may have a small proportion of oxygen in ester groups, as in the liquid glycerides, or even in keto or hydroxyl groups, as in many essential oils and castor oil.

Most vegetable and animal fats, the triglycerides, are at least partly liquid in their natural situation, where they are mainly stored in fairly massive layers as reserve energy foods or for thermal insulation. More complex oily substances are more widely distributed in living tissue, in the very thin membranes which keep appropriately separated the important biochemically active substances in the aqueous phases of cells. Oils are therefore by no means foreign to living organisms. Like many vital substances, they can, out of place, have lethal disorganizing effects.

If kerosine is spilt on almost any green leaf, the latter, if held up to the light, will be seen to become more transparent in patches. The oil displaces air from the spaces between the mesophyll cells, reducing the difference of refractive index. The process continues, not through further direct spread of oil but because the disorganized cells now release water, a further consequence of which is that the leaf wilts more rapidly than an untreated one, due to increased rate of evaporation of water.

Insects, particularly small ones, are even more vulnerable to applied oils, because they must rely on an even more efficient organization of native fatty molecules to prevent disastrous water loss. Oil can also flood the spiracles (breathing pores) of insects, as they do the stomata of leaves, and produce, in this case, rapid asphyxiation. A further effect is mechanical. In small-scale structures, surface tension of liquids becomes a very important force. Most insect cuticle is not wetted by water, a property necessary to prevent legs and other protruding organs becoming entangled by surface tension forces much greater than the muscle forces available. Oils do wet the cuticle.

The suffocating and entangling effects of oils are used together in the practice of spreading of oils on water surfaces in which mosquitoes breed. The larvae of these insects are free swimming but must take their oxygen directly from the air. The spiracles are united to an extended breathing tube at the posterior end which has a sharp and unwettable tip. When this is forced into the surface of clean water it connects with the atmosphere and anchors the insect in the surface so that it can remain for long periods at rest. Oil spread on the water surface blocks up the breathing tube and at the same time prevents the insect from anchoring itself in the surface. Other insects, such as Gerris spp., which skate upon water of normal high surface tension, are also incapacitated by an oil film. This method of destroying mosquito larvae in small water volumes (rain-butts, puddles, etc.) is still widely used. The oil must spread spontaneously on water. Refined mineral oils therefore require the addition of a small percentage of fatty acid or other oil-favourable surface-active compound. Waste engine oil from automobile maintenance is effective without adjuvants, being contaminated with oxidation products, and is widely used. Proprietary oils for mosquito larva control contain a chemically active insecticide to improve their action.

In control measures which rely on the gross "physical" toxicity of oils, it is the "oiliness" - the antithesis to water - which is important. Various oils differ, however, in their effectiveness and not solely because of differences in viscosity, volatility and surface behaviour. Associated chemical effects, not understood mechanistically, play an important part. Thus many plant leaves are undamaged by light applications of refined, saturated mineral oils, particularly when these are applied as emulsions. Unsaturated oils, aromatic oils and particularly those containing phenolic substances are much more damaging. Some essential oils, e.g. oil of citronella, have use as insect repellents. In this case, subtle and specific sensory responses to the vapour are called upon and the classification "oil" is almost accidental. Oils are, of course, widely used as solvents for more active pesticides, a subject dealt with in the chapter on formulation.

Vegetable oils had at one time a minor use in the control of powdery mildews. Soaps formed by their hydrolysis were moderately effective against some Botrytis fungi, the spores of which become lethally engorged with microscopically visible oil globules. Nearly all oils of pesticidal interest are, however, now derived from "mineral" sources - i.e. from distillation of the fossil fuels, coal and petroleum. The requirements of essentially paraffinic oils have come mainly from petroleum and of aromatic oils from coal tar. These distinctions, however, become increasingly less clear as both industries develop improved means of separation of valuable chemical intermediates from the complex mixtures of hydrocarbons present in the raw material and produced by pyrolysis or "cracking".

The largest tonnage of oils applied to the land is contributed by low-value, low-volatility by-products of petroleum refining which are used to bind soil to form primitive roadways in dry areas. The toxicity to vegetation of these heavy applications of crude oil provides a herbicidal bonus in this practice which is mainly confined, for economic reasons, to regions not far from oil-fields or refineries. More refined heavy petroleum oils are applied to the lanes between orchard trees in arid areas, and therefore mainly to citrus orchards, to reduce water loss directly from the soil, again with a herbicidal bonus. Low-grade oil fractions high in content of phenolic substances have been used for total weed suppression on gravel and earth roads on both domestic and industrial sites but this function is now taken over by much smaller dosages of less messy and objectionable but highly phytotoxic substances. One should perhaps include among herbicidal applications the use of oils as fuels in the flame gun to destroy top growth by heat. Research is actively carried out by the oil companies to make this process usefully selective by careful choice of flame temperature and time of exposure.

The next most massive use of oils as pesticides is that of coal-tar creosote for preservation of wood against fungal rots and insect attack. This application is confined to outside timber structures, farm buildings, telegraph poles, railway sleepers, etc., because staining and smell make it unsuitable for domestic timber. Its use is decreasing as mineral and synthetic materials take over some of these outside timber functions and as it becomes more worthwhile to work up crude coal-tar fractions for more valuable purified intermediates. The usage, however, is still enormous. Approximately 800 million litres of creosote, a substantial fraction imported, are used each year in the U.S.A. for timber preservation. The amount so treated is about 1% of the total weight of domestic production of sawn timber.

Creosote (a crude mixture of aromatic hydrocarbons with a minor proportion of phenol and naphthol homologues) is not very active in comparison with most synthetic pesticides, but it is cheap and can therefore be used in high dosage. The best results are obtained by pressure impregnation or heating and cooling in the liquid. Well-impregnated timber may contain several per cent of the oil. There is, incidentally, a good technical reason why timber preservation should be carried out with cheap, although not very active, chemicals in massive dose rather than with very active chemicals in much smaller dose. Adsorption on to the large internal surface of porous wood is high so that low doses have no possibility of deep penetration.

More active chemicals are mainly used in the impregnation of timber for domestic building, but are usually provided in oil solution since oils, because they do not swell the wood fibres themselves nor quickly penetrate their microstructure, spread throughout the gross porosity of the wood more quickly than aqueous solutions. Aqueous treatments, if heavy enough to be useful, also produce undesirable warping. Pentachlorophenol is widely used as a good fungicide very effective also against termites which are a major factor in wood deterioration in the subtropics. Organochlorine insecticides are added to kill wood-boring beetles.

The good fungicidal properties of copper are used in oil treatments by forming the oil-soluble copper naphthenates (naphthenic acids are mixtures of cyclic carboxylic acids derived from certain petroleum fractions by oxidation and alkali extraction). The green colour of such treatments is for some purposes objectionable. Zinc naphthenates provide a less active but colourless substitute, but they are now being displaced by triphenyltin derivatives. Proprietary oil formulations for treatment of timber may contain mixtures of these compounds.

Oil washes have been used for a long time to control some insect and spider-mite pests in orchards. Only light applications of emulsified paraffinic oils can safely be made to trees in leaf but these so-called "summer oil" treatments are effective against red spider. After a period of displacement by more active chemical acaricides, oils are now coming back into use because of the ability of this rapidly breeding pest to evolve strains resistant to chemical attack. Summer oil is now the most widely used acaricide in the U.S.A. It may be added that its rate of application may be reduced, with greater safety to the foliage, by dissolving polyisobutene in the oil. This solution leaves a permanent sticky deposit (the "active ingredient" of self-adhesive tapes and bandages) which immobilizes the pest.

While refined oils of limited activity must be used in summer, cruder and more aggressive oils are tolerated by the dormant trees of deciduous orchards in winter. Several important pests overwinter on the bark in the egg stage and "winter washes" are effective against these. Although tar oils with a high phenolic content are more effective against aphid eggs, the more paraffinic petroleum oils have a greater action on acarid (spider mite) and capsid eggs. Mixtures are preferred and addition of dinitrocresol extends the range and certainty of control. The action is certainly partly physical and best results are obtained when the emulsion used for spraying is rendered unstable by exposure so that a thin, continuous oil layer, able to penetrate into crevices, is left on the bark. Winter oil washes have tended to be displaced by systemic insecticides applied in the spring, but some orchardists, particularly on the European continent, are tending to return to the winter oils because of the development of resistance of insects to the alternative chemicals.

Another use where the physical property of "oiliness" is partly responsible for the toxic action is that of petroleum for selective weed control. A special fraction, less volatile than kerosine and containing a moderate proportion of unsaturated compounds, derived from a particular oil-field, gives good weed control in carrots. It is also used to a small extent, in other crops of the family Umbelliferae. Many farmers have been successful with a home-made emulsion of tractor vaporizer oil. It is not known why seedlings of the family Umbelliferae should be so much less vulnerable to oil damage than most other seedlings, but it may be associated with the fact that these plants are naturally rich in oils and their seeds particularly so. This use is declining as other, chemically active, herbicides have been discovered which are more reliable.

Liquid fatty acids in the C_6-C_{12} range are used as emulsions in the tobacco crop to inhibit the development of lateral shoots. Their action is not highly specific and they are sprayed at rates allowing accumulation in the leaf axils. The action may be essentially a physical one. Other oils have been used for this purpose although with more risk of damage to the leaf. The practice is known as chemical "anti-suckering". The methyl esters of acids in the C_8-C_{12} range are likewise used as a chemical "pinching" agent on ornamental plants such as azaleas.

An interesting recent hypothesis is that the sensitivity of various plants to certain soil-applied herbicides may depend on the amount of internal lipids in the plants and that this sensitivity, and therefore the selectivity of the herbicide, might be altered by adding suitable lipids around the roots of the crop seedlings, or by treating the seeds with lipids before sowing. This effect has been demonstrated with the herbicide trifluralin on barley, wheat and cotton seedlings.

Further Reading

American Chemical Society, Agricultural Application of Petroleum Products, (Washington, D.C., 1952)

Findlay, W.P.K., Timber Pests and Diseases, (Pergamon Press, Oxford, 1967)

Guthrie, V.B., (Ed), Petroleum Products Handbook, (McGraw Hill, London & New York, 1960)

Chapter 6

SYNTHETIC INSECTICIDES: MISCELLANEOUS AND ORGANOCHLORINES

Inorganic Compounds

The first synthetic insecticides were inorganic. In 1867 the pigment Paris green, a crystal compound of acetate and arsenite of copper, having approximate composition $Cu_4(CH_3CO_2)_2\ (AsO_2)_2$, was used successfully in the U.S.A. against the increasing population of Colorado beetle in the potato fields. It was later used against a wide variety of leaf-eating insects and against codling moth larvae on apples.

Paris green can make little claim to be specifically insecticidal. Rather is it a generally toxic compound which kills leaf-eating insects shortly after application but is relatively harmless to the eventual human consumer of the mature product. Selectivity depends upon timing and placement of the poison and on the feeding habits of the species attacked. The insect is much more voracious than man, consuming a much greater weight of fresh vegetable matter in relation to its body weight and its diet is restricted to a particular crop. The insect moreover eats the crop when it is freshly contaminated. Man eats only a portion of the crop, after the lapse of a safety interval. Tubers, roots and fruit have increased in size and were either not touched by the spray or have been subject to weathering since spraying.

Various other arsenites and arsenates were tested but many produced troublesome and unpredictable damage to the crop with symptoms of arsenic poisoning. Atmospheric carbon dioxide and exudates from the leaf cells are probably responsible for liberation of soluble arsenate from combination with the heavy metal. Lead arsenate, $PbHAsO_4$, first used against the gipsy moth in 1892, established itself as the compound safest to the crop but the addition of lead to the consumer hazard has never been popular among safety authorities. The use of lead arsenate, still the most effective product against codling moth, is nowhere prohibited, but a maximum level of residual contamination of the marketed produce is set in most countries. Washing of apples before marketing is widely practised to ensure conformity.

Sodium fluoride has been known, at least since 1842, to be highly toxic to insects. It has been used mainly in situations where a concentrated bait or barrier is effective - e.g. against cockroaches, earwigs, ants and other gregarious crawling species. It has limited value out of doors because it is too soluble (about 2%) in water to persist except in arid districts. The almost insoluble silicofluoride Na_3SiF_6, and native cryolite, Na_3AlF_6, are much more persistent and still sufficiently rapid in action to be useful. Silicofluoride is a by-product of the fertilizer (superphosphate) industry, the fluoride of the mineral apatite, $Ca_4F_2(PO_4)_2$, a component of all phosphate deposits, being released as fluosilicic acid during treatment of the phosphate rock with sulphuric acid.

Small terrestrial insects are very dependent on an organized fatty cuticle to prevent evaporation of body water. Many mineral dusts exert some insecticidal action by damaging this protective layer. The damaged cuticle is usually more permeable to insecticides, particularly if they are highly polar compounds. The formulation of fluorides with abrasive or adsorptive mineral dusts increases their effectiveness in dry situations and such dusts are used for protection of roof timbers against termite attack. A mixture of silica aerogel and ammonium fluosilicate is used as a dust or aerosol in agricultural premises and on farm animals.

Borax or boric acid is still widely used against cockroaches by dusting into their hiding places. It is much less active than many modern insecticides, judged on a milligram/kilogram basis, but the cockroach appears to be a very suspicious insect and is effectively repelled by most insecticides and even formulating substances. Boric acid is almost unique in not showing this repellent action.

Several other inorganic substances have been used for killing insects, including lead chromate and ammonium reineckate, $NH_4^+(Cr(NH_3)_2(SCN)_4)2H_2O^-$, but only those noted above still have any commercial importance.

Thiocyanates

The modern era of synthetic insecticides begins with the thiocyanates in the early 1930's. All the alkyl thiocyanates, $R - S - C{\equiv}N$, prepared by reaction of an alkyl halide with sodium thiocyanate, are insecticidal. In tests where the compound enters mainly from body contact, the activity is at its highest with R = dodecyl, but where fumigant effect is important, lower alkyl compounds are more effective. Lethane, introduced in the U.S.A. in 1936, consisted mainly of a compound with ether links in the alkyl chain, $C_4H_9OC_2H_4OC_2H_4SCN$. Such compounds, products of ethylene oxide technology, have advantage in production because the chlorine atom in $R - O - C_2H_4Cl$ is more reactive than that in $R - Cl$. They also provide a cheap route to compounds of 6-10 atom chain length, which are scarce among the natural fats. They have also a much higher water solubility than compounds having only CH_2 groups in the chain.

Lethane and similar compounds had some minor use in agriculture and were being considered for public health problems early in the Second World War when DDT arrived on the scene. Their development was, perhaps prematurely, arrested by the dramatic success of DDT. Their interesting properties as insecticides have remained largely unexploited. They are extremely rapid in action, in marked contrast to DDT, having "knock-down" properties almost as good as those of pyrethrum. They have, however, considerable irritant effect on human skin which would probably have stopped their use for impregnation of clothing against the body louse and for space-sprays against flying mosquitoes and flies. They tended also to damage the leaves of many crops.

The only compound of this series of present commercial importance is a more complex one, a thiocyanoacetic ester of a terpene alcohol, isoborneol (thanite).

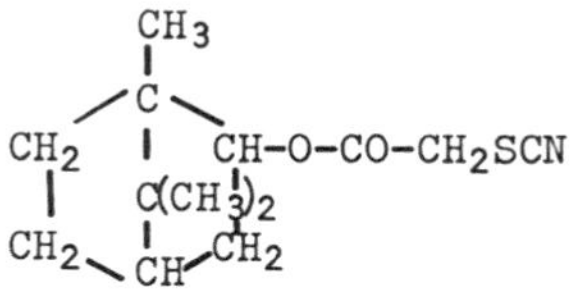

thanite

It is prepared by esterification of the alcohol with chloroacetylchloride, followed by reaction of the product with sodium thiocyanate. It is used mainly as a constituent of fly-sprays to protect livestock.

DDT

The insecticidal properties of the compound 1,1-bis(4-chlorophenyl)-2,2,2-trichloroethane were discovered in 1939. DDT, and compounds of similar action which followed it, had a profound effect on the whole subsequent history of pest control. DDT was produced in enormous quantities during the war and its use, under the direction of the U.S. Army, arrested a potential epidemic of typhus (Naples, 1943) for the first time in medical history. Its continued use in the next decade greatly reduced the enormous death-toll in India from another insect-borne disease, malaria, and it has been, and still is, the mainstay of the W.H.O. malaria control programme.

DDT was first prepared in 1874, by condensation of chloral and monochlorobenzene, agitated with some three times their combined weight of strong sulphuric acid (monohydrate), but its insecticidal properties remained undiscovered for 65 years.

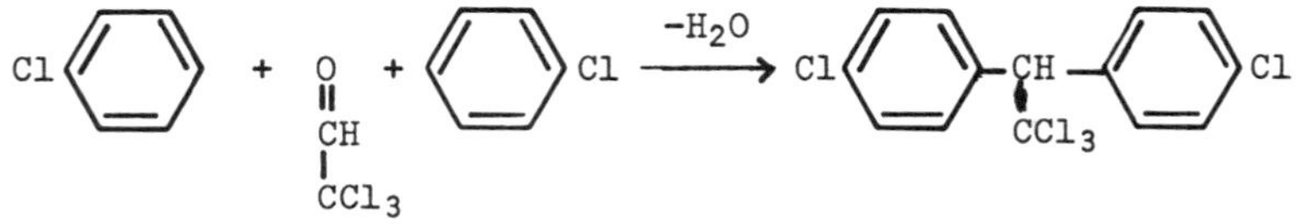

DDT

This reaction is the basis of the very economical commercial production of the compound. The mixture must be well agitated but needs no external heating, the temperature rising by heat of reaction to about 60°C when it is complete. After partial cooling, the mixture is poured, with further agitation, into excess water and the solid DDT, after washing on the filter, is sufficiently pure for most purposes.

As would be expected, chloral does not condense exclusively in the 4-positions of the benzene ring. In addition to the desired 4,4' compound, some 20-30% (according to conditions) of 2,4' compound is produced and a trace of the 2,2'. The isomers are also crystalline compounds, not of great insecticidal significance, but their formation involves some wastage of raw materials. Traces of more oily by-products, probably arising from reactions involving self-condensation of chloral and condensation with only one molecule of chlorobenzene, are more troublesome in the efficient production of wettable powder formulations. Improvement of the crude product can be achieved by a further hot-water washing or by washing with cold alcohol. Recrystallization from hot alcohol yields a substantially pure product but the increased cost on a very cheap commodity prohibits the use of purified material except for special purposes - e.g. in aerosol packs where the presence of traces of sticky impurities insoluble in the solvents used may completely block the fine orifice through which the liquid is discharged.

One of the advantages of DDT was that its manufacture could be carried out if necessary in easily constructed, or even improvised, plant. It has been manufactured on various scales in many countries throughout the world. Records of production are not available for most countries. In the U.S.A. which probably made nearly half the world total, about 15,000 tonnes was produced in 1945, rising to nearly 100,000 tonnes/year in the late 1950's, since when there has been a decline to 20,000 tonnes in 1971 under the impact of pressure to withdraw permission for some uses of these very persistent compounds.

The great stability of DDT seemed, at the time of its introduction, an advantage almost as important as its cheap and easy production. The only facile reaction it undergoes is loss of hydrogen chloride in the aliphatic centre of the molecule to give 1,1-bis(4-chlorophenyl)-2,2-dichloroethylene (DDE), a compound with no insecticidal action. This reaction, carried out under reflux in a solution of caustic soda in 95% alcohol, is used in control analysis. Pressure heating with an anhydrous solution of sodium alcoholate is necessary to remove the remaining chlorine atoms. The same dehydrochlorination reaction can also take place in anhydrous systems in the presence of ferric chloride and some other catalysts. This is the only reaction of significance in the storage of DDT formulations. Since hydrogen chloride is produced in the reaction the catalyst can form by reaction with the iron of a container, if exposed. Under normal conditions of storage, the risk is sufficiently reduced by ensuring that the DDT and other ingredients have negligible initial content of free acid.

A third advantage of DDT, which encouraged its rapid war-time and post-war development, was its very low toxicity to mammals and absence of skin irritancy. It seemed to have all the virtues and no vices at a time when there were pressing problems in the control of insect-borne diseases. It is now generally appreciated, in retrospect, that the low acute toxicity of DDT to mammals was accepted too easily as a guarantee of safety, but it must be remembered that DDT was developed when there was urgent need of it and most of the world was preoccupied with products far more dangerous. The atmosphere of struggle for immediate survival was not conductive to quiet thought about the possible long-term effects of synthetic insecticides.

DDT has a very wide spectrum of activity among the different families of insects and related organisms. Its properties gave it outstanding success as a residual deposit. It was active not only when ingested, but also against insects which only crawled on the deposit. Later compounds of the organochlorine class have equalled or even surpassed it in this respect and it is easy to forget how novel and important this property was when arsenicals requiring ingestion and the transient pyrethrum and nicotine dominated the scene.

DDT is not, of course, uniformly effective against all species. In general, aphids and spider mites are less susceptible than most. In the case of aphids the poor control is mainly due to the inability of the water-insoluble insecticide to reach colonies of stationary wingless aphids under, or within the folds of, leaves. A high proportion escape contact with the spray, but their predators, intrinsically more susceptible and making contact with residual deposits during their hunting activity, succumb. Use of a DDT spray against immobile aphids such as Brevicoryne on Brassica crops often therefore results in a drastic resurgence. Acyrthosiphon pisi on the pea crop, an unusually easily disturbed aphis, is effectively controlled.

The weakness of DDT against spider mites is more intrinsic, but the defect is again exaggerated in the field by the susceptibility of the predators, which are a very important factor in natural control. Extensive use of DDT in orchards, in the early years of its spectacular success, resulted in a very greatly increased red spider population which required other compounds for control.

Compounds Related to DDT

The 4,4' chlorine substituents in the DDT molecule can be replaced by several other groups yielding compounds of comparable activity. The unsubstituted compound is, however, of very low activity and the 4,4'-dihydroxy compound is quite inactive. The 4,4'-dibromo and dimethyl compounds have no advantages to justify their higher cost, but the even more expensive difluoro compound was fairly extensively used in Germany during the Second World War in preference to DDT for reasons which were obscure and presumably not valid, since it is no longer available. Its much lower melting point (19°C) makes it much more soluble than DDT so that a liquid residue can be left. On a non-porous surface, this is more rapidly active but it disappears more rapidly into the capillaries of a porous surface.

The only 4,4' analogue to have achieved an important commercial position is the dimethoxy compound (methoxychlor). It is made by the very facile condensation of anisole and chloral. It is not in general quite so active as DDT but is more vulnerable to biochemical attack so that it has much less tendency to accumulate in fat depots. The use of DDT to keep cow houses free of flies is now prohibited in most countries because of the significant transfer of DDT to the milk fat. Methoxychlor is permitted for this purpose as it is decomposed in the body before it reaches the milk.

In other compounds the aliphatic centre of the molecule is altered. The dichloroethylene derivative is, as noted above, inactive, but the dichloroethyl compound, TDE

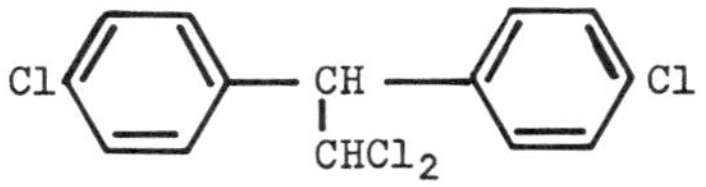

TDE

has found limited use in the U.S.A. on food crops, being significantly less toxic than DDT. It is prepared similarly to DDT, but requires dichloroacetaldehyde in place of chloral. This is obtained by controlled action of chlorine on ethyl alcohol at a temperature under 30°C and the product is condensed with monochlorobenzene immediately after preparation.

Lindane

This insecticide, with broadly similar biological potential to DDT, followed it very closely in development. By the action of elementary chlorine on benzene in the dark and with suitable catalysts, true chlorination (replacement of H by Cl) occurs in stages. By the action of elementary chlorine in the cold, without catalysts and in the presence of mercury arc light, an addition reaction occurs yielding benzene hexachloride or 1,2,3,4,5,6-hexachlorocyclohexane. It is formed without significant concentration of compounds intermediate between C_6H_6 and $C_6H_6Cl_6$, because, as soon as the aromatic ring becomes partly saturated, further addition of chlorine is almost instantaneous. The crude crystalline insecticidal product, with a strong and persistent musty smell, has been known at various times and places as BHC or HCH.

Each C atom in the compound is attached to two different other atoms and the linkage into a chain prevents free rotation. Several structural isomers are therefore possible as in inositol, $C_6H_6(OH)_6$, or in ring sugars. Only the γ-isomer, which has three adjacent axial chlorine atoms and three adjacent equatorial, has insecticidal activity.

The purified γ-isomer is now in most countries called lindane. The preparation of lindane will be seen to necessitate the wasteful production of some seven times its weight of unwanted isomers. The potential economic loss is partly offset by splitting the by-products by strong heating into hydrogen chloride and trichlorobenzene, $C_6H_6Cl_6 \rightarrow C_6H_3Cl_3 + 3HCl$. The same dehydrochlorination occurs in alcoholic alkali, except in the case of the β isomer. The trichlorobenzene is mainly the 1,2,4 compound which can readily be further chlorinated and led into the production line of higher chlorobenzenes.

Lindane has broadly similar action to DDT. Even the melting point (112°C) is close and both compounds leave a residue active via body contact. Lindane is in general rather more potent, the crude HCH being more nearly equivalent to DDT, but the response to both is variable with species and therefore some are more responsive to one than the other. Both can induce resistance but resistance developed to lindane is more likely to be coupled with resistance to the cyclodienes than to DDT. The biochemistry of its action is therefore probably basically different from that of DDT.

Lindane is considerably more volatile than DDT (9×10^{-6} mm at 20°C compared with 2×10^{-7} mm) but only slightly more volatile than aldrin (6×10^{-6} mm). It is significantly more soluble in water than either and a saturated solution will quickly provide a lethal dose to insects in contact with it. While its volatility gives it significant fumigant effect in a dry environment, its water solubility restricts its vapour phase movement in moist soil, but facilitates diffusion under the cuticle, and there is evidence of short-range systemic effect in plants. These properties make lindane a very effective seed dressing against soil insect attack.

Cyclodiene Insecticides

From 1945 onwards a number of very active contact insecticides were produced by the Diels-Alder reaction on hexachlorocyclopentadiene. The latter compound, unlike most dienes, will not easily react with itself and so it can be condensed with a wide range of adducts to give specific products in high yield. Development of a satisfactory manufacturing process for hexachlorocyclopentadiene presented some problems as the intermediate chlorination products self-polymerize explosively, but the problem was solved either by using a very large excess of chlorine in the gas phase or by using chlorine and alkali in an organic solvent.

Reaction of hexachlorocyclopentadiene with cyclopentadiene gave chlordene, which, on chlorination with chlorine in carbon tetrachloride, gave the first commercial insecticide of this type, chlordane.

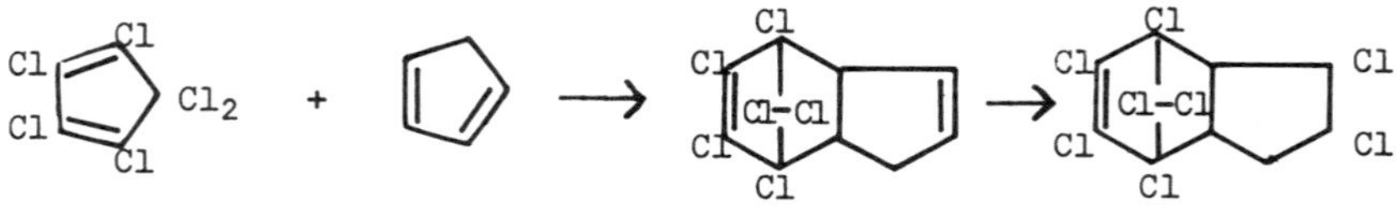

chlordene chlordane

Structures of this type exist in two forms, the "endo", which has the three five-membered rings in a "boat" form, and the "exo", which has the three rings in a "chair" form. Isomers also arise from different spatial arrangements of the chlorine atoms.

Chlorination of chlordene with sulphuryl chloride in carbon tetrachloride in the presence of catalytic amounts of benzoyl peroxide gave heptachlor, a more effective insecticide than chlordane. A related insecticide is chlorbicyclen.

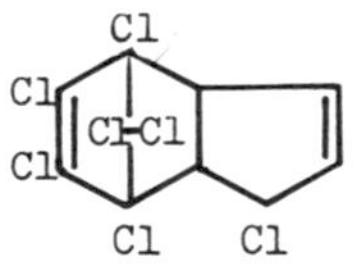

heptachlor

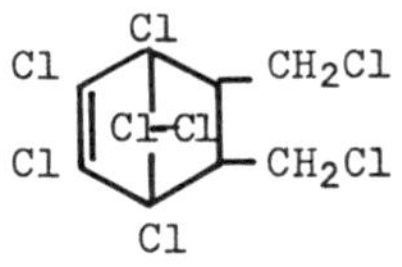

chlorbicyclen

The most important insecticides in this group contain four fused five-membered rings. Cyclopentadiene is reacted with acetylene to give dicycloheptadiene which is then condensed with hexachlorocyclopentadiene

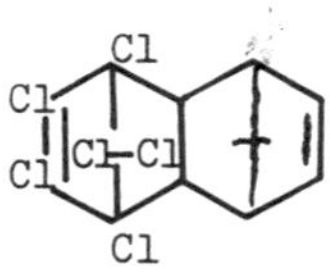

aldrin (exo-ends)
isodrin (exo-exo)

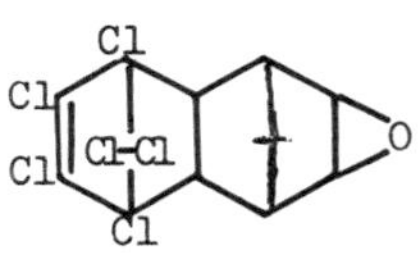

endrin (endo-endo)
dieldrin (endo-exo)

The exo-endo compound is aldrin and the exo-exo compound is isodrin. These can be readily oxidized with hydrogen peroxide to the epoxy compounds, endrin, the endo-endo form, and dieldrin, the endo-exo form.

Hexachloropentadiene can be dimerised by heating in the presence of aluminium chloride, yielding mirex, a stomach insecticide with little contact activity, used mainly against ants.

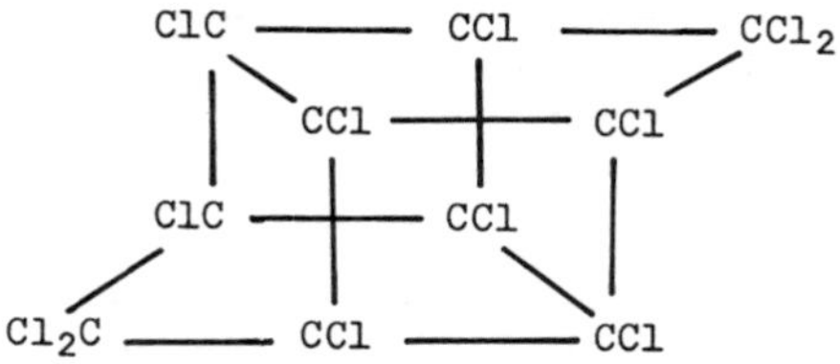

mirex

Most polychlorinated cyclodienes are chemically very stable and do not lose chlorine even when refluxed with alcoholic potassium hydroxide or treated with liquid ammonia. This stability has created problems because of their persistence in the environment and attempts have been made to find insecticidally effective compounds which are less stable. Some commercially successful products of this type are endosulfan and isobenzan which are oil-soluble compounds similar in activity to aldrin. However, although the spray residues are protected from reaction with water by their low solubilities, when the compounds penetrate lining tissue, they are broken down by hydrolysis.

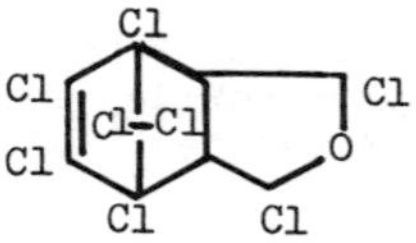

isobenzan

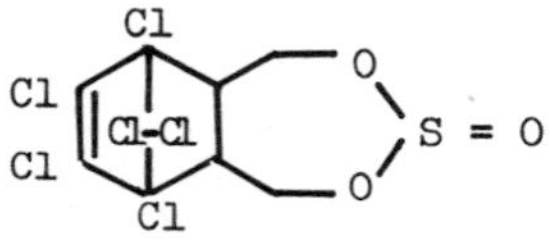

endosulfan

In the past decade the cyclodiene insecticides have come into disfavour because of their high mammalian toxicities and very ready absorption through skin, their extreme persistence in the environment which has resulted in their widespread accumulation in the body fats of animals and humans and their indiscriminate activity against beneficial insects as well as pests, and against birds and fish. Gross misuse of these products in the early days resulted in those incidents of widespread damage to domestic animals and wildlife which were dramatized by Rachel Carson in "Silent Spring" and which initiated a strong public reaction against these pesticides, a reaction which, unfortunately, has been indiscriminately extended by many groups to all pesticides. In most developed countries the cyclodiene insecticides have been withdrawn from many uses and greatly restricted for many others. It is now generally accepted by all who are concerned with pesticides that compounds of very high persistence are undesirable.

Aldrin, dieldrin and endrin are the most powerful general insecticides known. They are particularly effective where contact action and long persistence are required, but their lack of systemic action makes them of little use against sap-sucking species unless these are exposed to direct spray. The significant volatility of aldrin makes it the least persistent as an exposed deposit, but gives it an advantage for application in the soil against soil-dwelling insects. Dieldrin has been the most effective compound to date against ectoparasites (lice, ticks, blow flies, etc.) of sheep and cattle and was until recently used very widely in dips and sprays for this purpose. It gave a longer period of protection than any other compound. Its affinity for, and effectiveness on, animal hair extended this use to protection of woollen cloth and carpets from the ravages of moth and beetle, but in competition with other more specific protectants.

These compounds are significantly more toxic to mammals than DDT and lindane (approximate acute oral LD50 to rats in milligrams per kilogram body weight are DDT, 115; lindane, 125; aldrin, dieldrin, 50; endrin, 12). It is indeed remarkable that such a stable compound as dieldrin can be highly toxic to any species. It must undergo reactions in living tissues which have no parallel *in vitro*.

Chlorocamphene (toxaphene)

This product results from direct chlorination of camphene to a chlorine content of 67%. It is a mixture of several compounds and stereoisomers which are always formulated without separation. It has similar biological properties to lindane but is more soluble in petroleum hydrocarbons.

It was first introduced in 1948 but was overshadowed by DDT and the cyclodiene insecticides. However, in the last decade, when the latter products have come into disfavour and been either withdrawn or greatly restricted, chlorocamphene has emerged as the most acceptable compound. It now tops the sales of all insecticides in the USA where farmers use 17,000 tonnes annually, almost one quarter of the total synthetic organic insecticides used in the U.S.A. It has a reasonably low mammalian toxicity (oral LD50 90 mg/kg dermal LD50 1075 mg/kg) and it does not accumulate persistently in body fat, but is eliminated rapidly when intake is stopped. It is non-phytotoxic except to cucurbits.

Other Insecticides

Apart from organochlorine compounds, organophosphosphorus compounds and carbamates there have been no new classes of insecticides discovered until the recent development of the synthetic pyrethroids.

A number of compounds have been introduced but have made little commercial impact. They include:

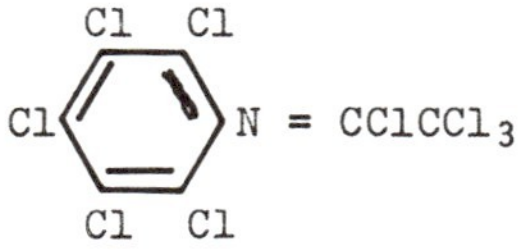

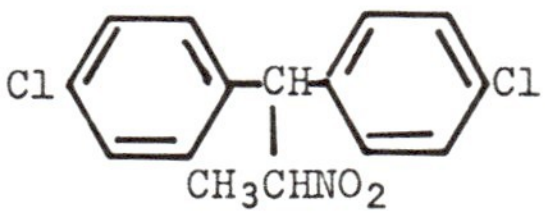

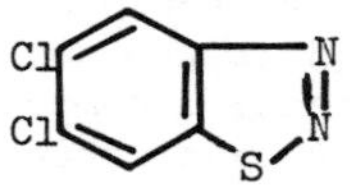

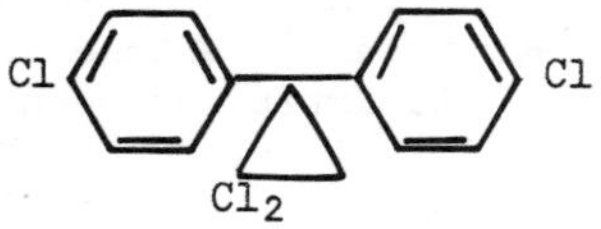

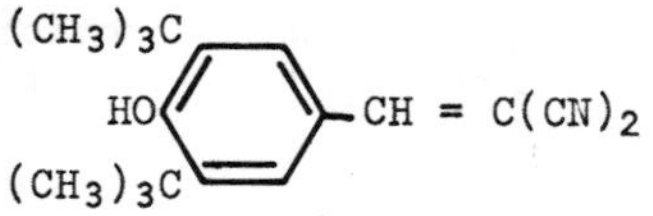

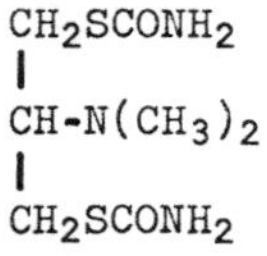

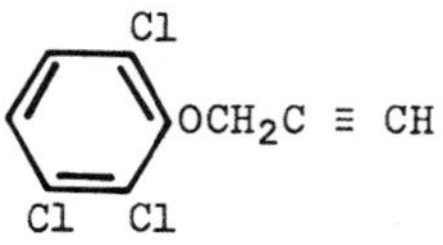

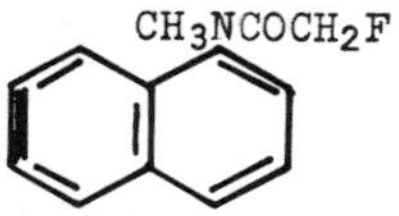

Further Reading

Brooks, G.T., Chlorinated Insecticides, Vols I & II, (CRC Press, Cleveland, Ohio, 1974)

Brown, A.W.A., Insect Control by Chemicals, (Chapman and Hall, London & Wiley, New York, 1951)

Metcalf, R.L., Organic Insecticides, (Interscience, London & New York, 1955)

Moriarty, F. (Ed)., Organochlorine Insecticides, (Academic Press, London & New York, 1975)

O'Brien, R.D., Insecticides, Action and Metabolism, (Academic Press, London, 1967)

Tahori, A.J., (Ed)., Insecticides, Proc. 2nd Int. IUPAC Cong. Pest. Chem. 2 volumes (Gordon and Breach, London & New York, 1972)

West, T.F. & Campbell, G.A., DDT and Newer Persistent Insecticides, (Chapman Hall, London, 1950)

World Health Organisation, The Place of DDT in Operations against Malaria, WHO Record No. 190, (Geneva 1971)

Chapter 7

SYNTHETIC INSECTICIDES: ORGANOPHOSPHORUS COMPOUNDS AND CARBAMATES

Research into toxic organophosphorus compounds was begun during the Second World War and was followed up in military research establishments in Germany, the U.K. and the U.S.A. The activities of competing investigation teams, as they followed the terminal advance of the allied armies, ensured greater publicity for the German work, while that in the U.K. and U.S.A. remained largely in secret files, but it is certainly true that the Germans had gone further towards practical development. Small stocks of two potential war gases had been charged into weapons and two other compounds were coming into use as insecticides.

The vapours of the war "gases", sarin and tabun, are much more lethal to insects than to man, but the public would not have tolerated the agricultural use of military weapons. Safer insecticidal compounds had to be developed, particularly since DDT and lindane had just become available and were, at the time, considered to be quite harmless to mammals. It was natural that in the immediate post-war years official and public attitudes over-emphasized the significance of acute toxicity and the dangers of inhalation and skin-contamination because of the analogy with war gases. Now that the very long persistence of the organochlorine insecticides such as DDT, aldrin and dieldrin, has been established, there is perhaps a tendency towards the opposite overemphasis.

The organophosphorus poisons are very numerous but form a well-recognised class. More is known about their mode of action, and the features of chemical structure and reactivity necessary for this action, than is the case with almost any other class of poison. The field is therefore one in which a great deal of chemical research towards new insecticides has been concentrated. Many thousand compounds must have been synthesized, mainly in industrial laboratories. It is probably true to say that most of these have proved to be insecticidal, a much higher proportion than when chemical synthesis explores wholly new fields. The objective of synthesis of organophosphorus compounds is not, however, to produce just another insecticide but to find some substantially improved selectivity, safety to mammals or desirable level of persistence.

The structure of most active compounds in this class can be described by the general formula

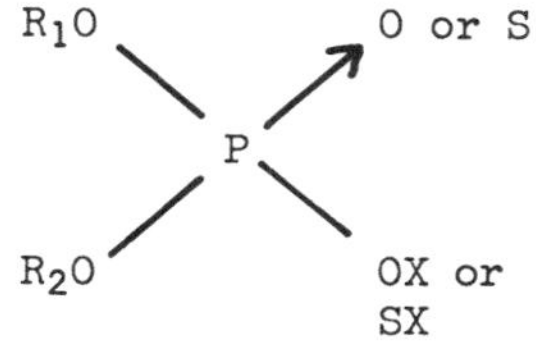

A few compounds have been made in which a higher group VI element, Se, or Te, replaces O or S as the recipient of the lone electron pair of the central phosphorus, but these are not economic. One could expect that the NH group could also act as acceptor but such compounds are subject to polymerization during attempted synthesis.

The stable alkoxy groups, R_1O and R_2O, are usually, for convenience of manufacture, the same and most frequently C_2H_5O- or CH_3O-. Propyl esters occasionally appear but esters of higher alcohols usually show much reduced activity. The group X which is linked to the phosphorus atom through oxygen or sulphur can be a very wide range of alkyl, aralkyl, aryl and heterocyclic groups with various substituents. Most of the ingenuity and energy of the organic chemist in his search for improved organophosphorus insecticides has gone into variations of this group, and the factors which affect choice of chemical structure are discussed later in this chapter.

A few early compounds had $(CH_3)_2N$ groups in place of R_1O and R_2O, and F in place of OX or SX. Two commercially used compounds were dimefox and mipafox, but they are little used nowadays because of their very high mammalian toxicities (oral LD50 5 mg/kg).

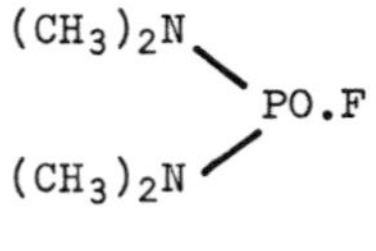

dimefox

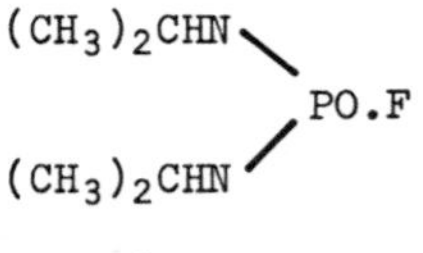

mipafox

The only other compound with $(CH_3)_2N$ groups instead of R_1O and R_2O which had any commercial utility was schradan, but this too is not now used because of its high mammalian toxicity (oral LD50 8 mg/kg).

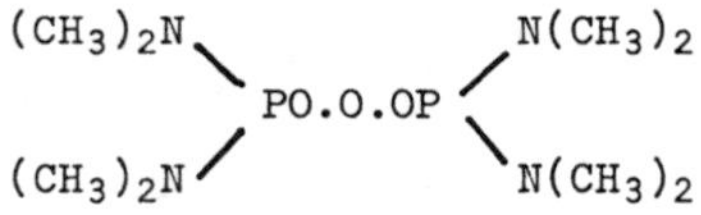

schradan

There are one or two commercially useful compounds in which the group X is directly linked to phosphorus, not through O or S, that is, they are phosphonates, not phosphates. The most important compound of this type is trichlorphon, manufactured from dimethyl phosphite and chloral (Eq. 1) or from phosphorus trichloride, chloral and methanol.

$$(CH_3O)_2.P.OH + CHO.CCl_3 \longrightarrow \underset{\text{trichlorphon}}{(CH_3O)_2.PO.CHOH.CCl_3} \qquad (1)$$

There are also some commercially-used phosphonates in which one of the alkoxy groups R_1O and R_2O is replaced by alkyl or aryl. The most important compounds of this type are fonofos, mercarphon, trichloronate, leptophos and cyanolate.

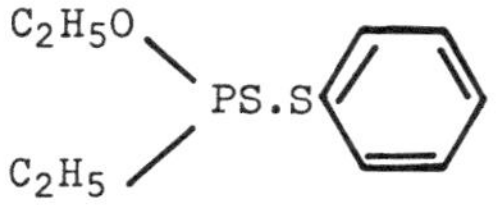

fonofos

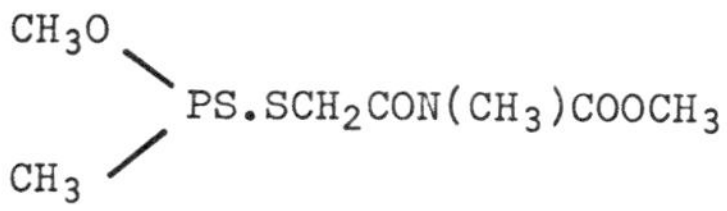

mercarphon

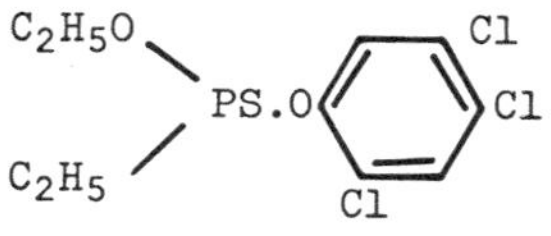

trichloronate

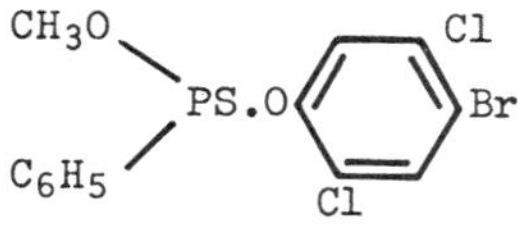

leptophos

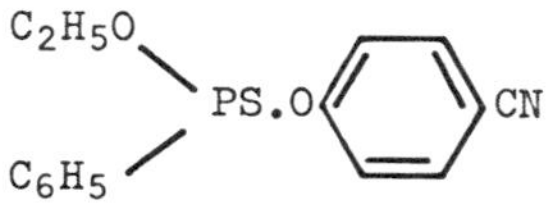

cyanolate

Recently there has been considerable interest in compounds in which the OX or SX group is replaced by NXY so that the compounds are phosphoroamidothioates. Representatives of this class are methamidophos, acephate, phosfolan and mephosfolan.

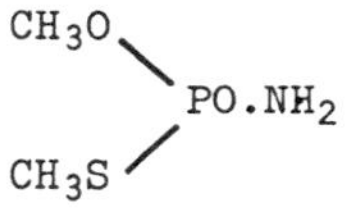

methamidophos

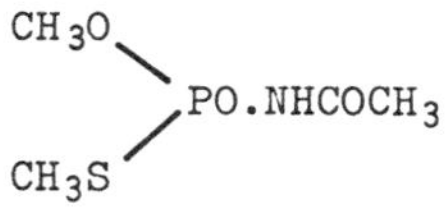

acephate

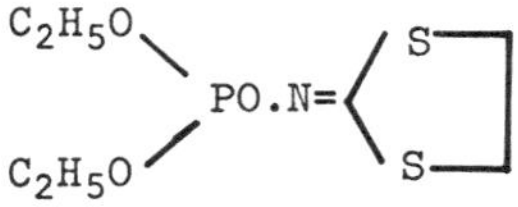

phosfolan

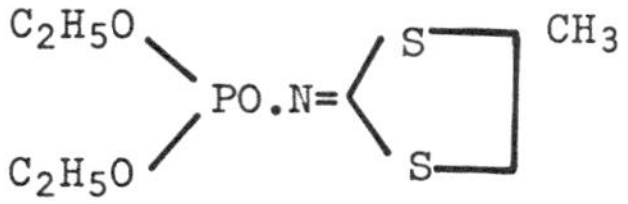

mephosfolan

Most of the commercially important organophosphorus insecticides are, however, derivatives of phosphoric or thiophosphoric acid. The earliest useful compounds of this type were phosphoric anhydrides such as TEPP and sulfotep, but these are little used now because of their high mammalian toxicities (oral LD50 5 mg/kg).

$$(C_2H_5O)_2.PO.O.OP.(OC_2H_5)_2$$

TEPP

$(C_2H_5O)_2.PS.O.SP.(OC_2H_5)_2$

sulfotep

Phosphates

Because phosphates tend to be unstable and to have high mammalian toxicities there are few useful insecticides of this type. Apart from paraoxon and fospirate, all the commercial products are dialkyl vinyl phosphates, and the most widely used of these is dichlorvos.

Dimethyl phosphates X	$(CH_3O)_2.PO.OX$ common name
2,2-dichlorovinyl	dichlorvos
1,2-dibromo-2,2-dichloroethyl	naled
2-methoxycarbonyl-1-methylvinyl	mevinphos
1-methyl-2-methylcarbamoylvinyl	monocrotophos
1-methyl-2-dimethylcarbamoylvinyl	dicrotophos
2-chloro-2-diethylcarbamoyl-1-methylvinyl	phosphamidon
1-methyl-2-(1-phenylethoxycarbonyl)vinyl	crotoxyphos
2-chloro-1-(2,4,5-trichlorophenyl)vinyl	tetrachlorvinphos
3,5,6-trichloro-2-pyridyl	fospirate

Diethyl phosphates X	$(C_2H_5O)_2.PO.OX$ common name
4-nitrophenyl	paraoxon
2-chloro-1-(2,4-dichlorophenyl)vinyl	chlorfenvinphos
2-bromo-1-(2,4-dichlorophenyl)vinyl	bromfenvinphos

Phosphates can be manufactured by reaction of a dialkyl phosphorochloridate with an hydroxy compound, generally in an organic solvent in presence of sodium carbonate or an organic base. Thus, paraoxon is prepared by heating diethyl phosphorochloridate with sodium 4-nitrophenoxide in acetonitrile for 2 hours at 75° (Eq. 2), and mevinphos similarly from dimethyl phosphorochloridate and the sodium enolate of methyl acetoacetate (Eq. 3).

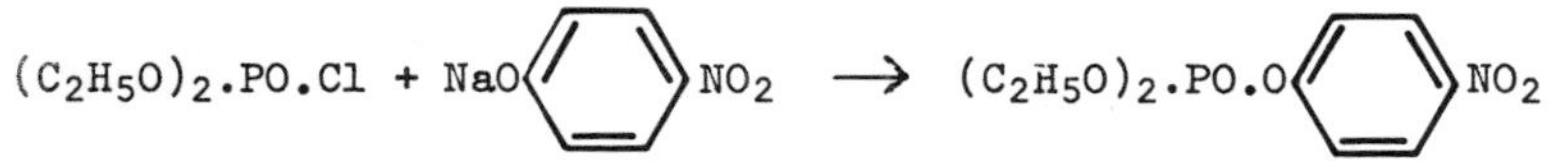

(2)

paraoxon

$$(CH_3O)_2.PO.Cl + \frac{CH_3}{NaO}\!\!>C{=}CHCOOCH_3 \longrightarrow (CH_3O)_2.PO.O\!\!\frac{CH_3}{}\!\!>C{=}CHCOOCH_3 \qquad (3)$$

mevinphos (95% trans)

The dialkyl phosphorochloridates can be easily manufactured from phosphorus oxychloride and an alcohol (Eq. 4) or by reaction of dialkyl phosphites either with chlorine (Eq. 5) or with carbon tetrachloride in presence of a tertiary amine (Eq. 6). Dialkyl phosphites are made from phosphorus trichloride and an alcohol; in the absence of added base the reaction stops at the dialkyl phosphite stage (Eq. 7). More economically, a mixture of alcohol and water may be used (Eq. 8).

$$POCl_3 + 2ROH \longrightarrow (RO)_2.PO.Cl + 2HCl \qquad (4)$$

$$(RO)_2.P.OH + Cl_2 \longrightarrow (RO)_2.PO.Cl + RCl \qquad (5)$$

$$(RO)_2.P.OH + CCl_4 \longrightarrow (RO)_2.PO.Cl + CHCl_3 \qquad (6)$$

$$PCl_3 + 3ROH \longrightarrow (RO)_2.P.OH + RCl + 2HCl \qquad (7)$$

$$PCl_3 + 2ROH + H_2O \longrightarrow (RO)_2.P.OH + 3HCl \qquad (8)$$

Phosphates can also be manufactured by the Perkow reaction of a trialkyl phosphite with an α-halogenated carbonyl compound and this is the process used for all the vinyl phosphates - mevinphos, monocrotophos, dicrotophos, phosphamidon, crotoxyphos, tetrachlorvinphos, chlorfenvinphos and bromfenvinphos. For example, trimethyl phosphite and methyl α-chloroacetoacetate give mevinphos (Eq. 9). The product from this route is 2:1 cis:trans whereas the alternative route from dimethyl phosphorochloridate (Eq. 3) gives 95% trans. Dichlorvos is similarly manufactured by heating trimethyl phosphite and chloral in benzene at 60-70° (Eq. 10).

$$(CH_3O)_3P + CH_3COCHClCOOCH_3 \longrightarrow (CH_3O)_2.PO.O-\overset{CH_3}{\overset{|}{C}}{=}CHCOOCH_3 \qquad (9)$$

mevinphos (2:1 cis:trans)

$$(CH_3O)_3P + CCl_3CHO \longrightarrow (CH_3O)_2.PO.O.CH{=}CCl_2 + CH_3Cl \qquad (10)$$

Trialkyl phosphites are manufactured from phosphorus trichloride and an alcohol in presence of base (Eq. 11).

$$PCl_3 + 3ROH \longrightarrow (RO)_3.P + 3HCl \qquad (11)$$

The only commercially-used phosphate which is not a dimethyl or diethyl compound is di-n-propyl 4-methylthiophenyl phosphate (propaphos).

O-Phosphorothioates

The O-phosphorothioates are, in general, more stable than the phosphates, and less toxic to men and animals, so they are more widely used. The main commercial products in this class are (1) 2-alkylthioethyl phosphorothioates (2) dialkyl aryl phosphorothioates. The most widely-used are parathion, parathion-methyl, diazinon and demeton.

Dimethyl O-phosphorothioates X	$(CH_3O)_2.PS.OX$ common name
2,2-dichlorovinyl	thiophosvin
2-(methylthio)ethyl	demephion-O
2-(ethylthio)ethyl	demeton-O-methyl
2-methoxycarbonyl-1-methylvinyl	methacrifos
4-nitrophenyl	parathion-methyl
2-chloro-4-nitrophenyl	phosnichlor
3-methyl-4-nitrophenyl	fenitrothion
4-cyanophenyl	cyanophos
2,4,5-trichlorophenyl	fenchlorphos
4-bromo-2,5-dichlorophenyl	bromophos
3-methyl-4-methylthiophenyl	fenthion
4-(4-chlorophenylazo)phenyl	azothoate
2,3,5-trichloro-6-pyridyl	fospirate
2-diethylamino-6-methylpyrimidin-4-yl	pyrimiphos-methyl
2-isopropyl-6-ethoxypyrimidin-4-yl	lirimphate
2-ethyl-6-ethoxypyrimidin-4-yl	ethoxypyrithion
2-quinoxalyl	merquinphos

Diethyl O-phosphorothioates X	$(C_2H_5O)_2.PS.OX$ common name
2-(ethylthio)ethyl	demeton-O
4-nitrophenyl	parathion
4-methylsulphinylphenyl	fensulfothion
2,4-dichlorophenyl	dichlofenthion
4-bromo-2,5-dichlorophenyl	bromophos-ethyl
cyanobenzylideneamino	phoxim
3,5,6-trichloro-2-pyridyl	chlorpyrifos-ethyl
5-phenyl-3-isoxazolyl	isoxathion
2-dimethylamino-6-methylpyrimidin-4-yl	pyrimitate
2-diethylamino-6-methylpyrimidin-4-yl	pirimiphos-ethyl
2-ethylacetamido-6-methylpyrimidin-4-yl	piracetaphos
2-isopropyl-6-methylpyrimidin-4-yl	diazinon
6-ethoxy-2-ethylpyrimidin-4-yl	etrimfos
3-chloro-4-methyl-7-coumarinyl	coumaphos
1-isopropyl-5-chloro-1,2,4-triazol-3-yl	isazophos
3,4-benzo-7-coumarinyl	coumithioate

O-Phosphorothioates can isomerize to S-phosphorothioates, so the commercial products are often a mixture of both, especially in the case of the 2-alkylthioethyl compounds. They are manufactured by reaction of an hydroxy compound with a dialkyl phosphorochloridothionate in the presence of bases and often, also , of copper powder. Careful control of temperature is essential, especially with O-methyl esters, to prevent isomerisation into S-methyl esters and to restrict decomposition. For example, fenitrothion is manufactured by heating 3-methyl-4-nitrophenol with dimethyl phosphorochloridothionate at 60-80° in methyl isobutyl ketone in presence of potassium carbonate (Eq. 12).

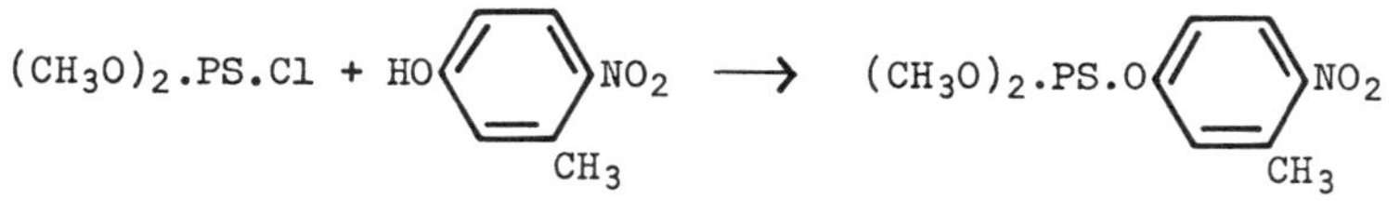 (12)

Dialkyl phosphorochloridothionates can be prepared by reaction of phosphorus thiochloride with an alcohol in presence of a base (Eq. 13) but a more economical manufacturing process is to chlorinate the dialkyl phosphorodithioic acids made by reaction of phosphorus pentasulphide with an alcohol (Eq. 14 and 15).

$$PSCl_3 + 2ROH \longrightarrow (RO)_2.PS.Cl + 2HCl \quad (13)$$

$$P_2S_5 + 4ROH \longrightarrow 2(RO)_2.PS.SH + H_2S \quad (14)$$

$$2(RO)_2.PS.SH + 3Cl_2 \longrightarrow 2(RO)_2.PS.Cl + 2HCl + S_2Cl_2 \quad (15)$$

S-Phosphorothioates

In general, the S-phosphorothioates have greater mammalian toxicities and smaller insecticidal activities than the O-phosphorothioates so are much less used.

Dimethyl S-phosphorothioates X	$(CH_3O)_2.PO.SX$ common name
2-(methylthio)ethyl	demephion-S
2-(ethylthio)ethyl	demeton-S-methyl
2-(ethylsulphinyl)ethyl	oxydemeton-methyl
methylcarbamoyl methyl	omethoate
2-(1-methylcarbamoylethylthio)ethyl	vamidothion
5-methoxy-4-pyron-2-ylmethyl	endothion

Diethyl S-phosphorothioates X	$(C_2H_5O)_2.PO.SX$ common name
2-(ethylthio)ethyl	demeton-S
2-diethylaminoethyl	amiton
ethoxycarbonylmethyl	acetofos
1-cyano-1-methylethylcarbamoylmethyl	cyanthoate
6-chloro-4-thiochromanyl	thiochrophos

Reaction of dialkyl phosphorochloridate with a mercaptan is not a suitable manufacturing process for S-phosphorothioates because attack takes place on the O-alkyl group as well as on the chlorine atom (Eq. 16).

$$(RO)_2.PO.Cl + NaSX \longrightarrow R.S.X + \begin{matrix} RO \\ NaO \end{matrix}\!\!>PO.Cl \quad (16)$$

However, demeton-S is manufactured from diethyl phosphorochloridate and the sodium salt of 2-ethylthioethylthiol (Eq. 17).

$$(C_2H_5O)_2.PO.Cl + NaSCH_2CH_2SC_2H_5 \longrightarrow (C_2H_5O)_2.PO.SCH_2CH_2SC_2H_5 + NaCl \quad (17)$$

demeton-S

Generally, S-phosphorothioates are manufactured by reaction of a dialkyl phosphorothioate with a halo compound. Thus, acetofos is made from ammonium diethyl phosphorothioate and ethyl chloroacetate (Eq. 18).

$$(C_2H_5O)_2.PO.SNH_4 + ClCH_2COOC_2H_5 \longrightarrow (C_2H_5O)_2.PO.SCH_2COOC_2H_5 \quad (18)$$

acetofos

The dialkylphosphorothioates are manufactured by treatment of dialkyl phosphorochloridothionates with alkali (Eq. 19), by reaction of dialkyl phosphorochloridates with sulphides (Eq. 20), or by reaction of dialkyl phosphites with sulphur and ammonia at room temperature (Eq. 21).

$$(RO)_2.PS.Cl + KOH \longrightarrow (RO)_2.PO.SK + KCl \quad (19)$$

$$(RO)_2.PO.Cl + K_2S \longrightarrow (RO)_2.PO.SK + KCl \quad (20)$$

$$(RO)_2.P.OH + S + NH_3 \longrightarrow (RO)_2.PO.SNH_4 \quad (21)$$

Phosphorodithioates

This is the most commercially important class of phosphorus insecticides. In contrast to the phosphates and phosphorothioates, aryl esters are unimportant. Nearly all useful compounds have a methyl group attached to the -S, and this methyl group carries an ester, amide, carbamoyl, sulphide or heterocyclic grouping. The most widely used compounds of this class are disulfoton, malathion, phorate, azinophos methyl, ethion and dimethoate.

Dimethyl phosphorodithioates X	$(CH_3O)_2.PS.SX$ common name
2-acetamidoethyl	amiphos
2-(ethylthio)-ethyl	thiometon
methylcarbamoylmethyl	dimethoate
ethylcarbamoylmethyl	ethoate-methyl
2-methoxyethylcarbamoylmethyl	amidithion
1,2-di(ethoxycarbonyl)ethyl	malathion
ethoxycarbonylbenzyl	phenthoate
morpholinocarbonylmethyl	morphothion
4,6-diamino-1,3,5-triazin-2-ylmethyl	menazon
3,4-dihydro-4-oxobenzo[d]-1,2,3-triazin-3-ylmethyl	azinphos-methyl
formylmethylcarbamoylmethyl	formothion
mercaptomethylphthalimide	phosmet
2-methoxy-1,3,4-thiadiazol-5-(4H)-onyl-4-methyl	methodathion

Diethyl phosphorodithioates X	$(C_2H_5O)_2.PS.SX$ common name
ethylthiomethyl	phorate
tert butylthiomethyl	terbufos

Diethyl phosphorodithioates X	$(C_2H_5O)_2.PS.SX$ common name
2-(ethylthio)ethyl	disulfoton
2-(ethylsulphinyl)ethyl	oxydisulfoton
ethoxycarbonylmethyl	acethion
isopropylcarbamoylmethyl	prothoate
ethoxycarbonylmethylcarbamoylmethyl	mecarbam
methylene-bis-	ethion
4-chlorophenylthiomethyl	carbophenothion
1,4-dioxan-2,3-ylidene-bis-	dioxathion
6-chloro-4-thiochromanyl	dithiochrophos
3,4-dihydro-4-oxobenzo[d]-1,2,3-triazin-3-ylmethyl	azinphos-ethyl
2-chloro-1-phthalimidoethyl	dialifos
6-chloro-2-oxybenzoxazolin-3-ylmethyl	phosalone

They are manufactured from dialkyl phosphorodithioic acids by reaction with a halo-compound (Eq. 22), by addition to an aldehyde (Eq. 23), or by addition to an olefin (Eq. 24).

$$(CH_3O)_2.PS.SNa + ClCH_2CH_2C_2H_5 \longrightarrow \underset{\text{thiometon}}{(CH_3O)_2.PS.SCH_2CH_2C_2H_5} + NaCl \qquad (22)$$

$$(C_2H_5O)_2.PS.SH + CH_2O + C_2H_5SH \longrightarrow \underset{\text{phorate}}{(C_2H_5O)_2.PS.SCH_2C_2H_5} \qquad (23)$$

$$(CH_3O)_2.PS.SH + \begin{array}{l} CH.COOC_2H_5 \\ \| \\ CH.COOC_2H_5 \end{array} \longrightarrow \begin{array}{l} (CH_3O)_2.PS.S.CH.COOC_2H_5 \\ \qquad\qquad\quad | \\ \qquad\qquad\; CH_2.COOC_2H_5 \\ \qquad\qquad\quad \text{malathion} \end{array} \qquad (24)$$

Direct formation of a P-S bond by reaction of dialkyl phosphorochloridothioates with sodium derivatives of thiols is not a suitable manufacturing process because of a side-reaction (Eq. 25).

$$(RO)_2.PS.Cl + NaSX \longrightarrow R.S.X + \begin{array}{l} RO \searrow \\ \qquad PS.Cl \\ NaO \nearrow \end{array} \qquad (25)$$

The required dialkyl phosphorodithioic acids can, as described previously (Eq. 14), be easily manufactured by reaction of phosphorus pentasulphide with an alcohol.

Manufacture of Organophosphorus Insecticides

Dialkyl phosphorochloridates and dialkyl phosphorochloridothionates react readily with water in the presence of bases to give phosphoric anhydrides (Eq. 26).

$$(CH_3O)_2.PO.Cl + H_2O \longrightarrow [(CH_3O)_2.PO]_2O + 2HCl \qquad (26)$$

These anhydrides can, therefore, easily be formed as impurities in the manufacture of organophosphorus insecticides and, because they have very high mammalian toxicities, give final products with toxicities exceeding the permitted levels. Toxic impurities can also arise during manufacture of these products as a result of disproportionation reactions of the phosphoric anhydrides. (Eq. 27).

$$2(RO)_2.PO.O.OP.(OR)_2 \longrightarrow (RO)_3PO + (RO)_2.PO.O.\underset{\displaystyle OR}{\underset{|}{P}}O.O.OP.(OR)_2 \qquad (27)$$

Very careful choice of reaction conditions and very strict control of manufacturing operations are essential, not only because impurities may increase toxicity but also because they may decrease stability. Most of the products are oily liquids which cannot be distilled without decomposition, even under reduced pressure, so very little purification of the final product is possible. The most that can usually be done is to dissolve the product in a solvent such as chloroform and to remove acidic impurities by extraction with water. Frequently the final stage of the manufacturing process is carried out in a stirred two-phase system. All operations are strictly controlled to give an acceptably pure technical product directly from the manufacturing process.

Mode of Action

All of these compounds inhibit the action of several ester-splitting enzymes in living organisms. They are particularly effective against cholinesterase, which hydrolyses the acetylcholine generated in myoneural junctions during the transmission of motor-commands. In the absence of effective cholinesterase, the acetylcholine accumulates and interferes with the co-ordination of muscular response. Such interference in the muscles of the vital organs produces serious symptoms and eventually death.

The mechanism of inhibition of cholinesterase by the organophosphorus compounds is fairly well understood but only by indirect evidence. More direct evidence is available on the much better characterized enzyme chymotrypsin. It has been shown that, when paraoxon is presented to this enzyme in dilute solution, the phosphorus of one molecule of the inhibitor becomes locked up in every molecule of enzyme inhibited. At the same time one free nitrophenate ion appears. The active site of the enzyme evidently splits the molecule of the inhibitor, but is then unable to release the phosphoryl moiety. This blocks further action whereas the unchanged enzyme can attack, split and release some thousands of its normal substrate molecules per second.

Actual behaviour in the living organism, or even on enzymes *in vitro*, is, of course, more complex than this. The blockage of the enzyme by the substituted phosphate group is not necessarily quite permanent and the phosphate ester may be hydrolyzed *in situ*. Thus there may be reactivation of the attacked enzyme, although most recovery from partial intoxication must await replacement of the enzyme. The compound itself is in some cases not directly active and requires chemical alteration before it becomes so. The compound, or its derivatives, may be subject to other reactions which compete with the toxic one. The disposition of groups other than those directly concerned in the inhibition reaction may assist adsorption of the toxic molecule on to the enzyme in a suitable posture for reaction to occur.

Biochemical Reactions Affecting Selectivity

The enzymic hydrolysis of liberated acetylcholine seems to be essential to nerve-muscle relations in all animals. The organophosphorus poisons may therefore seem at first to be wholly unsuitable for selective action. Selectivity, however, arises from the important complex of other reactions outlined in the last paragraph. A compound may be intrinsically toxic to all animals but, of similar doses externally administered, the fraction which can reach the active site sufficiently quickly can vary widely between species. Some of the ways in which variation can arise can be illustrated by reference to the compounds exemplified.

TEPP and paraoxon are direct inhibitors of cholinesterase, i.e. they react rapidly in this respect in sterile solutions. None of the other compounds (much more useful as insecticides) react so effectively in vitro. They are also more slowly hydrolyzed in alkaline solution. There is a general positive correlation between speed of enzyme inhibition in vitro and speed of alkaline hydrolysis. Dimefox and schradan, particularly the latter, are so inactive against cholinesterase in vitro that one would not expect them to be toxic. After treatment with oxidizing agents for a suitable time, however, their solutions become strongly inhibitory. Excessive oxidation, or prolonged storage in water solution after mild oxidation, destroys this new activity. It has been established that both insect tissues and mammalian liver produce the same transient active products.

The oxidation products are not only toxic but very much less stable, particularly to hydrolysis, than the original compound. They are produced in plant tissue also, but so slowly that there is very little accumulation, the main products identified being the harmless hydrolysis products. All organophosphorus compounds tested in which there is a free acidic function on the phosphorus atom are non-toxic.

It is evident that further oxidative demethylation reactions could occur, through intermediates still more unstable and probably intrinsically more toxic. The important characteristic is that toxicity is directly due to the reaction of an unstable intermediate. In theory a species could be quite unaffected if the applied compound underwent no reaction at all in its tissues, although no example is known. It could also be unaffected if oxidation in the tissues were too vigorous or occurred in places too remote from the site of action. Selectivity depends on the balance of production and destruction of the active toxicant.

P→S compounds are much less easily hydrolyzed at the P-X link than the corresponding P→O compounds. Parathion, in a pure state, is a very feeble inhibitor in vitro, but is converted by oxidation into the very active paraoxon. This conversion takes place in living tissues and there is evidence that it occurs to a greater extent in insects than in mammals.

Demeton methyl is a compound in which activity is changed by oxidation in the X group of the general formula - i.e. in the group which is liberated during reaction with the enzyme. The thioether substituent becomes progressively changed to the sulphoxide and sulphone. The influence on rates of enzyme inhibition and hydrolysis is not in this case so great as in the other examples, but there is an important effect on the solubility balance of the compound which becomes much more hydrophilic. The compound is a very effective systemic aphicide. The lipophilic nature of the applied compound is probably useful in helping penetration of the leaf cuticle and the hydrophilic oxidation product formed in the leaf cells is then transported in the vascular system.

In dimethoate, as in parathion, there is need for oxidation of P→S to P→O before inhibition of cholinesterase is effective, but in this case another type of reaction is important. The P-S link is the one subject to fission during reaction with the enzyme and is also subject to hydrolysis, but the amide, -CO-NH-, link can also be hydrolyzed and this reaction appears to be much more favoured by enzyme systems of some species than of others. Amide hydrolysis can occur to a significant extent in mammals before activating oxidation of P→S takes place and the resulting thiophosphoryl acetate ion is rapidly excreted in the urine.

Similar, but greater possibility of inactivating hydrolysis is present in malathion where two carboxylic ester groups are present. Malathion is a feebler insecticide than many other organophosphorus compounds in use but its toxicity to mammals is reduced much more than its toxicity to insects.

Transport Processes Affecting Selectivity

This brief account of the reactivity of the organophosphorus insecticides and the part it can play in their selective action is, of course, very incomplete. It is not the purpose of this book to go at all deeply into the biochemistry of toxicity. Longer and more specialist books must be consulted for full accounts of the present knowledge of biochemical mechanisms. Although much is known, and this knowledge can be applied to alter structures in ways that will probably affect selectivity in a desirable direction, much more is still unknown. Synthetic work cannot make useful progress without the guidance at every stage of biological experiment. The study of toxicants discovered fortuitously contributes much more to knowledge of normal biochemistry than can the latter, as yet, contribute to the design of new toxicants.

The organophosphorus compounds have been used to outline the principles of competition between toxicant-producing and toxicant-destroying reactions, because these principles have to date contributed more to this field than to any other. This is partly because the toxic mechanism is much better known and partly because the toxic reaction requires a rather reactive molecule.

All present insecticides act, finally, on some very general biochemical mechanism and rely for their selectivity on quantitative rather than qualitative differences. Most biochemical mechanisms are remarkably universal. Essentially quantitative differences in basically the same biochemical processes account for the difference in shape, life history and behaviour between man and his louse. Perhaps they can fairly be relied on to make a compound lethal to one but safe for the other.

The significant competition is also not confined to that between parallel or consecutive reactions. Physical transport processes are always involved - diffusion through cuticle, cells and cell membranes; flow in blood, lymph, sap or phloem. If decomposition goes on during transport, the fraction of applied compound which can reach its destination unchanged depends on the relative speeds of transport and reaction. The transport processes are not all favourable. An externally applied compound can be lost by evaporation or by rain-washing. These rates are in competition with rate of entry.

If all the applied substance were to reach the vital site it might seem not to matter how long it would take to do so. All toxic reactions, however, destroy or inactivate some chemical which is so essential that it must be continually replaced. The rate of arrival of the toxicant is therefore in competition with the rate of replacement of the vital chemical attacked. Wide differences in the balance between rates of destruction and replacement are responsible for some poisons - e.g. lead - having a long-term chronic or cumulative effect, while, at the other extreme, oxalate and cyanide have acute significance only.

Rates of diffusion through tissues and of transport by flow of liquid in conducting vessels are determined by the size and shape of molecules and by partition between water and oily phases. In these respects, the organophosphorus insecticides cover a wide range - much wider than most pesticides. At one extreme, schradan is so strongly water-favourable that only the exceptionally good solvent, chloroform, is useful for extracting it from water. At the other, parathion and many similar compounds partition favourably into vegetable oils from water.

With so wide a range of physical properties in compounds of considerable reactivity, it is not surprising that a wide range of toxicity is found among different species and different modes of application. It is generally found that the more direct means of administration to a living organism - such as injection into the blood stream of higher animals or presentation of volatile compounds in the ambient air - show less difference between species than less direct means. The latter, of course, include most of the practical ways of using insecticides, such as leaving on vegetation a deposit which will be ingested or picked up in walking or applying a spray which directly contaminates the body cuticle.

Carbamate Insecticides

The carbamates are closely related in biological action and in resistance development to the organophosphorus insecticides and, like them, inhibit cholinesterase. The activity of different members of the class is rather more dependent on substituent positions and on stereoisomerism than in the case with organophosphorus compounds. The three classes which have yielded commercially useful products are (1) the N-methylcarbamates of phenols, the most widely-used of which are carbaryl and metalkamate (2) the N-methylcarbamates of oximes, the most widely-used of which is methomyl (3) the N-methylcarbamates and N,N-dimethylcarbamates of hydroxy heterocyclic compounds, the most widely-used of which is carbofuran.

Methylcarbamates of phenols R	$RO.CONHCH_3$ common name
1-naphthyl	carbaryl
3-methylphenyl	MTMC
2-isopropylphenyl	isoprocarb
3-dimethylethylphenyl	metalkamate
3-methyl-5-isopropylphenyl	promecarb

Methylcarbamates of phenols R	$RO.CONHCH_3$ common name
2-chloro-4,5-dimethylphenyl	carbanolate
2-isopropoxyphenyl	propoxur
3,5-dimethyl-4-methylthiophenyl	methiocarb
2-ethylthiomethylphenyl	ethiophencarb
4-dimethylamino-3-methylphenyl	aminocarb
4-dimethylamino-3,5-dimethylphenyl	mexacarbate
4-diallylamino-3,5-dimethylphenyl	allyxycarb

These are all manufactured by reaction of the appropriate sodium phenoxide with phosgene at 10-50°, generally by adding a toluene solution of phosgene to an alkaline aqueous solution of the phenoxide (Eq. 28), and then reacting the intermediate chloroformate thus formed with an aqueous solution of methylamine (Eq. 29). The products are usually crystalline solids which can be filtered off.

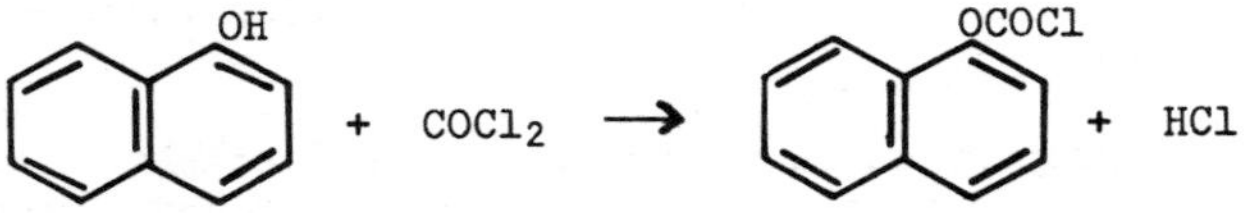

(28)

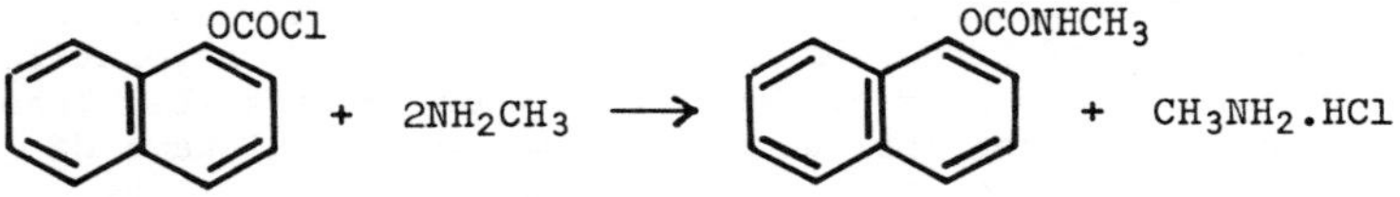

(29)

Methylcarbamates of oximes parent aldehyde	$R{=}NO.CONHCH_3$ common name
2-methyl-2(methylthio)propionaldehyde	aldicarb
1-(methylthio)acetaldehyde	methomyl
Dimethylcarbamoyl-methylthioformaldehyde	thioxamyl
1-(2-cyanoethylthio)acetaldehyde	thiocarboxime

Aldicarb is manufactured by reaction of isobutene with nitrosyl chloride to give the dimer of 2-chloro-2-methyl-1-nitrosopropane (Eq. 30) which is then treated with methanethiol in aqueous sodium hydroxide to form 2-methyl-2-methylthiopropionaldehyde oxime (Eq. 31) which is converted to aldicarb by reaction with methyl isocyanate in methylene chloride. (Eq. 32)

$$(CH_3)_2C{=}CH_2 + NOCl \longrightarrow (CH_3)_2.CCl.CH_2.NO \qquad (30)$$

$$(CH_3)_2.CCl.CH_2.NO + CH_3SH \longrightarrow (CH_3)_2.C(SCH_3).CH{=}NOH \qquad (31)$$

$$(CH_3)_2.C(SCH_3).CH{=}NOH + CH_3NCO \longrightarrow (CH_3)_2.C(SCH_3).CH{=}NOCONHCH_3 \qquad (32)$$

Methomyl is manufactured by preparing acetaldoxime from acetaldehyde and hydroxylamine (Eq. 33) and chlorinating this in water at -10° to give 1-chloroacetaldoxime (Eq. 34). This is then reacted with methanethiol in aqueous sodium hydroxide and the 1-(methylthio)acetaldoxime (Eq. 35) thus formed is refluxed with methyl isocyanate in methylene chloride (Eq. 36).

$$CH_3CHO + NH_2OH \longrightarrow CH_3CH{=}NOH \qquad (33)$$

$$CH_3CH{=}NOH + Cl_2 \longrightarrow CH_3CCl{=}NOH \qquad (34)$$

$$CH_3CCl{=}NOH + CH_3SH \longrightarrow CH_3C(SCH_3){=}NOH \qquad (35)$$

$$CH_3C(SCH_3){=}NOH + CH_3NCO \longrightarrow CH_3C(SCH_3){=}NOCONHCH_3 \qquad (36)$$

The other thioimidates, thioxamyl and thiocarboxime are manufactured similarly.

Two more complex oxime carbamates which have recently been introduced are tazimcarb and thiophanox.

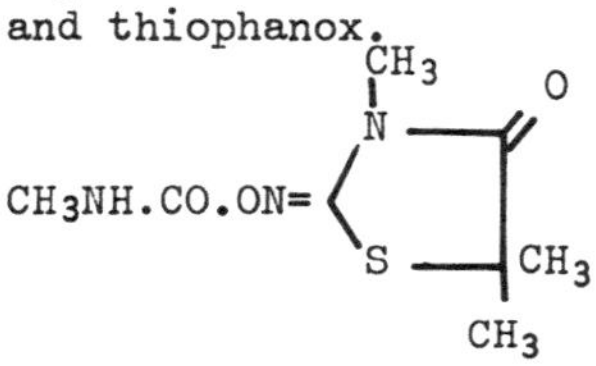

tazimcarb

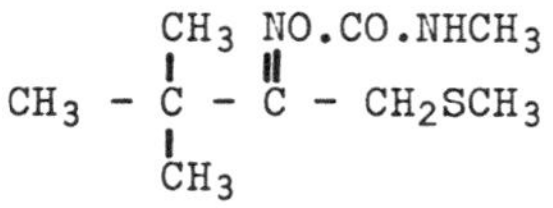

thiophanox

Methylcarbamates of heterocycles R	$RO.CO.NHCH_3$ common name
2,3-dihydro-2-methylbenzofuran-7-yl	decarbofuran
2,3-dihydro-2,2-dimethylbenzofuran-7-yl	carbofuran
2,2-dimethylbenzo-1,3-dioxo-4-yl	bendiocarb
2-(1,3-dioxolane-2-yl)phenyl	dioxacarb

Dimethylcarbamates of heterocycles R	$RO.CO.N(CH_3)_2$ common name
5,6-dimethyl-2-dimethylaminopyrimidin-4-yl	pirimicarb
1-isopropyl-3-methylpyrazol-5-yl	-
1-dimethylcarbamoyl-5-methylpyrazol-3-yl	dimetilan

The methylcarbamates are made by reaction of the hydroxy heterocyclic compound with methyl isocyanate. This is exemplified by the most widely-used compound, carbofuran.

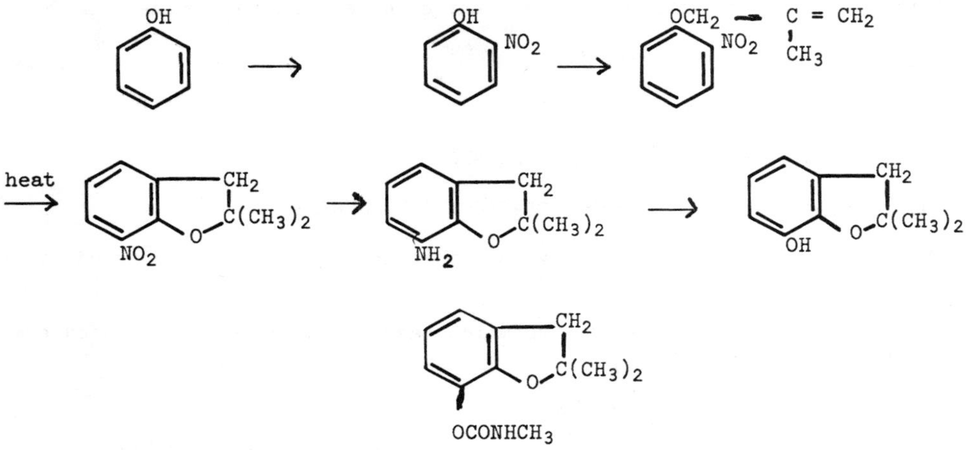

carbofuran

The dimethylcarbamates are made by reaction of the hydroxy compound with a toluene solution of dimethylcarbamoyl chloride in presence of a base such as potassium carbonate. The toluene solution of dimethylcarbamoyl chloride is made by reaction of a toluene solution of phosgene with an aqueous solution of dimethylamine and the toluene layer is separated and used without further purification. Because dimethylcarbamoyl chloride has come under suspicion as a possible carcinogen alternative processes of reacting the hydroxy heterocycle with phosgene to obtain an intermediate chloroformate which is then treated with dimethylamine, have been developed for products of this type.

Further Reading

Corbett, J.R., The Biochemical Mode of Action of Pesticides, (Academic Press, London & New York, 1974)

Eto, M., Organophosphorus Pesticides: Organic and Biological Chemistry, (CRC Press, Cleveland, Ohio, 1974)

Heath, D.F., Organophosphorus Poisons, (Pergamon Press, Oxford, 1961)

O'Brien, R.D., Insecticides, Action and Metabolism, (Academic Press, London & New York, 1967)

O'Brien, R.D., Toxic Phosphorus Esters, (Academic Press, London & New York, 1960)

Street, J.C., (Ed), Pesticide Selectivity, (Marcel Dekker, New York, 1975)

Chapter 8

PESTICIDES OF NATURAL ORIGIN AND SYNTHETIC PYRETHROIDS

Natural Protectants

The struggle between prey and predators, hosts and parasites is as old as life itself. It is not surprising that the natural chemical weapon has evolved along with others. Probably it was at first an offensive weapon used by less mobile animals to prey upon more mobile ones - as in the poisoned tips of the tentacles of the primitive sea anemones. It is now, however, just as often used in defence, or even as a deterrent. Thus the plant deadly nightshade is highly toxic to most animals, but these have learned (in the evolutionary sense) to avoid it.

Examples could be quoted, from every major class of living organisms, of species which in nature poison other species. Some of the most toxic of known chemicals are of natural origin - strychnine, curare, snake-venoms and many others - a fact often forgotten by those who condemn the chemical pesticides industry for "unnatural" practices. Some vegetable extracts are still used to poison pest mammals - e.g. red squill for rats and strychnine for moles.

The biggest potential contribution of natural chemicals for pest control might appear to be as fungicides, because, in most cases where there has been extensive investigation of the mechanism of immunity of some strains of species of higher plant to parasitic fungi which attack related species, it has been found that a special endogenous fungitoxic chemical is responsible. An example is wyerone isolated by Wain from broad bean tissue. This field has, however, so far failed to produce any commercially useful products. The compounds have been too unstable or too non-systemic to be useful when extracted and reapplied. Their effectiveness in the originating species depends upon their continued synthesis at all vulnerable sites. Active research is in progress at many centres and may eventually yield compounds of commercial utility or suggest synthetic possibilities for such compounds. The active quinone, juglone, which was isolated from walnut juice, led to the examination of other quinones. The fungicides dichlone and chloranil which are still used in seed-dressing and some horticultural applications, can be considered as industrial products suggested by natural ones.

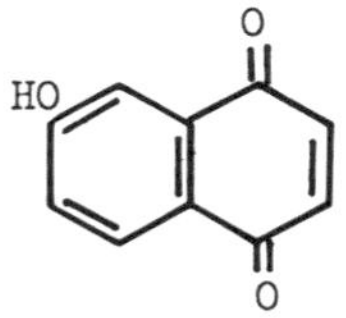

Natural juglone

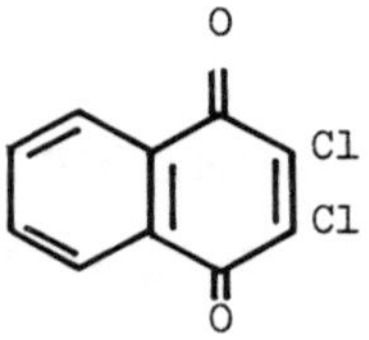

dichlone

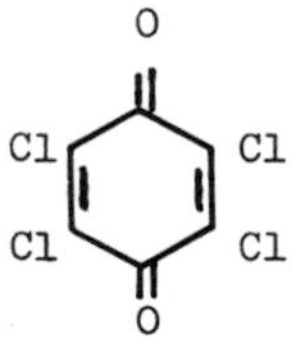

chloranil

To date, by far the most important contributions of natural products to commercial pesticides have been made by insecticides of vegetable origin. For a long time the only effective insecticides were vegetable products. The "industrial insecticide revolution" really only started, in a modest way, with alkylthiocyanates ("lethanes") in the 1930's, then quickly rose to a great pace after DDT was taken up by military authorities in 1942.

The chief natural products are nicotine, derris, pyrethrum and the newer product, ryania. Crude extracts of the plants producing these products, or dried and powdered plant tissue, have a long history of use in South-east Asia as fish poisons or insecticides. They are still used commercially to some extent as insecticides although they have been largely replaced by the synthetic organic insecticides.

None of these natural products is a universal insecticide. Tobacco leaf may contain several per cent of nicotine and yet the plant is subject to attack by several insects. The root from which derris is obtained is subject to attack in storage by the larvae of a species of beetle. The crop from the flowers of which the very active pyrethrins are extracted needs protection in the field by synthetic insecticides. The breeding of resistant species is not a problem exclusively of the synthetics, but, in the case of the natural compounds, a stable "stalemate" has long been reached.

An interesting recent discovery is azadirachtin, a naturally-occurring insect repellent, which may provide Indian and South American peasant farmers with a cheap or even free insect pest control. Azadirachtin is a compound produced by the Neem tree and by the Persian lilac or Chinaberry. It has two separate effects on several common insect pests, including caterpillars and locusts. It firstly repels them from eating the leaves or seeds of the plant. Then, if insects do get past the repellent barrier, azadirachtin upsets the cycle of moults by which they grow, slows down their growth and eventually kills them.

Neem trees and Chinaberries are widespread in parts of India and South America. Peasant farmers in these areas could cut down food loss due to insect pests simply by crushing up and powdering seeds or leaves of these trees and sprinkling them over their crops. Azadirachtin may also be developed as a cheap insecticide but more tests will be needed first to make sure that it is totally non-toxic to humans - at present there is a distinct risk that in high concentrations it might be toxic. The mode of action of azadirachtin is not yet clear, although its chemical structure is fairly close to that of the insect moulting hormones, the so-called ecdysones. Such products might provide indigenous insecticides for those developing countries which cannot afford to buy expensive modern synthetics.

Nicotine

The use of tobacco as an insecticide is of very long standing, dating from the eighteenth century. The active principle, nicotine, is the main alkaloid of the tobacco plant and is present in the leaves at concentrations of 1-8% as a salt of malic acid and citric acid. Until recent years it had a large commercial use as a constituent of insecticidal formulations, but has now been largely replaced by the organophosphorus compounds, although around 500 t/year are still used in the U.S.A.

Derris and Lonchocarpus

It was the practice in tropical countries over centuries to obtain fish by introducing into the water vegetable products from certain plant roots to stupefy them and cause them to float on the surface. In 1911 the first application was made for a patent for the use of derris in the manufacture of insecticides, sheep dips and the like. Up to the 1940's derris and lonchocarpus were used widely and both owe their insecticidal properties to a group of compounds known as rotenoids of which the most important is the compound rotenone.

The main virtues of the rotenoid insecticides were their low toxicities to man and to most mammals except pigs and their short persistence. They are still used in controlling some pests in home gardens, and as powders for keeping dogs and other domestic animals free from fleas and other ectoparasites, but they have been replaced in sheep and cattle dips by organophosphorus compounds. However, it must be remembered that these compounds were originally used as powerful fish poisons and that they are very toxic indeed to fish so their use near ponds, lakes and rivers must be carefully controlled.

Ryania

Substances present in the stems and roots of *Ryania speciosa* - a shrub indigenous to tropical South America - are highly toxic to some insects, particularly caterpillars. The main constituent is the neutral substance, ryanodine, having the formula $C_{25}H_{35}NO_9$. The ground stems are marketed in the U.S.A. for use as a contact and stomach poison. Products containing ryania have not yet been marketed in Europe. They are much in favour in the orchards of eastern Canada.

Pyrethrum

In 1851 Koch stated, for what appears to be the first time, that a peasant-produced insecticide, well-known in Persia and the Caucasus, consisted of a mixture of the powdered, dried flowers of *Chrysanthemum*, *roseum* and

C. carneum. The discovery and early commercial development of powders based on C. cinerariaefolium from which a later "Dalmatian powder" was prepared - is equally obscure. The term "pyrethrum" is now applied to cinerariaefolium since the other pyrethrin-containing plants are not of commercial interest. Serious cultivation of C. cinerariaefolium began in 1886 and extended until the 1914-1918 war, when the supply from Europe was cut off and Japan became the principal world source. In about 1927 the growing of pyrethrum was taken up in the highlands of Kenya and later in other parts of Africa. For many years African production has provided the main source of the world's supply. During the past few years the commercial growing of pyrethrum flowers has been developing in Ecuador. In addition the Administration of the Territory of Papua and New Guinea is now promoting production of pyrethrum in the Western Highlands of New Guinea and an extraction plant has been installed at Mount Hagen.

Processing of Flowers

The flowers are harvested and dried in the country of origin and ground in various types of mill depending upon the use for which the product is intended. Nowadays the bulk of the flowers is subjected to an extractive process with an organic solvent of low boiling point in factories located close to the flower-growing areas. The active insecticides - the pyrethrins - are soluble in a wide range of organic solvents such as ethylene dichloride and petroleum fractions. After removing the solvent these extracts are dark green in colour and are used in the preparation of synergized mixtures and other formulated products. Considerable quantities are subjected to further treatment with selective solvents or to molecular distillation to give pale-coloured extracts. In some cases a solution of the extract in the synergist piperonyl butoxide is subjected to vacuum distillation.

Uses

Today the main use of pyrethrum is as a constituent of space sprays against flying insects, e.g. flies and mosquitoes, in the form of oil solutions applied through a suitable spraying device or by aerosol dispensers. Pyrethrum has such a rapid action as a contact insecticide that the question whether it also has stomach poison activity is irrelevant. The pyrethrins rapidly lose activity in the presence of oxygen and light but despite this weakness, the use of pyrethrum, unlike that of derris and nicotine, has grown with the increase in the use of synthetic chemicals. The pyrethrins owe their importance to their outstandingly rapid knock-down action and to their very low mammalian toxicity, which is largely due to the ease with which they are metabolized into non-toxic products.

Structure of the active principles

The structures of the active esters of pyrethrum flowers are shown below and the composition of the active principles of a typical extract of pyrethrum is pyrethrin I 35%, pyrethrin II 32%, cinerin I 10%, cinerin II 14%, jasmolin I 5% and jasmolin II 4%.

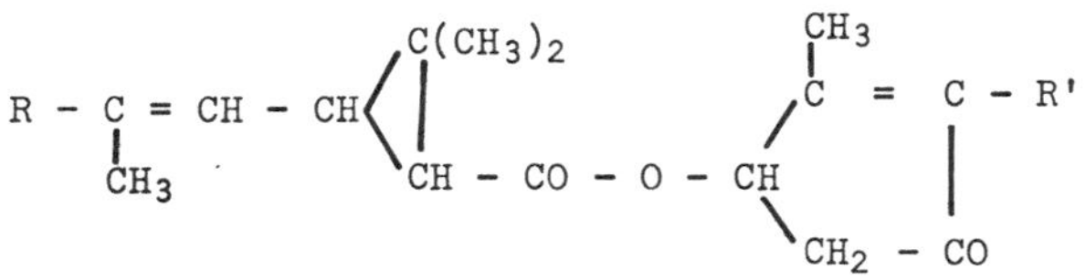

for pyrethrin I	R =	CH_3; R' = $CH_2.CH{:}CH.CH{:}CH_2$
for pyrethrin II	R =	$COOCH_3$;
	R' =	$CH_2.CH{:}CH.CH{:}CH_2$
for cinerin I	R =	CH_3;R' = $CH_2.CH{:}CH.CH_3$
for cinerin II	R =	$COOCH_3$; R' = $CH_2.CH{:}CH.CH_3$
for jasmolin I	R =	CH_3; R' = $CH_2.CH{:}CH.CH_2.CH_3$
for jasmolin II	R =	$COOCH_3$;
	R' =	$CH_2.CH{:}CH.CH_2.CH_3$

Pyrethrins and related compounds

The pyrethrum extracts of commerce contain 25% of the active principles.

Pyrethrum synergists

An examination of the constituents of sesame oil led to the development of a number of compounds containing the methylene-dioxyphenyl group as pyrethrum synergists. The most widely used synergist for the pyrethrins is an ether of propylpiperonyl alcohol always known by the pseudo systematic name of piperonyl butoxide.

Synthetic pyrethroids

Pyrethrin I is the most important natural constituent lethal to insects while pyrethrin II provides much of the rapid knock-down effect. The jasmolins are lower in activity than the four main insecticidal constituents. The insecticidal activity of pyrethrin I is thought to depend upon the overall structure of the ester; in particular on methyl groups at C-2 on the cyclopropane ring maintained in a definite stereochemical disposition with respect to an unsaturated side chain at C-3 and the ester link at C-1. Without steric constraint, the ester probably takes an S-trans conformation and, supported by the near planar cyclopentenolone ring, the *cis*-pentadienyl side chain can adopt only certain orientations with respect to the features of the acid structure mentioned above. High insecticidal potency is probably related to the ability of the molecule to adopt at the site of action an appropriate shape or conformation which will be influenced by the absolute configurations of the asymmetric centres at C-1 of the cyclopropane ring and at C-4 of the cyclopentenone ring. In pyrethrin I, these configurations are respectively R and S and inverting either diminishes or eliminates insecticidal activity. This outline is taken from the review papers by Elliott listed under further reading. These papers should be consulted for information on (i) the spectrum of activity of synthetic pyrethroids, (ii) significance of stereochemical structure (geometrical and optical) in relation to insecticidal activity, (iii) mammalian toxicity and metabolism, (iv) relationship of structure to lethal and knock-down properties, (v) stability in light and oxygen, (vi) relationship of polarity to biological activity and (vii) mode of action.

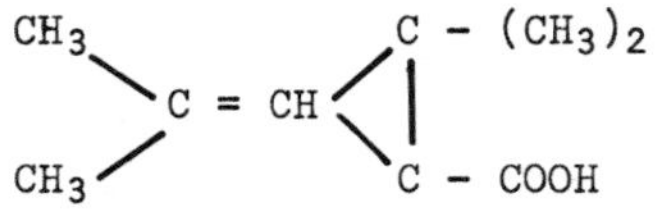

chrysanthemic acid

In 1949 LaForge and Schechter announced the synthesis of allethrin which has $R{=}CH_3$ and $R^1{=}CH_2{-}CH{=}CH_2$, that is, cinerin I with the 2-butenyl group replaced by allyl. The structure of allethrin permits geometrical isomerism about the cyclopropane ring and each geometrical isomer has optical isomers. In allethrin all the isomers were present but allethrolone, the alcoholic component, is now resolved on a commercial scale. The (±)- form esterfied with synthetic (+)- trans chrysanthemic acid gives bioallethrin and esterified with the (+)- trans acid, s-bioallethrin. The latter corresponds to pyrethrin I in stereochemical form and displays an outstandingly rapid knock-down action against some species of insects. Allethrin is less effective than the pyrethrins against a variety of insects and when synergized the disparity in toxicity between allethrin and the pyrethrins becomes even more marked. Because of its volatility and thermal stability much allethrin is used in mosquito coils.

Tetramethrin, reported in 1964, was the next synthetic pyrethroid to be produced commercially. The alcoholic component differs from that in other synthetic pyrethroids and, although tetramethrin knocks down insects rapidly, it is not generally a good lethal agent.

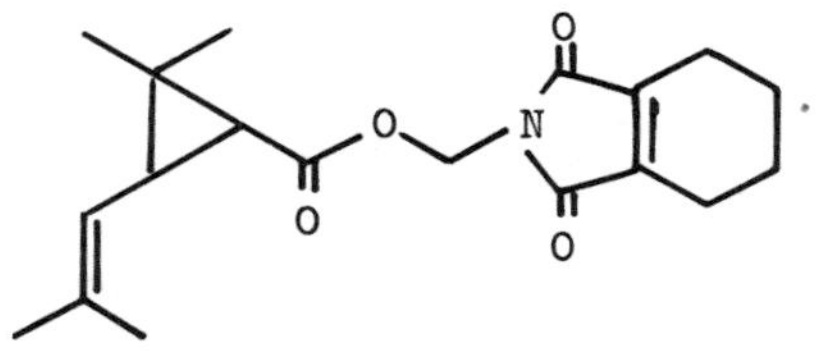

tetramethrin

In 1966 esters of 5-benzyl furyl-3-methanol were shown by Elliott and co-workers to possess properties superior to those of the natural esters containing pyrethrolone as the alcoholic component. The furylmethyl esters were low in mammalian toxicity and warranted commercial production of resmethrin and bioresmethrin. The latter prepared from (+)- trans chrysanthemic acid had especially high insecticidal activity and very low mammalian toxicity. Other pyrethroids derived from furan alcohols include K-othrin, kadethrin, proparthrin and prothrin.

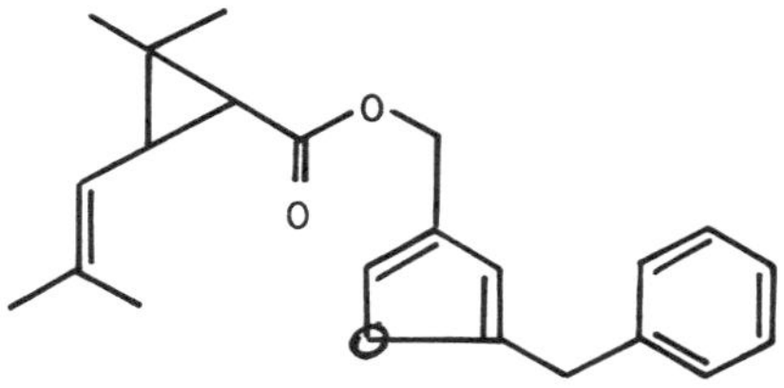

Bioresmethrin

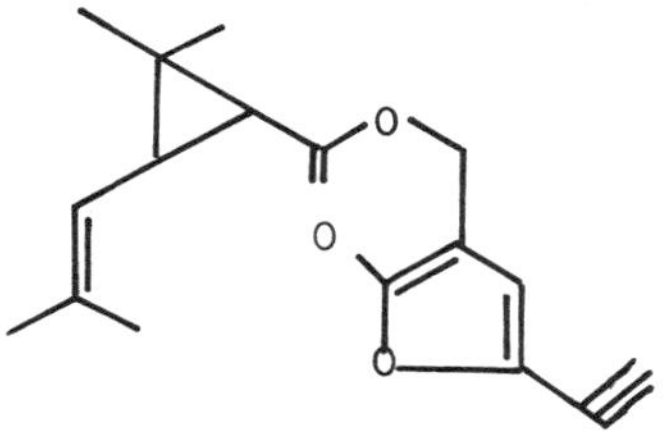

Proparthrin

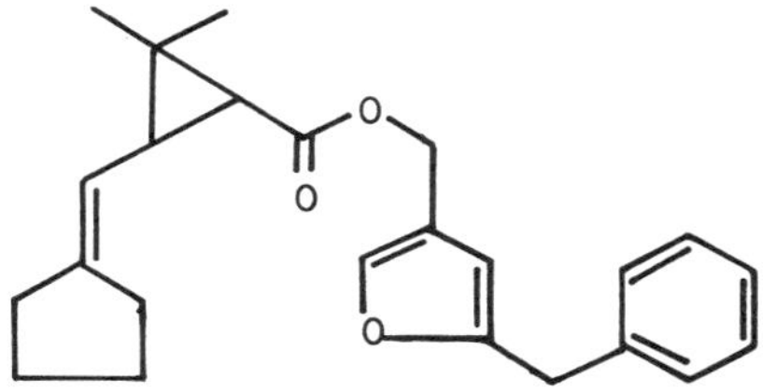

K-othrin

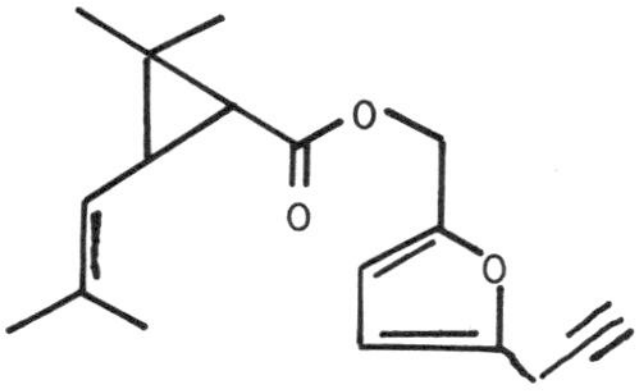

Prothrin

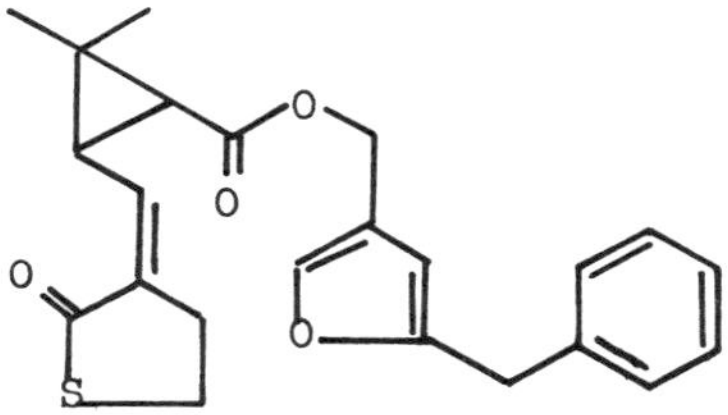

Kadethrin

The pyrethroid K-othrin has an even greater toxicity against some insect species than bioresmethrin but the mammalian toxicity is also higher. Kadethrin knocks down houseflies more rapidly than any other compound yet reported. Proparthrin and prothrin have found a use in Japan as constituents of aerosol sprays.

Workers in Japan and Great Britain next discovered, independently, that the 5-benzylfuryl-3-methanol could be replaced by 3-phenoxybenzyl alcohol, thus providing more accessible and less expensive esters such as phenothrin.

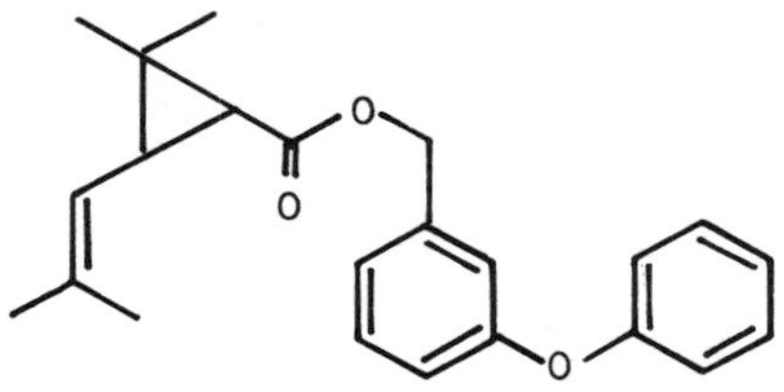

Phenothrin

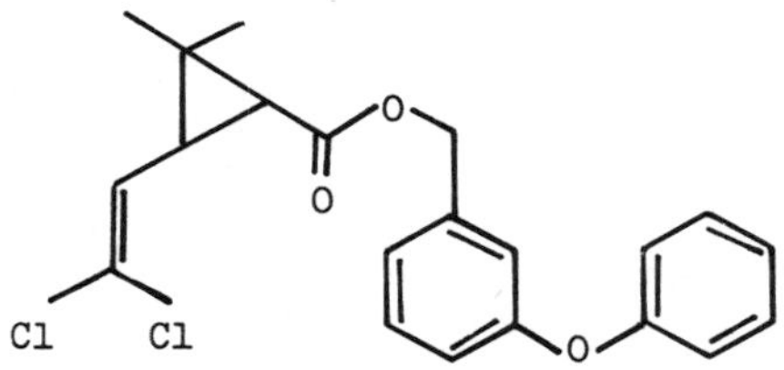

Permethrin

The synthetic pyrethroids mentioned so far failed to extend the range of application established over many years for natural pyrethrum, because they were all unstable to light and air. At Rothamsted Experimental Station in 1972 an exceptionally valuable combination of properties was found in the esters (of which permethrin is an example) of 3-phenoxybenzyl alcohol with cis- and trans- dichlorovinyl analogues of chrysanthemic acid having chlorine atoms in place of the methyl groups in the isobutenyl side chain. Not only was permethrin active against a number of insect species but it was very much more stable in oxygen and light than the other pyrethroids and exerted a prolonged residual action. This opened up for the first time the possibility of using synthetic pyrethroids for control of pests of plants and animals in the field. As the speed of knockdown of insects is less than that of the natural pyrethrins, permethrin exerts both contact and stomach action.

A further outstanding advance was disclosed by Japanese workers in 1974 when Ohno and co-workers showed that highly active esters (e.g. S-5439) could be obtained when various α-isopropylarylacetic acids replaced chrysanthemic acid. Further increase of insecticidal potency was generally obtained by substituting an α-cyano group into 3-phenoxybenzyl esters (e.g. fenvalerate). In the chrysanthemic series, esters derived from α-cyano-3-phenoxybenzyl alcohol were two to three times more active than those from 3-phenoxybenzyl alcohol. The dichlorovinylcyclopropane ester (cypermethrin) is of particular interest.

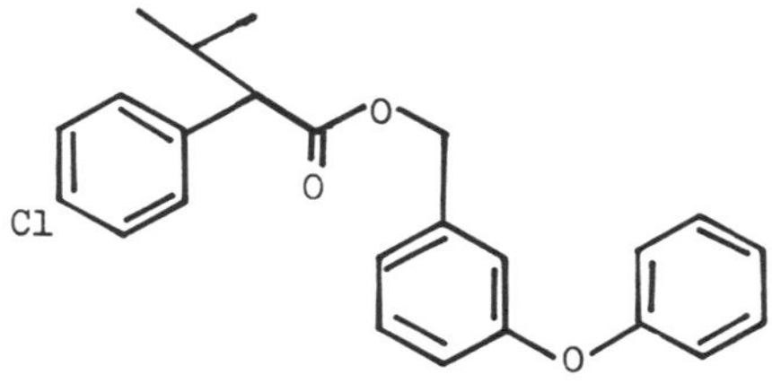

S-5439

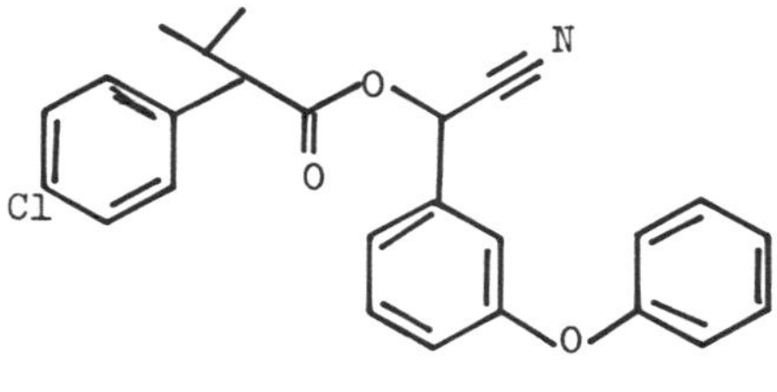

fenvalerate

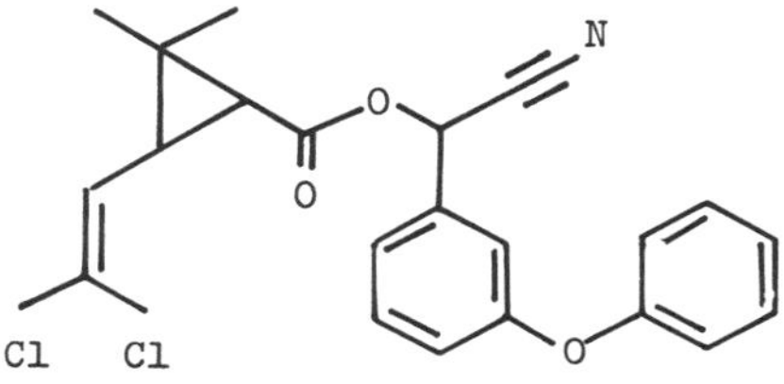

cypermethrin

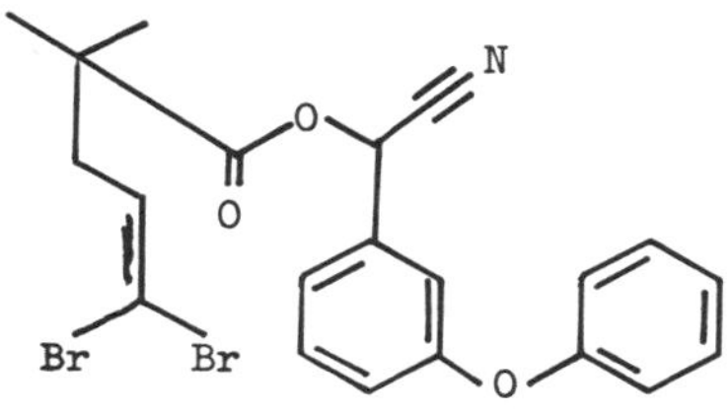

decamethrin

The ester formed between (±)-α-cyano-3-phenoxybenzyl alcohol and (IR, cis)-3-(2,2-dibromovinyl)-2,2-dimethylcyclopropane carboxylic acid gave an active compound from which a single isomer (decamethrin) was separated. This crystalline compound reported by Elliott in 1974 has an insecticidal activity (on a molar basis) approximately 1700 times that of pyrethrin I to houseflies. It is (S)-α-cyano-3-phenoxybenzyl (IR, 3R)-3-(2,2-dibromovinyl)-2,2-dimethylcyclopropane carboxylate.

Insecticidal properties and applications of synthetic pyrethoids

The high insecticidal activity and low mammalian toxicity of pyrethroids are especially significant now that compounds stable to light and oxygen are potentially available. Lepidoptera, especially the larval stages, are serious economic pests of crops such as cotton, maize, tobacco, rice, sugar beet and cane sugar. The natural pyrethrins are effective against lepidopterous larvae but, in the past, instability has precluded their use in the field. The recently synthesized pyrethroids are also very potent and the stability of some of them has opened up the possibility of large scale use in crop protection. This illustrates how just one particular property, in this case stability to air and light, can make the difference between a very limited commercial utility and a potential major world-wide use in horticulture and agriculture.

The synthetic pyrethroids have very low toxicities to mammals and to birds but their toxicity to fish is rather high and could create a need for caution in application near watercourses, rivers and lakes. They are rapidly degraded in soil and have no detectable ill-effects on soil microflora and microfauna. As their insecticidal activity is not specific they are likely to prove hazardous to bees and to other beneficial insects so care will have to be taken in the way they are used. They are also not active against mites so the problems of resurgence of mites which arise with DDT will have to be guarded against. There is some evidence that resistant species of insects may develop when they are used on a large scale under field conditions.

Further Reading

American Chemical Society, Proc. 172nd Annual Meeting of Division of Pesticide Chemistry, (Washington, D.C., 1976)

American Chemical Society, Symposium on Natural Pest Control Agents, Advances in Chemistry Series, No. 53, (Washington, D.C., 1966)

Casida, J.E., (Ed) Pyrethrum, (Academic Press, London & New York, 1973)

Crosby, D.G., Natural Pest Control Agents, (American Chemical Society, Washington, D.C., 1976)

Elliott, M. et al, Environmental Health Perspective, March 1976, (London)

Elliott, M., Insecticides for the Future: Needs and Prospects, (Wiley, London & New York, 1977)

Elliott, M. et al, Pesticide Science, June 1976, (London)

United States Department of Agriculture, Insecticides from Plants, USDA Handbook, No. 154, (Washington, D.C., 1954)

Chapter 9

REPELLENTS, ATTRACTANTS AND OTHER BEHAVIOUR-CONTROLLING COMPOUNDS

Repellents and attractants might seem to be compounds evoking directly opposite responses. In fact neither response is simple. Repulsion is the less complex behaviour. It is much more frequently evoked by synthetic compounds and is much less specific than attraction. It is partly for this reason that repellents are already made in large amounts and widely used with good effect, while attractants are still in the development stage. There are, however, two other good reasons for the much more advanced state of repellents.

The most important is that repulsion, if efficient enough, is itself a useful means of pest control, which need not be identified with pest destruction. If we could stop the mosquito from biting us, the carrot fly from laying eggs near our carrots or the deer from stripping bark from our forest trees, we should have done all we need in our local interest. If the pests are far-ranging, like pigeons or deer, the use of an effective repellent, were one known, could be strategically better than killing, unless killing were carried out over a wide area. An attractant, on the other hand, can serve no useful control purpose unless the pests attracted are killed or sterilized by some other agent. An attractant must be a part of a combined operation. A repellent can be useful in itself.

Another important difference is that a repellent need not be effective over a great distance from its site of application. It would be preferable to drive the mosquitoes out of range of sight and hearing but the only really necessary effect is to stop them biting. Only a very small proportion of the population of mosquitoes within flight range may come near enough to be attracted to blood or repelled by repellent. An attractant, to be useful, must lure a high proportion of the population to its destruction, and therefore must be effective over a long range. It has a more difficult task to perform.

Insect Repellents for Man and Livestock

Coal-fire soot scattered along the row of seedlings is a very effective deterrent to slugs. Its effect is probably mainly mechanical, providing a very unsuitable surface for the crawling process of the animal. Pepper, as a repellent for cats, is effective only when it is disturbed by scraping. Evil-tasting oily products smeared on tree-boles, in so far as they are effective in preventing damage by deer and rodents, are distasteful to tongue and lips and even this may be partly mechanical.

At the other extreme, citronella oil is a powerful repellent of flying mosquitoes. If objects contaminated with it are interspersed in a large chamber, or out of doors, with similar but absorbing objects (for example, gauze cylinders, some filled with cotton wool moistened with the oil, others with active charcoal), mosquitoes do not merely not stay on the oil-scented surfaces, but clearly avoid them before they are within alighting distance. It will be seen below that the mosquito repellents now most widely used appear to act at a distance by cancelling the signals of attraction.

There are many species of biting insects. At best they are a nuisance, sometimes almost intolerable, but many also carry diseases. The best known example is the transmission of malaria by anopheline mosquitoes, but there are other serious diseases, prevalent in the tropics, which are carried by biting insects. In the case of an obligate parasite which lives as well as feeds upon its host, for example the body louse (the carrier of typhus), attention throughout the community to personal hygiene is the best treatment. It has been said that the most important public health insecticide is soap and water. With many free-living flies, however, especially when man is only one of many host species, personal hygiene is much less effective.

It is against the free-living biting insects outside urban areas that repellents are mainly used. The problem became particularly acute for military personnel in jungle warfare. At one period during the Second World War it was estimated that the probability of a soldier being incapacitated by disease before he engaged the human enemy was over 90%. Most of the casualties were through insect-borne diseases.

Oil of citronella has been mentioned as an effective repellent. Its main defects are that it is short-lived, being too volatile, and that many people do not like its aggressive smell. During the war the sources of supply were largely cut off. Intensive search for less objectionable, longer lasting but still effective repellents was undertaken. Very important further requirements were low toxicity and non-irritancy to the human skin. Dimethyl phthalate, easily and cheaply produced from phthalic anhydride and methanol, was soon found to be very effective and was very widely used.

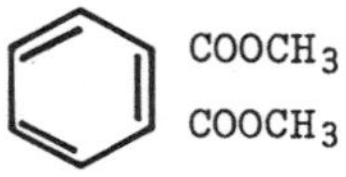

dimethyl phthalate

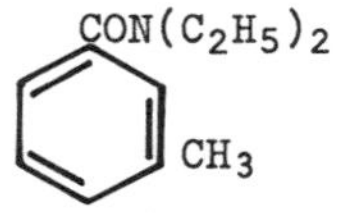

DEET

Dimethyl phthalate was not, however, effective against all inimical species and the search went on and thousands of compounds were tested. Two became widely used in admixture with dimethyl phthalate to produce a broader spectrum of action. These were 6-butoxycarbonyl-2,3-dihydro-2,2-dimethyl pyran-4-one (butopyronoxyl), made from mesityl oxide and dibutyl oxalate, and 3-hydroxymethylheptan-4-ol (ethohexadiol), made by hydrogenation of butyraldol. A third compound, the most generally effective of all and the one most used today, came too late for war-time use. It is N,N-diethyl-3-methylbenzamide (DEET), manufactured from 3-methylbenzoyl chloride and diethylamine. The 2- and 4- methyl compounds are also active, but less so than the 3-methyl isomer. Dibutyl succinate, dipropyl isocinchomeronate and butoxypolypropylene glycol have also found some use as insect repellents.

These compounds do not steer the mosquitoes away in flight as does oil or citronella. Although a tangible smear produces a departure reaction in an alighting insect, particularly in the case of DEET, their most important effect seems to be to cancel out the attractant properties of the animal body. It has been demonstrated that they prevent mosquitoes from detecting the presence of a nearby warm, wet surface. If flat copper bars are laid on the floor of a cage and heated at one end and cooled at the other, they are ignored by the mosquitoes as long as they are dry. However, when covered with dampened cloth (black was found to be better than white), mosquitoes alighted very much more frequently on the warm part of the bar; in fact they seldom landed on the cold part. When the air was permeated with molecules of a repellent by hanging a sheet of filter paper moistened with the particular repellent in the upper part of the cage, the behaviour of the mosquitoes was completely changed. The number alighting on the bar was very greatly reduced and instead of being concentrated near the warm end they appeared to be distributed along its length.

Repellents are formulated as oils, creams or gels for hand application, or put up in self-dispensing aerosol packs. Creams and aerosols incorporating a sun screen with the repellents are marketed.

Pyrethrum deposits have considerable repellent effect, as well as being insecticidal. It is a matter of everyday observation that after spraying a room with a kerosine solution of pyrethrum (fly spray) or use of a pyrethrum aerosol, flies tend to avoid this part of the house for some time. The burning of pellets or coils containing pyrethrum is practised in various parts of the world to repel mosquitoes from rooms, particularly bedrooms. Although a sufficient dose of pyrethrum rapidly renders an insect immobile, the first effect of slight pick-up is to produce a state of great excitement. The mechanism of repellence in this case may be a direct consequence. There is some parallel to the behaviour of gases according to the kinetic theory. Concentration decreases in a part of the gas volume which is heated. The effect of a low dosage of pyrethrum on flying insects is somewhat equivalent to several thousand degrees temperature rise on gas molecules.

Pyrethrum preparations are often used for spraying milking sheds, and cattle coming in, in order to reduce the fly nuisance. Here the immediate excitation on first contact is the most important effect. The population of flying insects in the shed is reduced, but the number of insects alighting long enough to bite is much further reduced.

The market for insect repellents for livestock, both domestic and range-fed, is potentially very great. These animals suffer from many ectoparasites, such as ticks, fleas and lice, which are carriers of disease. Also the nuisance insects often pester them in much larger numbers than they do humans. The fly *Musca autumnalis*, a nectar feeder in its native Europe, was introduced into North America where it adopted a habit of sucking facial liquids on cattle. It has become the major nuisance face-fly and animals have been known to die through exhaustion under the intolerable discomfort.

Many repellents are known, including those used successfully to protect man, which also effectively protect cattle, but they are too transient and expensive to be economic for the latter purpose. This is a very active field of research, but insecticides seem to be making more progress than repellents. For both, in the case of range animals, automatic means of application have had to be devised such as oil-smeared rubbing posts, and dust-filled sacks suspended below back level over entrances to shelters or drinking stations.

Repellents in Agriculture

Insect repellents have found little use in modern agriculture for protection of growing crops. The large areas involved would probably make their use uneconomical even if it could be made effective. Their use is also incompatible with monoculture in large fields, within which there can be no choice of food for the insect. Under these conditions, only a repellent which also completely inhibited feeding, could be effective.

Some use is made in agriculture of chemical repellents for birds and small mammals which cause tremendous damage to growing crops. Although public sentiment views destruction of insects, or even of rodents, with equanimity, killing of birds and other "pleasant" animals provokes violent public reaction.

9,10-Anthraquinone is used as a seed dressing to deter birds. An interesting compound is 4-aminopyridine which does not kill birds at the concentrations used for repellence but causes those which eat bait containing it to behave in a manner which alarms the whole flock and makes them leave and not return. In some way, not understood, this compound creates a disturbance in birds and produces in them fear of treated areas.

The fungicides tetramethylthiuram disulphide and zinc dimethyldithiocarbamate-cyclohexylamine complex are used to repel deer and rabbits, methyl nonyl ketone is used to repel cats and dogs, and tertbutylsulphenyl dimethyldithiocarbamate is incorporated into rubber and plastic insulation for electric cables to repel rodents from gnawing them.

Attractants

Traps and poison baits are perhaps the most ancient of all pest control devices. Many insect traps do not make use of chemicals, e.g. light traps. The earliest "chemical" traps were based entirely on natural attractants, e.g. laying down slices of old potato to collect wireworms or slugs. This technique came into the province of the chemical industry when chemical means of killing the trapped insects were used or when chemical poisons were mixed with the natural baits. For example, it used to be the practice, in glasshouses built on old pasture, to sow lines of wheat between tomatoes. This "trap crop" attracted most of the wireworms, which were then killed by trickling cyanide solution along the rows. As an example of use of chemical poisons mixed with natural baits widespread distribution of a mixture of moist bran bait with an organochlorine insecticide over the non-agricultural breeding grounds of the desert locust has been extensively used to stop populations increasing to swarming densities. Pellets of bran, metaldehyde and a binder are the most widely used means of controlling slugs.

An evident disadvantage of a natural bait is that, in many situations, it may be necessary to use more bait than is saved in food. If, for example, a pest attacks a high proportion of fruits in an orchard it might be necessary to lay down poisoned fruits in greater quantity than the fruits it is hoped to harvest, in order to make any significant reduction in pest populations. Such a practice would be hopelessly uneconomical.

An obvious alternative would be to find cheap synthetic attractants which could be used as baits. Insects appear to be attracted to sources of food by specific chemical stimuli - the phytomones. Very little is known about the chemical natures of these, and those synthetic attractants which have been discovered have been found by empirical testing not by constructive thought. Two effective compounds which have been developed are sec butyl 2-methylbenzoate, which attracts the Mediterranean fruit fly, and methyl eugenol, which attracts the tropical fruit fly.

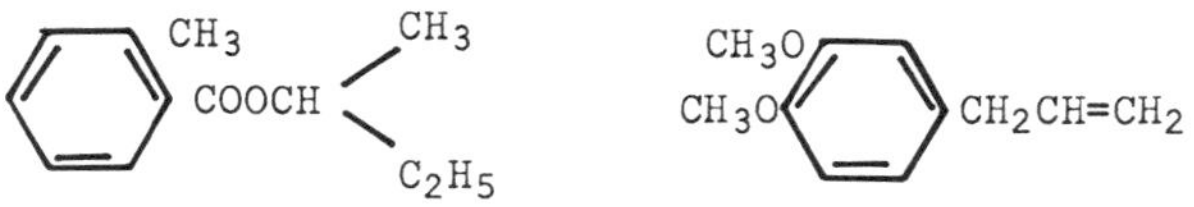

sec butyl 2-methylbenzoate   methyl eugenol

Given the right circumstances, use of such attractants can be effective. A mixture of methyl eugenol and insecticide was impregnated into porous wafers and scattered from the air over a Caribbean island and proved so efficient in attracting and killing the tropical fruit fly that this insect was practically exterminated in that limited environment.

Insect Sex Attractants

It has long been known that the females of some species of insect can excite an approaching response in males at a distance, in favourable cases, of miles. It is clear that volatile substances of extremely high physiological activity must be responsible. Now that something is known about these sex "pheromones", it has been estimated that the threshold quantity for response can be as low as about 30 molecules or 10^{-14} μg. The molecules are picked up by the antennae of the males where they produce electrical impulses in the nerves which initiate a muscular response. This response is produced only by the specific pheromone.

Interest in the sex-pheromones has recently been intensified by two developments. Firstly, the exploitation of modern microtechniques - chromatographic separation and nuclear-magnetic-resonance, infrared, and mass-spectrograph analysis - has made possible the chemical identification of some of the compounds responsible, available only in minute quantities. Secondly, public pressure for more selective methods of pest control and the increasing importance of insect resistance, have directed attention to this most specific of all responses and one without which reproduction would appear impossible.

Many of the natural insect sex pheromones have now been isolated, identified and synthesised. In general, they tend to be long chain unsaturated aliphatic alcohols or ketones or related compounds, but activity for a particular species is critically dependent on precise structure and configuration. For example, only the 10-trans-12-cis isomer of the silkworm moth attractant shown below is effective. The insect can distinguish geometrical isomers and enantiomers.

$CH_3(CH_2)_2CH{=}CHCH{=}CH(CH_2)_8CH_2OH$
natural silkworm moth attractant

$CH_3(CH_2)_5CH(OOCCH_3)CH_2CH{=}CH(CH_2)_5CH_2OH$
natural gypsy moth attractant

$CH_3(CH_2)_5CH(OOCCH_3)CH_2CH{=}CH(CH_2)_7CH_2OH$
synthetic gypsy moth attractant

Nevertheless, provided the structural and steric features essential for activity are retained it is possible to produce synthetic compounds which have attractive properties similar to the natural pheromones. For example, the synthetic compound 12-acetoxy-1-hydroxy-9-octadecene, which can be manufactured economically from ricinoleic acid obtained from castor oil, has similar properties to the natural gypsy moth attractant.

The sex pheromones are not always produced by the female. In the case of the cotton boll weevil it is the male which produces four pheromones which attract the female.

Nor is a sexual response the only one which can be produced by pheromones. Pheromones are, in general terms, any chemicals secreted by an animal to stimulate some form of physiological or behavioural response from another member of the same species. Ants, for example, secrete pheromones as a trail to guide other members of the colony back to the nest. "Alarm pheromones" are emitted by some insect species to warn other members of danger. Tree bark beetles secrete "aggregation pheromones" to bring them together for a concerted onslaught on a pine-tree.

It appears, therefore, that much of the behaviour of insects is mediated by chemical stimuli and it is an attractive thought to synthesize natural stimuli or active analogues and to use these to interfere with natural behaviour of insect pests in a way which could be used to control them. Such a method of control would be environmentally safe, specific to a particular pest and free from the possibility of toxic residues. This is, however much more difficult than it sounds because we understand so little about the complexities of insect behaviour that we have no real idea how best to use natural or synthetic pheromones for insect control.

Chemical Control of Insect Behaviour

The behaviour of insects, particularly of the social insects such as bees and ants, has long been a challenge to human understanding. Much of the behaviour makes no sense at all in terms of human motives, emotions or intelligence. The directionally accurate homing flight of a honey-bee from a distant foraging expedition arouses wonder and admiration and yet, if the hive is moved a few feet during the bee's absence, the behaviour of the returning bee appears incomprehensibly stupid. It will fly about to exhaustion near the old location of the alighting board, apparently unable to detect the conspicuous hive in the immediate vicinity.

Insect behaviour requires completely impartial and objective study: no interpretation should ever be based on human behaviour. It is being studied very deeply by the knowledgeable few, but their work is very little known to the intelligent layman or the worker in related fields. Certainly until recently little interest was taken by practical agricultural entomologists. The natural-product chemist confines his resourcefulness to identifying characteristically olfactory substances available in extremely minute quantities. Yet, without presentation of basic knowledge of insect behaviour to the agricultural research worker, the discoveries of the chemist are not likely to become of much practical value.

Attraction and repulsion are essentially vector (directional) behaviour patterns. Concentration of a chemical substance is essentially a scalar (non-directional) quantity. No substance can attract or repel per se. A very steep concentration gradient, which is met only close to some source, as at the edges of a small thermal up-current of air, or in extreme form at an interface between liquid and air, has been shown to produce a vector response, but this cannot explain attraction from great distances. In the only cases where the mechanism of such attraction has been clearly demonstrated, it was found that the chemical "command" is "fly upwind". The vector part of the response thus depends on wind direction, not on concentration gradient. The insect flies in such direction that, judged by visual marking on the ground, it makes slowest progress. If it cannot make progress at all in the slowest direction (i.e. the wind is too strong), it does not fly. This response is perfectly suited to natural conditions but, by suitable tricks with moving light patterns on the floor of a wind tunnel, the behaviour of the insect can be made totally inappropriate and it is by such experiments that the detailed response can be analyzed.

The command may be "fly upwind". It may be "turn sharply". It may be "fly faster". It may, as with a contact repellent, be simply "fly" or, as on contact with a suitable host surface, "don't fly". More than one command may be involved and one stimulus may cancel another. Many of the commands responsible for feeding control and other activities are chemical in origin. Their definite "stop-go" nature has been shown by many experiments in which insects have been caused to eat, suicidally, non-food which tastes right, to starve in the presence of food or even to engorge to the point of actual bursting.

Practical Use of Attractants

It is clear from the preceding discussion that isolation and identification of a pheromone by the organic chemist does not mean that it can be immediately used as an alternative method to conventional pesticides for insect control. Results in the field have so far been disappointing. In order to be able to make effective use of pheromones, it will be essential to accumulate much more detailed knowledge of the ecology and behaviour of the species we wish to control than we have at present.

For example, it was thought that, if a whole area were sprayed with a synthetic sex pheromone, the concentration level would be everywhere so high that it would completely mask the natural emission of the female insect and render it undetectable to the male. This is the so-called "confusion technique". However, it underestimated the resourcefulness of nature and of the male insect which seemed to have other means of locating the female in a population dense enough to be a threat to a crop. Possibly the pheromone may be of significance only in a sparse and scattered population.

It is also doubtful whether such a use of sex pheromones would be economical since relatively large amounts would need to be used for the confusion technique and the compounds are extremely expensive compared with conventional insecticides. It is unlikely that their manufacture and sale would give profits large enough to justify the full range of toxicological and residue studies which are required for registration.

Field work on the use of pheromones is going on and possibly the most promising results are being obtained with the cotton boll weevil. However, their only practical use at the moment is to bait traps to give indications of insect populations so that application of conventional insecticides can be timed to produce the greatest effect with the minimum amount of insecticide. So, although insect pheromones provide interesting topics for natural product chemists they will be of little value for practical crop protection and pest control until a lot more work has been done by the biologists.

Anti-feeding Compounds

The feeding of insects is initiated by chemoreception in the appropriate organ. The initiating compound is characteristic of the host range of species and is usually a token without significant nutritional value. The token substance applied to paper may even initiate feeding on this totally unsuitable food. The feeding initiator is not usually identical with the compound which leads the insect to the host. The signal to cease feeding is more complex. "Surgical" procedures can cancel this signal so that the insect feeds until it bursts, and at least one compound, methyl eugenol, is known which has the same effect without surgery in the tropical fruit fly.

Inappropriate or excessive feeding has not so far been considered outside the laboratory, but compounds which appear to cancel the signal to initiate feeding are known and research in this field to find better "anti-feeding" compounds has been pursued. No repellent action need be associated with the anti-feeding action. In the presence of the anti-feeding compound the insect may starve to death while remaining on its host plant.

Possible application in agriculture seems to be opened up by the discovery that 4-(dimethyl-triazeno)-acetanilide (DTA)

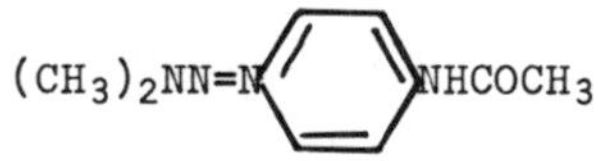

DTA

inhibited the feeding of southern army worms when applied at dosage rates in the range of commercial insecticides. The screening tests showed that the compound was not effective against aphids or mites, but was effective against army worms and Mexican bean-beetle larvae. The non-toxic compound left the predators unaffected and their appetites did not seem to be influenced. It was not highly toxic to honey-bees and did not inhibit them from feeding on treated sugar solutions. It showed some systemic behaviour in the host plant.

A limited field test was organized mainly to ascertain the effectiveness and possibilities of the anti-feeding concept as a method of insect control. Although the compound was ineffective against chewing insects which fed inside the fruit, such as the plum curculio, corn earworm and cornborer, the first season's results were promising in that most of the leaf-feeding caterpillars and beetles were controlled by the material. However, more extensive field tests on some cotton insects, boll weevil, bollworm and leaf-worm, did not give any useful results. It appeared that when low insect population was present the effect of the deterrent compound was dominant, but under heavy pressure from an increasing population the deterrent ceased to be effective. No further compounds of this type have been developed, and this is no longer considered a very promising area of research.

Oviposition Inhibitors

It has been found that, after the female boll weevil has deposited her egg in a cotton bud, she cements the hole with her frass. This frass releases volatile compounds, which appear to be specific monoterpene alcohols or aldehydes, which deter other females from depositing an egg in the same bud. If the deterrent compounds could be identified and synthesized or if synthetic analogues which were effective could be discovered, spraying a cotton crop with these might be an effective method of controlling the pest.

Further Reading

Beroza, M., Pest Management with Insect Sex Attractants, (American Chemical Society, Washington, D.C., 1976)

Wood, D.C., Silverstein, R.M. & Nakajima, M., (Eds)., Control of Insect Behaviour by Natural Products, (Academic Press, London & New York, 1970)

Symposium on insect pheromones, Pesticide Science, December 1976 (London)

Chapter 10

CHEMICALS USED AGAINST OTHER INVERTEBRATES

Acaricides

The mites belong to the class Arachnida and differ from insects in having unsegmented bodies, eight legs and no antennae. They also differ in their biochemistry, consequently, many insecticides do not affect them. This has caused problems in the past when application of general insecticides in orchards destroyed the insect predators of the mites and converted them from minor into major pests. The dangers of upsetting ecological balances are nowadays well appreciated and precautions are taken to avoid them.

Mites are commercially important pests of many fruit and vegetable crops and of stored grain and flour. They are a special problem in glasshouses which provide a favourable environment. Compounds called acaricides have been developed to control them. Mites have a great ability to develop resistant strains quickly so the useful life of any acaricide tends to be short - 10 years or so.

A productive chemical class for acaricides has been the bridged diphenyl compounds, of which some of the more important examples are illustrated below. The earliest diphenyl compound used, mainly as a smoke in greenhouses, was azobenzene.

X	name
$-OCH_2O-$	DCPM
-SS-	DDDS
$-CH_2S-$	chlorbenside
$-OSO_2-$	chlorfenson
OH -C - CH_3	chlorfenethol
OH -C - CCl_3	dicofol
OH -C - $COOC_2H_5$	chlorobenzilate

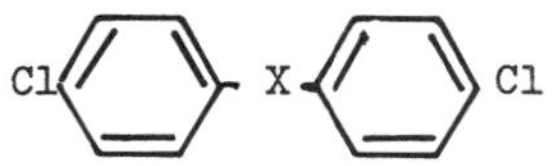

A similar compound, which was the most commercially useful acaricide for fruit trees for many years until resistance developed, is tetradifon.

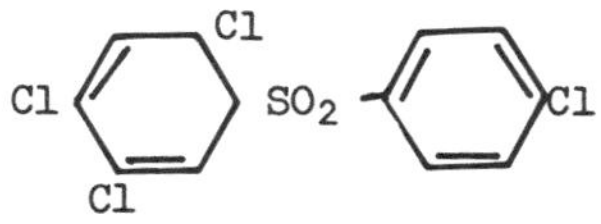

tetradifon

This was the most active of a large number of chlorinated diphenyl sulphones tested. It is manufactured by reaction of 1,2,4-trichlorobenzene with chlorosulphonic acid to give 2,4,5-trichlorobenzene sulphonyl chloride which is then condensed with chlorobenzene.

The most successful and widely used acaricide in recent years has been an organo-tin compound, cyhexatin. A similar, more recent compound, is neostanox.

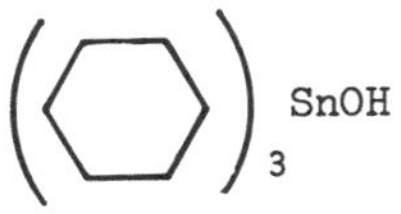

cyhexatin

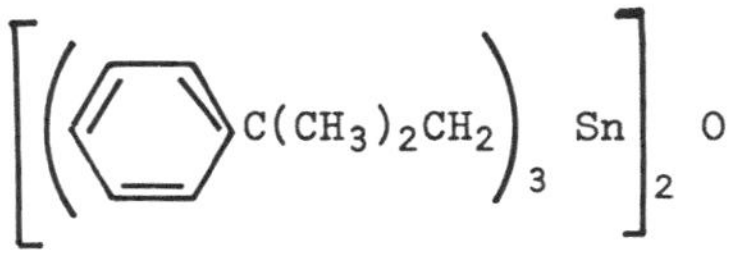

neostanox

Some nitrophenol derivatives have been commercially used.

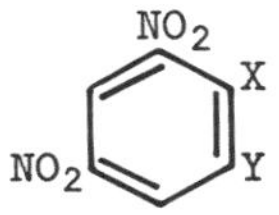

X	Y	name
OH	CH_3	DNOC
$CH_3CH{=}CHCOO$	C_8H_{17}	dinocap-6
$(CH_3)_2C{=}CHCOO$	C_4H_9	binapacryl
CH_3OCOO	C_8H_{17}	dinocton-6
$(CH_3)_2CHOCOO$	C_4H_9	dinobuton
OH	C_6H_5	dinex

Miscellaneous acaricides include:

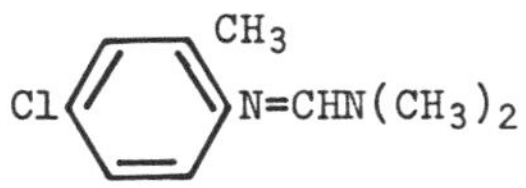

chlordimeform

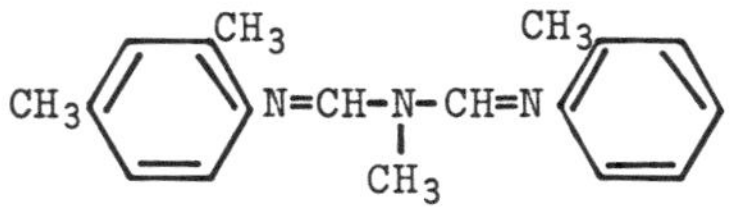

amitraz

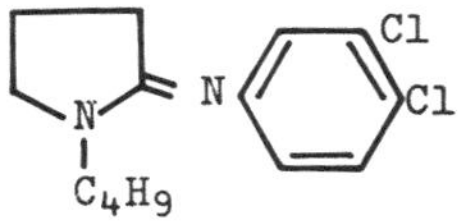

clenpyrin

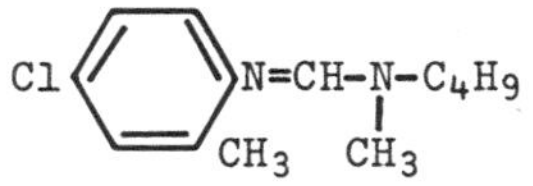

chloromebuform

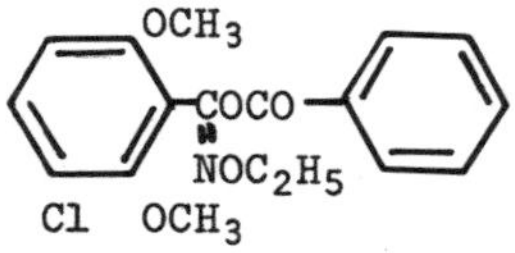

benzoximate

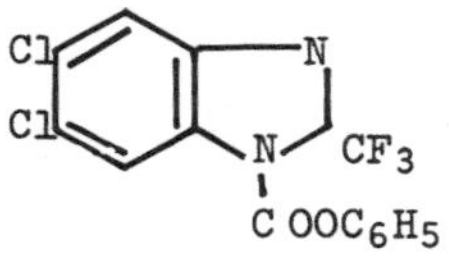

fenazaflor

Spider mites rasp the tissues of leaves and suck up the cell contents. They take up in this way a relatively large volume of cell sap and consequently tend to be vulnerable to the water-soluble systemic organophosphorus aphicides as well as to the specific acaricides mentioned, which have only a very limited systemic action.

It should be noted that other members of the Arachnida, especially the cattle tick and sheep scab mite, do extensive damage to agricultural animals. Various chemicals are used in their control but a discussion of animal health products is outside the scope of this book and the reader should consult the books indicated at the end of this chapter for further information.

Nematicides

Nematodes are small worm-like organisms, generally about 1 mm long, which either live on plant roots as ectoparasites or enter the plant tissues via the roots and become endoparasites in leaves and stems. There are very many species which are ubiquitous and do vast damage to crops, not only by feeding on the plants and causing them to become stunted, unyielding and less resistant to diseases, but also by actively transmitting virus diseases and allowing entry of fungi and bacteria through the damaged roots. Many produce galls. Their eggs remain dormant in the soil for long periods and often hatch out only under the influence of chemical substances secreted from the roots of the growing plants. The full nematode fauna of soil and the precise damage they do are imperfectly known as yet.

Because the eggs are often protected by cysts they are very invulnerable to attack of any kind and so chemical control is a difficult problem. The principal method of control is still crop rotation but this has its limitations, and the modern tendency towards monoculture has exacerbated the problem. An attractive method of control would be chemicals which would cause the eggs to hatch in the absence of host plants, but this has not yet been achieved. The natural hatching factors are complex oxidized sugars, quite unsuitable for commercial production.

The most widely used methods for controlling nematodes involve rather drastic and expensive treatment of the soil before sowing or planting. They are application of heat, generally in the form of live steam, or volatile chemicals having a wide spectrum of toxicity. In both cases almost complete sterilization of the treated soil results, but it is virtually impossible to achieve this in the whole depth of soil, even in glasshouse practice. However, most soil nematodes are not deep-dwelling and are usually eliminated, so the results in practice are better than might be expected. Some nematodes have seasonal migratory habits and it is important in these cases that treatment is carried out when they are not in the deep layers of the soil. The damage done by nematodes to crops is much more serious if they are present during the early life of the transplant or seedling so elimination from the initial rooting zone may give the most satisfactory results.

All the fumigants used to control nematodes are toxic to plants and must be applied from one week to several months before planting. The rate of application is usually 200-450 kg/ha. The compounds most widely used are methyl bromide, ethylene dibromide, 1,3-dichloropropene, methyl isothiocyanate, formaldehyde, carbon disulphide and chloropicrin. The first four are generally used when nematodes are the primary target and the last three when control of soil fungi is also required, but they are all effective general sterilants in sufficiently high dosage. 1,2-Dibromo-3-chloropropane has sufficiently low phytotoxicity to permit its use under favourable and controlled conditions for control of nematodes in citrus orchards. The toxicant is introduced into the soil along one side of each row of trees in one year and along the other side in the next in order to reduce direct chemical damage.

All these compounds disperse through the soil in the vapour phase and, if good nematode control is to be achieved, distribution within the soil must be very good.

If a non-volatile nematicide is to be used this distribution must be achieved mechanically or by the use of heavy irrigation following treatment with a water-soluble compound. The first method involves introduction of a dust formulation into the soil and stirring to the necessary depth with a powerful rotary hoe. This method is effective only on easily-worked, friable soils, and is expensive. It has been applied to good potato land using an inorganic mercury salt as the nematicide. The second method can be applied only to good-structured, free-draining soils in which the compound can be washed down quickly and uniformly. If slow downward leaching under natural rain is to be relied on, the compound must not be significantly adsorbed on soil colloids. This imposes an almost impossible condition for complex non-ionized substances.

Metham sodium is commonly used as a drench and acts by breaking down into methyl isothiocyanate. A number of organophosphorus compounds are also very useful for this type of application, for example,

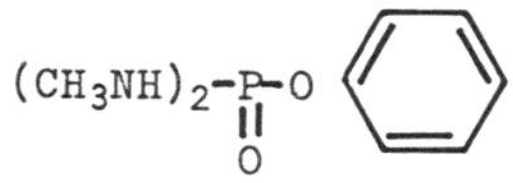

diamidfos

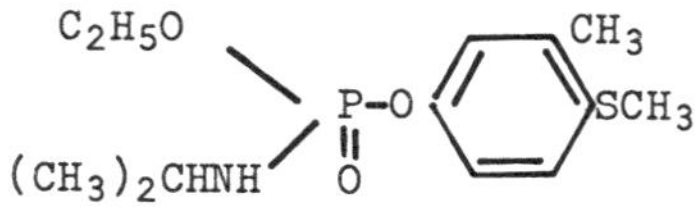

One of the most effective compounds for incorporation into topsoil by rotary cultivation is dazomet. However, applying a nematicide to soil in the autumn creates difficulties and non-phytotoxic granular materials are being sought which can be applied to seedbeds in the spring. Some organophosphorus compounds are promising but the most effective substances are oxime carbamates such as aldicarb. At 7-14 kg/ha this compound prevents nematode injury to crops and stops them multiplying.

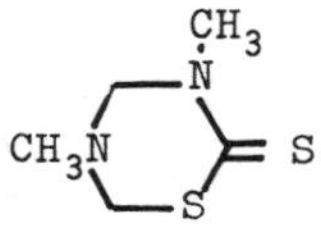

dazomet

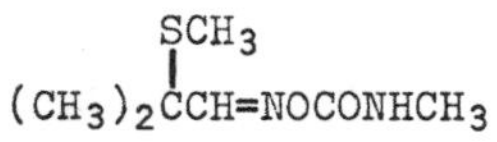

aldicarb

The leaf and stem-dwelling nematodes may be controlled by several organophosphorus pesticides. Parathion is effective on chrysanthemums. Thionazin is used on bulbs and corms. Diclofenthion is less poisonous to men and animals.

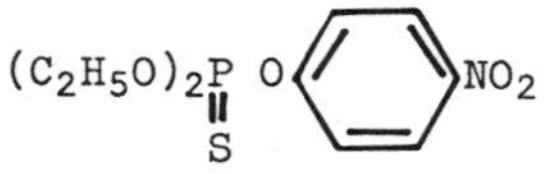

parathion

thionazin

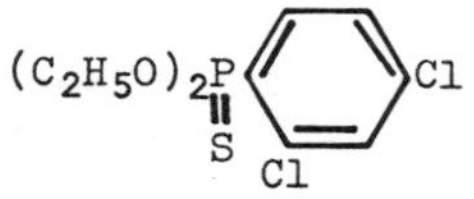

diclofenthion

Molluscicides

Slugs and snails cause serious damage to crops, especially to potatoes. In a wet spring they can reduce the market value of many green and root vegetables, and, in the case of wheat, may devour much autumn-sown seed before emergence. No really satisfactory method of chemical control has yet been found. The most effective chemical known, much used in bait form among garden crops, is a tetramer of acetaldehyde known as metaldehyde or simply "meta". It is made by polymerization of acetaldehyde in solution in alcohol below 30°C in the presence of sulphuric acid, which catalyses the interconversion of the monomer and the various polymers. Meta is considered to be predominantly one stereoisomer of the eight-membered ring formed from four molecules of acetaldehyde. A poison bait of metaldehyde mixed with bran is broadcast thinly or placed in small heaps in the evening when the soil is moist. Ingestion of the poison by the mollusc induces increased production of slime and causes death by desiccation.

Because meta does not give fully satisfactory control under field conditions many other compounds have been tried. Amongst those which are reasonably effective are carbaryl, methiocarb, phorate, aldicarb and thiocarboxime. However, because of the toxicity of these compounds to man, animals and wildlife, their use is restricted and there is a need for discovery of safer compounds. It is interesting to note that beer and cider are very attractive to slugs but also very lethal to them.

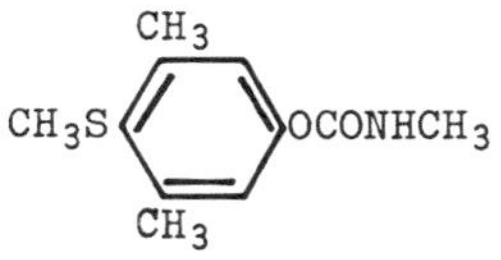

methiocarb

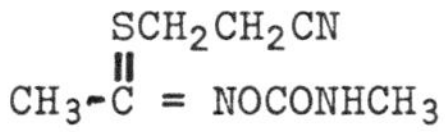

thiocarboxime

Poison baits are curative treatments applied after damage has been done. Preventative methods would be better but those which are currently available kill too few of the pests and those deep in the soil survive and rapidly restore the population in the surface layers. Much more needs to be known about behaviour and population dynamics and about the mechanism of slime secretion.

Water snails are a problem because they harbour a phase in the life cycle of the Schistosoma organism which is responsible for one of the most debilitating and widespread tropical diseases in man, bilharzia.

Aquatic species of snail, like all aquatic animals, must be adapted to make good molecular contact with a great volume of water in order to obtain sufficient oxygen. It is therefore understandable that they are more vulnerable to lipophilic substances dissolved at only low concentration. Most of the compounds effective against aquatic snails are lipophilic (or are salts of lipophilic acids) with the exception of copper sulphate which is still used in large quantity, and is the only common agent for both land and water species. Water containing 2-3 ppm of cupric ion kills nearly all snails after 24 hours' exposure. Even at present copper prices, few organic molluscicides can compete for cost but several factors operate against the successful use of copper. It is toxic also to aquatic vegetation, particularly algae, and can therefore upset the complex ecology of streams and ponds. It is also toxic to fish. It is rapidly inactivated in alkaline hard waters by precipitation as the basic carbonate and in muddy waters by adsorption on base-exchanging clay particles.

Pentachlorophenol (PCP), introduced into the water as a concentrated solution of the sodium salt, is also effective at a few parts per million and has been extensively used. It is much less strongly adsorbed than copper and is also more toxic to snail eggs, but is just as toxic to fish and some vegetation. Handling the concentrate is a considerable toxic hazard to man. Acrolein (acrylic aldehyde, $CH_2=CH-CHO$) has also achieved some success as a poison for river-dwelling snails. It is sufficiently water soluble not to evaporate excessively over the period required for action, but its lachrymatory property makes handling the concentrate unpleasant.

While the newer, more active organic compounds are not adsorbed by clays, they are adsorbed by organic constituents of mud and therefore share with copper the difficulty of adequate penetration of stagnant water along vegetated river margins. None is yet known which has a satisfactorily low toxicity to fish. Their chief advantage over copper and PCP is their lower phytotoxicity. A compound which has had some success is 5,2'-dichloro-4'-nitrosalicylanilide (niclosamide), prepared by reaction of 5-chlorosalicyloyl chloride with 2-chloro-4-nitroaniline.

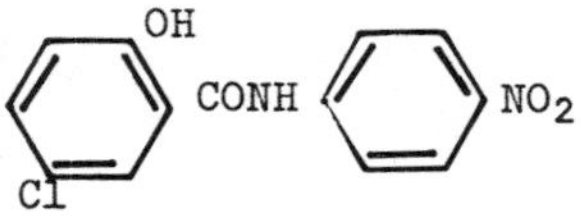

niclosamide

A related aquatic problem is the economic loss caused in marine transport by barnacles which attach themselves to the hulls of ships and seriously increase power to speed ratio. Special antifouling paints are used to combat these organisms. These paints generally contain sparingly soluble copper compounds, especially cuprous oxide, but some more complex organic compounds are now being introduced as antifouling paint constituents.

Control of Other Invertebrates

Woodlice cause some damage but are easily controlled with DDT or lindane. Millipedes cause sporadic and unpredictable damage to root crops but can be controlled with lindane, methiocarb or aldicarb. Symphylids cause damage to young vegetable seedlings and are best controlled with aldicarb.

Further Reading

Dickson, D.W., Nematode Control Guide, (University of Florida, 1969)

Jones, F.G.W. & Jones, M.G., Pests of Field Crops, (Arnold, London, 1974)

Chapter 11

CHEMICALS USED AGAINST VERTEBRATES

While there are many voices raised against the possible long-term or side-effects of the use of insecticides, few are raised against the actual killing of harmful insects. There is, however, much damage done to human interests by creatures which are closer to man in the scale of evolution. Birds and small mammals are, unlike insects, warm-blooded so they can cause losses throughout the year. Also, because of their highly developed central nervous systems, they can be more purposeful, and their greater mobility makes them less dependent on environmental conditions. Agriculture has greatly modified their way of life and provided them with an abundance of food. Some birds, like the sparrow, have become completely dependent on man.

In the countries with the highest standard of living, much opposition is met to killing these animals, especially by poison. Gin traps were made illegal in England and Wales by the Pests Act of 1954 but even humane traps are completely unselective. In a musk-rat campaign in Scotland, 1000 musk-rats were trapped together with 6000 other animals. Certain poisons such as red squill and phosphorus have been prohibited by the Cruel Poisons Act of 1962. As a result of the Protection of Animals Act of 1911 only rats, mice, voles, shrews and moles may be poisoned in England, although it is legal to gas rabbits, and there is somewhat more freedom in Scotland. Under the Agriculture Act of 1972, poisons may be used against squirrels and coypus in certain circumstances. The Protection of Birds Act of 1954 protects wild birds generally but specifies those that may be killed, including wood-pigeon, sparrow, rook, jackdaw, crow, magpie, jay and bullfinch. In all, there are about 30 Acts in the UK which govern destruction of birds and small mammals, and any control measures must be in accord not only with these, but also with public sentiment. This sentiment arises from greater leisure in developed countries for altruistic interest in animal life and particularly from the remoteness of most of the population from the hard facts of agricultural life. The housewife buying plums in a city shop does not appreciate the problems of the grower who has to shoot bullfinches in order to provide them.

This type of public sentiment does not occur in countries where most of the population is engaged in, or closely in contact with, agriculture and where food is short. In parts of Africa, flocks of weaver birds, running into millions, completely devastate vast areas of cereals. There are few weapons, however indiscriminate, which have not been tried against this pest, but they have had remarkably little success.

Rats and Mice

In developed countries, public sentiment varies widely from species to species, possibly as a result of conditioning by childhood stories and toys - many babies have a cuddly bunny but few have a cuddly rat. Rats are so universally loathed by man that few communities object to attacking them by poison. The rat devours growing and stored crops, it often fouls more food that it eats, it carries diseases such as bubonic plague, which is transmitted by the rat flea, and Weil's disease, its excreta contains bacteria which cause food poisoning, and it is destructive to buildings, embankments, drains, cables, etc. The rat population of the UK is estimated to be the same as the human population - 55 million. The field mouse and domestic mouse do similar damage and are also generally disliked, especially by women. Field mice dig up newly-planted seeds such as peas and sugar beet and feed on soft fruits and cereals.

In closed situations such as warehouses and ships' holds rats and mice can be controlled by sealing and fumigating with self-dispersing poisons such as sulphur dioxide, hydrogen cyanide or methyl bromide.

In farms and urban buildings rats and mice are controlled by poison baits. Most of these are fairly toxic to other animals and to humans but restriction of their action to rats and mice is usually achieved by placement of baits containing a general poison in situations frequented by the pests but inaccessible to domestic animals. Ideally a rodenticide should not be unpalatable to rats and mice, should not arouse their suspicions and cause them to become "bait shy" if more than one feed is required to give a lethal dose, should cause them to seek the open air before dying so as to avoid hygiene and odour problems, should not act so rapidly that the rodent perceives warning symptoms before a lethal dose has been ingested and should be much less toxic to domestic animals, especially to cats and dogs which may catch and devour affected rats or mice. No known rodenticide meets all these requirements.

For a long time zinc phosphide, arsenious oxide and a natural plant extract, red squill, were the most common rodenticides. Red squill is safe because it is rapidly decomposed in the body of the rat and, if taken directly by most other animals, it rapidly acts as an emetic; rats, however, cannot vomit. The odour of zinc phosphide is repulsive to most animals but is accepted by the rat.

The first synthetic organic rat poison of commercial importance was 1-naphthyl thiourea (antu), manufactured by reaction of 1-naphthylamine with ammonium thiocyanate. Another earlier organic rodenticide was 2-chloro-4-dimethylamino-6-methylpyrimidine (crimidine).

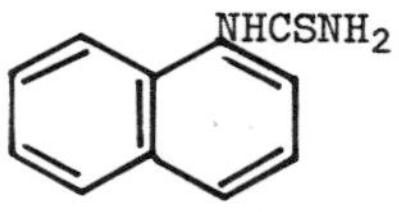

antu

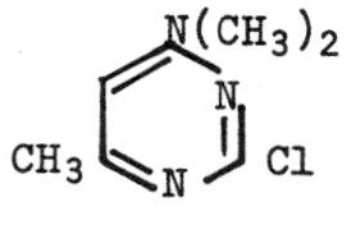

crimidine

A particularly successful and widely-used compound, especially against mice was chloralose; manufactured by reaction of chloral with glucose.

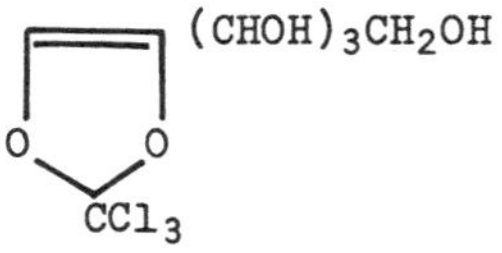

chloralose

Fluoroacetic acid, prepared by high temperature reaction of chloroacetic acid and sodium fluoride, was used either as sodium salt, the amide or the anilide. Its use is mainly restricted to ships and sewers because of its high mammalian toxicity.

All the above compounds are accepted by the rat but are also highly toxic to other animals. They are acute poisons intended to be lethal in a single dose and to kill quickly. As rats may not take sufficient of an acute poison bait immediately it is usual to "prebait" by offering unpoisoned bait until the rodents are feeding freely.

Chronic poisons, which are used at concentrations which kill only after several doses, were introduced around 1950. The most widely-used and successful has been 3-(1-phenyl-2-acetylethyl)-4-hydroxycoumarin (warfarin), which is manufactured by reaction of hydroxycoumarin with benzalacetone.

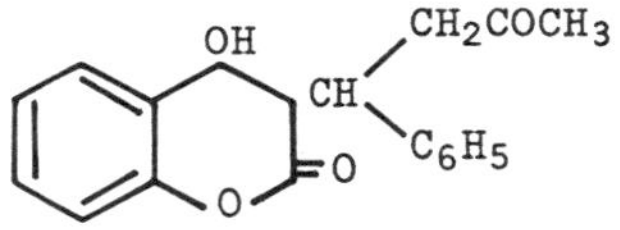

warfarin

Warfarin has the effect of arresting coagulation of blood and, with strict control of dosage, is used for this purpose in human medicine. Non-coagulation appears to be induced more easily in rats and mice than in most other animals and to have more drastic effects, but safe use is mainly dependent on placement of baits. The rodents die of internal haemorrhage after coming into the open in search of water. Warfarin has no tendency to produce poison-shyness.

Unfortunately warfarin-resistant strains of rats and mice have spread alarmingly throughout the UK and Europe so that some of the older poisons have had to be brought back into use. A number of other coumarins and related indanediones have been introduced, such as coumachlor, fumarin, coumatetralyl, bromadiolone, difenacoum, pindone and diphacinone, in order to try to control warfarin-resistant rodents.

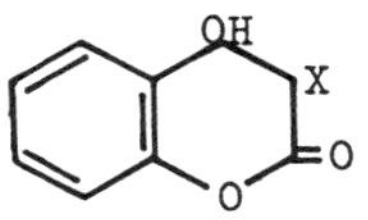

X = 1-(4-chlorophenyl)-2-acetylethyl	coumachlor
1-(2-furyl)-2-acetylethyl	fumarin
1,2,3,4-tetrahydronaphth-1-yl	coumatetralyl
3-(4,4'-diphenyl)-1,2,3,4-tetrahydro-naphth-1-yl	difenacoum
3-(4'-bromodiphenyl)-3-hydroxy-1-phenyl propyl	bromadiolone

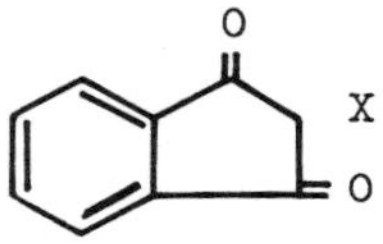

X = pivaloyl — pindone
diphenylacetyl — diphacinone

There is need for new types of chemicals which have the properties for a successful rodenticide. Two recent developments are an organophosphorus compound, phosazetim, and a very interesting novel type of pesticide, a silicon compound, silatrane.

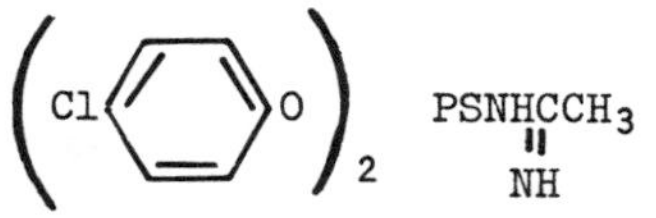

phosazetim

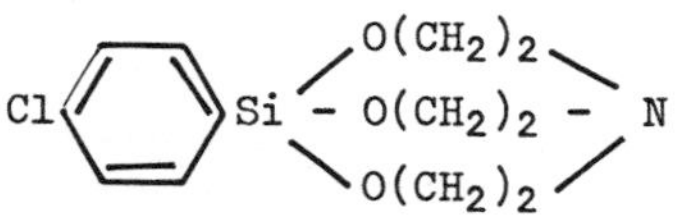

silatrane

However, economical, effective and safe use of rodenticides demands much more knowledge of the habits and behaviour of rats and mice.

Rabbits

Control of rabbits has been studied most extensively in Australia and New Zealand where they are major pests. Control by the myxamatosis virus has been very successful but the virus is becoming attenuated and resistant strains of rabbits are emerging. In the UK, deliberate spread of this disease was prohibited by the Pest Act of 1954 as the public found the spectacle of dying rabbits distasteful. Nevertheless, it must be realized that rabbit damage in the UK amounts to about £100M per year.

Poison baits containing rodenticides are used against rabbits in Australia and New Zealand but, in the UK poisoning of rabbits is illegal, although they are still shot, snared, ferreted and trapped. Chemical control is generally by hydrogen cyanide introduced into burrows. Contractors may employ anhydrous hydrogen cyanide with special equipment and precautions but, for general use, a powdered water-soluble cyanide is available. A suitable powdered formulation with good storage properties under reasonably dry conditions is a mixture of sodium cyanide, magnesium carbonate and anhydrous magnesium sulphate. It is placed in burrows with a long spoon or blown in with a dust gun. Chloropicrin is used instead of hydrogen cyanide in New Zealand but not in the UK.

Moles

Moles eat insects and aerate soil but they can destroy seedlings by burrowing under them and their hills can damage machinery and degrade pastures. Moles are controlled by trapping and by gassing either by hydrogen cyanide or more primitively, but cheaply, by fitting a flexible pipe adaptor to a tractor exhaust and leading the gas into openings made in the runs. Baiting is also used and the poison employed is invariably strychnine which is injected into the bodies of earthworms which are then dropped back into the runs.

Squirrels

Squirrels do extensive damage to field crops and trees. In the UK, use of rodenticide baits is now permitted in areas where there are no viable populations of red squirrels.

Coypu and Mink

These two animals were introduced into closed farms in the UK for fur-raising but escaped and colonised. The coypu has established itself in East Anglia where it eats sugar beet and cereals and damages river banks. Rodenticides such as warfarin may now be used against them.

Other Mammals

Control of game animals such as hares, deer, etc is a problem because of legal restrictions, public sentiment and private interests. An interesting new chemical approach is based on the fact that animals have an acute sense of smell and use chemical secretions - pheromones - from exterior glands to convey various messages to other members of their species. One of the purposes is to demarcate the individual's territory, and other members of the species will not cross such a "scent barrier". Some of the natural pheromones have been isolated, identified and synthesised, and others will follow. The hope is that they can be used for artificial marking to protect cultivated lands, but a considerable knowledge of the behaviour of the species will be needed if they are to be used effectively.

Birds

Public sentiment is strongly against poisoning birds, even those which it is legal to kill and even when they are seen to do extensive damage. In the UK the wood-pigeon is a major pest. It causes about £5M per year agricultural loss. One approach to chemical control is to use a bait containing a non-lethal narcotic. For this purpose chloralose is almost exclusively used. Shortly after feeding on the treated grains, birds become stupid and sleepy. If left alone, they recover completely. The farmer can kill all the wood-pigeons and put the pheasants in a safe place to recover. Use as a bait of large grain, such as field beans, which smaller birds cannot swallow helps to make the operation automatically selective. The wood-pigeon is, however, a very mobile pest and destruction on one isolated farm is of little value. Labour-costly use of chloralose can therefore be effective only in a co-operative campaign carried out by all farmers in an extensive area. Consequently it has been little used.

Chemical repellents would be useful. Sprays that, while harmless, would make buds unpalatable to bullfinches, seeds unpalatable to rooks and pigeons, bark unpalatable to rabbits and deer, would satisfy every interest. Unfortunately there has been little progress along these lines. Products based on impure anthraquinone are used successfully in seed dressing to reduce consumption by rooks, but they are not effective with most bird species.

It is salutary to realize that by far the most significant aspect of the bird pest problem nowadays is the danger they present to jet aircraft and consequently to life and safety of passengers, a risk which grows as aircraft get bigger and more numerous. Methods of dispersing birds from airfields are badly needed.

General

Control of vertebrates is an example of the need to develop systems of total pest management, in which chemicals will play merely one part. An essential prerequisite for developing such systems is a detailed knowledge of the ecology of the pest species, and of its population dynamics. Most control methods currently used depress numbers for a short time only, which has the effect of reducing competition and increasing breeding success, so the population rapidly reestablishes itself. Shooting and trapping in particular effect only minor local improvements. Effective control requires understanding of habits and behaviour, for example, it may be much more effective to shoot or trap when numbers are low instead of waiting, as is customary, until numbers are high. There is obviously scope for "second-generation" chemicals which do not kill directly but provide control by affecting some aspect of behaviour. If the pest can be deterred from the crop there may be no need actually to kill it.

Further Reading

Davis, R.A., Control of Rats and Mice, Ministry of Agriculture, Fisheries and Food Bulletin, No. 181, (H.M.S.O., London, 1961)

Murton, R.K. & Wright, E.N., (Ed)., The Problems of Birds as Pests, Institute of Biology Symposium, No. 17, (Academic Press, London & New York, 1968)

Chapter 12

FUNGAL DISEASES AND PROTECTION BY HEAVY METAL COMPOUNDS

Fungal Diseases

The fungi form a very large group of primitive plants which cover a wide range of size and habit, from the field mushroom and its larger relatives to unicellular yeasts. Most fungi have a complex life cycle. They are propagated by means of asexually produced spores which, in many species, are of more than one type. A sexual stage is also known in most species, but many, including potato blight, can persist through thousands of generations without it. Some spores are very short-lived and must quickly reach a favourable environment to survive. Other spores, usually from the sexual stage, can remain viable in a dormant condition for a long period in a dry or cold environment. This resting stage is often essential for survival of the species from one season to the next. Some species go through many cycles in one season. Others, like those living inside mature trees and forming fruiting "brackets" outside the bark, take many years from one fruiting stage to the next.

Fungi are devoid of chlorophyll. They must therefore obtain their energy-food from other organic matter. The saprophytes live on dead matter and are largely responsible for the breakdown of animal and plant remains in the soil. In this role they are essential in agriculture but they can also attack structural materials of natural origin - timber, and fabrics based on vegetable or animal fibres. Such materials are more at risk in an agricultural than in an urban situation - fencing posts more than roof timbers, stack covers more than household linen. Protection of timber against fungi has been touched on in the chapter on Oils.

A few fungi, for example the several species causing "ring-worm", are parasitic on animals, including man. Within the chemical industry the manufacture of compounds for the treatment of these diseases is accepted as belonging to the pharmaceutical branch and will not be dealt with in this book. The fungi causing plant diseases are accepted as targets for the pesticides industry. These fungi are the most effective competitors of man for the products of his agricultural labour. Chemical control and selection of immune varieties are the only alternatives to indirect practices of good husbandry - crop rotation, general hygienic measures and evasive tactics. Often these indirect practices are in conflict with efficient business organization of simple cropping systems.

A few fungal species, notably of the genus Botrytis, can be either saprophytic or parasitic. The soft rots of strawberry and raspberry fruits, for example, overwinter on ground litter and usually spread on to the fruits from adhering dead leaf or calyx fragments. Many parasitic fungi are, however, highly specialized, and can develop only on living tissues of a single, or few, host species.

Often there are two quite different hosts, each harbouring a different phase of the complex life-cycle. The rust disease of wheat has the barberry as an alternate host and the rust of white pine the currant. Drastic manual weeding-out of barberry in the wheat areas of France, and later of the U.S.A., led to a considerable reduction in the incidence of rust. Control of pine rust by eradication of currant species in forest areas of U.S.A. is not yet so successful, probably because there are too many small wild plants. Herbicides are now being used in this indirect attack on a fungus. In the U.K. where currant is abundant and cultivated, the planting of white pine has been abandoned. One side-result is a minor export trade in blackcurrant juice from the U.K. to the U.S.A.

Many other examples could be quoted where the prevalence of a parasitic fungus has prevented the cultivation of a crop in an area otherwise suitable. Fungus attack is more damaging in large-scale monocultures than in a natural mixed stand where sensitive host individuals are widely spaced. Increasing level of parasite population is the chief reason for the adoption of crop-rotation practices. For these reasons it is very difficult to assess the economic damage done by parasitic fungi, much of this damage being hidden and indirect. Even the direct damage is enormous.

Host and Parasite

The big problem in the chemical attack on parasitic fungi is that of toxicity to the host. Host and parasite are too close biochemically. The biggest difference is the presence of a photosynthetic mechanism in the higher plants. Photosynthesis is localized in the small chloroplasts inside the leaf cells. These produce the sucrose, which the rest of the plant uses, by similar biochemical reactions to those operating in a fungus, for metabolic energy and as the source of combined carbon. A higher plant can, in fact, almost be considered, from the biochemical point of view, as a fungus parasitic on its own chloroplasts. This is not far from the truth even from the anatomical viewpoint. The primitive green lichens consist of a non-green fungus species enveloping a separately reproduced population of unicellular green algae, living in intimate and necessary association (symbiosis). Most higher plants show a relic of this behaviour in that the chlorophyll is carried in the cytoplasm of the seed-germ and is not genetically controlled by the chromosomes.

Several important herbicides inactivate an essential step in photosynthesis and are lethal to higher plants without much affecting fungi. The fact that this major biochemical difference is the wrong way round makes the search for safe fungicides more difficult. Once the parasitic fungus has gained entry to the host tissues, the discrimination must be biochemical and the compound must be systemic. However, many practical "fungicides" act only prophylactically. They must be applied before the fungus invades so that they can lie on the outside of the tough leaf cuticle of the host ready to attack the fungus spore at the germinating stage, when it is most vulnerable.

Disinfection of Seed

Some diseases overwinter as dormant spores on the seed of the host. The host and parasite germinate together and the parasite lives internally in the host without killing it. Its spores are either produced alongside the seeds of the host or replace these seeds and then infect healthy seed through the air. The most important of such diseases are the smuts and bunt of wheat, but many others are in this class. They are controlled by chemicals applied to the crop seed before sowing. The spores are not affected while in the dormant state but are killed as they germinate. The germ tube of the parasite must move over the host seed surface seeking entry, while the host radicle grows more rapidly and downwards, away from the seed. Selective toxicity to the parasite is largely dependent on this geometrical difference.

The "dressing" of seed protects the future crop not only against seed-borne diseases but also against soil-borne fungal diseases and voracious insects. Its effectiveness is necessarily limited because of the very localized application and the need for biochemical selectivity. The convenience and economy of seed-dressing makes it, however, a clear choice when the disease can be controlled this way.

Protective Fungicides for Growing Crops

If a purely protective, rather than systemic, fungicide is to be effective it must stop the infection from "taking" on the crop. At this stage the fungus spore is exposed and vulnerable and a compound applied to the leaf surface has the best chance of being taken up into the germ tube of the spore while not penetrating significantly into the tissue of the mature leaf. Once the fungal hyphae have penetrated into the internal cells of the host a protective fungicide ceases to be effective. It is significant that the only fungi which are fairly effectively dealt with after establishment - the powdery mildews - penetrate only into the superficial leaf cells. These diseases are caused by a very high population of very small individuals.

Much the largest tonnage of protective fungicides used in agriculture consists of inorganic compounds of copper of low solubility and either elementary sulphur or organic compounds containing sulphur. Some of these products are only a little younger than the knowledge that most plant diseases are caused by fungi.

Potato blight hit Europe, particularly Ireland, in 1842. About 10 years later, due mainly to the work of a great amateur in the U.K., the Rev M. J. Berkeley, and a great professional in Germany, Anton de Bary, it was becoming recognized that a parasitic fungus could be responsible. It took another 10 years at least, after much academic polemics, for this idea to be generally accepted. Sulphur dust had already been used successfully against the powdery mildew of vines, and first salt, then later copper sulphate dried off with lime, had been found to give a cleaner crop of wheat after application to the seed. The use of salt was suggested by the clean crop resulting from seed salvaged from a shipwreck.

It was not till 1882 that Millardet in France saw the significance of another accidental discovery. A paste of lime and copper sulphate had been painted on vines by the roadside in the Bordeaux district solely to discourage local boys from stealing the fruit. Millardet realized that this "Bordeaux mixture" had discouraged a far more deadly enemy than schoolboys.

The downy mildew which had been building up to disastrous levels in France was not attacking the roadside vines.

There have, of course, been developments in the inorganic copper and sulphur products made available by the industry, but the changes have been more concerned with cost, packaging and convenience factors than with biological efficiency. Many plant pathologists consider that the original Bordeaux mixture made up in the field is frequently found to be more effective in comparative tests and never less efficient than the more convenient modern products.

It is very difficult to decide whether one product is generally a more efficient protective fungicide than another. The comments in the next section refer directly to copper fungicides and potato blight but are relevant to protective applications of "insoluble" fungicides in general against diseases which infect the foliage from the air.

Efficacy of Protective Treatment

The reason for the difficulty is that the biological and climatic variables may have an over-riding influence. These include local intensity of source of infection, many aspects of climatic conditions including weather after treatment, nutritional status of the crop, predisposing damage to the crop by wind and insects. In the case of potato blight, if the sprayings have been perfectly timed, almost any copper product will be effective. At the other extreme, if spraying is left until symptoms are already present, no practical field treatment is of any real value. Unfortunately, choice of timing cannot be made entirely on facts known at the time of choice, since future weather, and therefore "luck", is important.

Intermediate conditions can occur where, in well laid-out trials, formulation X can be shown to be significantly more effective than formulation Y, but other conditions can occur where the reverse is true. It is important for the chemist to realize this kind of limitation on the improvements he can make at the chemical end, and some brief explanation is therefore justified of how these differences can arise. It is possible to state the significance of some factors clearly, while admitting that there are others whose significance may be as great but not known. Only the influence of solubility and adhesion will be discussed here.

At one extreme, some copper compounds are too soluble. If a solution of the sulphate is sprayed and the weather remains dry, the crop will be severely damaged or even killed. If sufficient rain follows soon enough to save the crop, there will not be an effective residual deposit to prevent the germination of incoming spores. Copper sulphate cannot, therefore, be used alone as a fungicide. At the other extreme, some copper compounds, e.g. cuprous iodide or cupric ferrocyanide, are so insoluble that they have no effect on either host or parasite. An intermediate solubility, like that of the hydroxide or many basic salts, and some degree of adhesion of the solid to the leaf surface are necessary. These factors enable surface moisture (without which the spore cannot germinate) to let the copper migrate to the spore but prevent total loss in the first shower of rain.

It is evident that there is no absolute optimum either for solubility or adhesiveness. If heavy rain is going to occur a few days after spraying, the only product which may give any degree of residual control is one of very low solubility and high adhesion. If only dew or drizzle is experienced after spraying, a better residual deposit will be obtained from a more soluble product. It is true that a deposit which will survive heavy rain will be at least as active after light drizzle, but the pressure of infection is likely to be much greater in the second case, since the heavy rain washes off most of the spores it brings. There is thus a different optimum set of physical properties for the two weather conditions.

One answer to this problem is to use formulations with a range of properties so that the behaviour is at least adequate under a wide range of weather conditions. This accounts for the success of products which include other fungicides (dithiocarbamates and organotin compounds, see below) along with low-soluble copper compounds.

By careful addition of aqueous ammonia (3 equivalents) to well-stirred aqueous cupric chloride (4 equivalents) and subsequent dialysis through cellulose, an indefinitely stable colloidal solution of the oxychloride can be obtained. It has been claimed that products of this type can produce a more complete and uniform cover by spraying. While this is true under idealized laboratory conditions, the property confers doubtful advantage in the field. As the suspension concentrates by evaporation after spraying, coagulation of the colloidal matter occurs, particularly if hard water has been used to dilute the concentrate. Moreover, the fine structure of the deposit within any small area is usually of less significance than the gross pattern which results from the wetting and drainage behaviour of the diluted suspension as a whole and its redistribution by the action of dew and rain.

In the case of potato blight, the conclusions drawn from laboratory and field experiments seem widely at variance. In the laboratory a very uniform coverage of finely divided fungicide is necessary to give efficient protection against a heavy artificial inoculum. In the field, a relatively poor deposit, even that produced by low-volume spraying from the air, appears to give as good a protection as that obtained by high-volume spraying. No spraying is of much value once the disease is active within the crop. The last sentence gives a useful clue to the general explanation of the discrepancy. The field spraying protects the crop only against the incoming inoculum from distant sources. This test is not made in the laboratory. After "ideal" application in the laboratory, which cannot even be approached in the field, protection can be secured against a dense inoculum of fresh spores.

Why the high population of fresh spores is so much more resistant is not clearly understood, but the reason why the spores which have travelled for a distance are killed by a not very uniform deposit is probably that they need liquid water to germinate. The infection therefore starts within the portion of the leaf covered by the pendant drops at the lower margins. The water forming these drops has collected both copper and spores.

"Fixed Coppers"

Bordeaux mixture is prepared by mixing a slurry of lime in water into a solution of copper sulphate in water. This is often done in the actual spray tank. Originally, quicklime was employed but it is now possible to supply slaked lime packaged so as to avoid carbonating. Calcium carbide has been used as a source of lime slurry, but explosion, probably initiated by cuprous acetylide, has occurred. The lime is added in considerable molecular excess, about equal in weight to the hydrated "blue vitriol". The final product contains about 12.75% of metallic copper. It is very safe to use because its toxicity to men and animals is low. The rather gelatinous nature of the fresh precipitate (and Bordeaux mixture should be made immediately before use) helps to increase the deposit by delaying drainage from high-volume application. The carbonation of the excess lime during exposure on the leaf probably assists the adhesion of the final deposit.

Bordeaux mixture was rapidly followed by Burgundy mixture, in which the lime is replaced by washing soda. In eau celeste excess ammonia was used to hold the copper in solution in the sprayer but liberate it on exposure. Much more risk of plant damage was incurred by its use, not only because the copper is transiently more available but because a high concentration of free ammonia is itself phytotoxic. Neither of these two products is used today. They have been replaced by the single package "fixed coppers" in which a finely divided low-soluble copper compound is packed in a form stable in storage and readily dispersible in water only. The copper compound is usually a basic cupric salt.

The solubility of these compounds is not easily defined nor is its significance simple. The free cupric ion concentration present in a suspension in distilled water should be less than about 5 ppm. The solubility is, however, inevitably dependent on pH. While water on a leaf surface is likely to be slightly alkaline, potassium carbonate being excreted more freely than organic acids, the germinating fungus spores often excrete other compounds, of which the simplest is glycine, which form soluble chelates with cupric ions. The fungus spore therefore generally accumulates copper within itself at a much higher concentration than that of free ionic copper in a water suspension of the fungicide.

Nearly all the copper is precipitated from aqueous solutions of cupric salts by less alkali than corresponds to a simple stoichiometric equation - e.g. $CuCl_2 + 2NaOH \rightarrow Cu(OH)_2 + 2NaCl$. Precipitation is usually complete with about three-quarters of this amount of alkali, and the existence of some definite crystalline compounds has been demonstrated by X-ray analysis. The most widely used compound is the oxychloride, of approximate composition $3Cu\ (OH)_2.CuCl_2$. It is usually prepared by reaction of hydrochloric acid on scrap metallic copper in the presence of blown air.

The oxychloride forms as a very fine suspension which is filtered off, dried and mixed with a small proportion of wetting agent and a sticker such as starch or glue. The particle size in the final suspension remains finer if the concentrated product is marketed without drying, but packaging of suspensions is troublesome and their behaviour on storage unreliable. It is generally sold as a 50% wettable powder or 10%-25% dusts.

Other copper compounds of a similar type which have been used include basic copper sulphate, $CuSO_4.3Cu(OH)_2.H_2O$, basic copper carbonate, $Cu(OH)_2.CuCO_3$, tetracopper calcium oxychloride, $4Cu(OH)_2.CaCl_2$, and basic cupric zinc sulphate complex. All these compounds, like the oxychloride, have low toxicities to men and animals.

Products containing yellow cuprous oxide can be used in place of the many basic cupric salts available. This compound is obtained as a fine precipitate during reduction of cupric salts in buffered solution. It is best known to most chemists as the solid produced in Fehling's solution by reducing sugars. Such processes are, of course, far too costly for a commodity product to be used in agriculture, and, industrially, an electrolytic oxidation is used. Cuprous oxide is therefore mainly produced in countries with cheap water power. Cuprous oxide is much less soluble in distilled water, in the absence of air, than are the basic cupric salts but it is slowly brought into solution by oxidation on exposure. As in the case of the cupric compounds, the reaction with leaf and spore exudates is important.

There is no clear evidence of advantage for either cuprous oxide or basic cupric salts for control of any one fungus disease. Preference depends more often on local prejudices and the impact of effective salesmanship than on technical performance. Even colour has had an influence on the choice. The blue-green cupric products leave a less noticeable deposit on foliage (an advantage perhaps to the seller of the crop but a disadvantage for the spray operator) but the yellow colour of cuprous oxide has the merit in Buddhist countries of a holy and healing significance. One of the authors has been asked in all seriousness during his technical career to colour cuprous oxide green and to colour cupric oxychloride yellow.

Substantially different copper formulations are those in which an organic salt is formed and applied in oil solution. The mixed naphthenic acids (carboxylic acids of alicyclic hydrocarbons) obtained from petroleum oxidation are the most widely used but salts of fatty acids, particularly linoleic and oleic, are also oil soluble and, in this state, very fungicidal. These compounds have been mentioned under "oils" for wood preservation. They have been tried in agriculture but are too phytotoxic to be acceptable for most applications.

Other Inorganic Fungicides

Mercury

Many heavy metals are highly toxic, especially to microorganisms. The decoration with silver of some foodstuffs available to the wealthy minorities in the Indian subcontinent probably has its justification in protection against microparasites of the gut. This metal has also been used in the treatment of ringworm. The metals other than copper which have found fungicidal use in agriculture are zinc, chromium and nickel used in inorganic forms, mercury, used both in inorganic and organic compounds and tin, used only in organic compounds. Germanium and lead compounds analogous to those of tin have also been tried but without showing any advantage. Cadmium succinate is used as a fungicide on turf.

Zinc and copper complex chromates have been found useful against potato blight but have not survived long comparison with the standard copper products and later developments. In practice, therefore, other inorganic usage is confined to mercury and nickel salts. Of the former, the almost insoluble mercurous chloride is used by soil application, particularly against club-root disease of the cabbage tribe. It is also effective against fungi which damage fine turf. Mercury salts are toxic to most organisms and these dressings also control the cabbage-root-fly, other soil insects and nematodes. For effective control of the latter, however, the toxicant must be dispersed through several inches depth of soil. Although it has been claimed that the vapour of elementary mercury, to which the salt may be reduced in the soil, is able to diffuse in the soil air-space, too high a dosage is required to be economic.

Mercury is an inherently toxic element and some concern has been felt about the possible environmental effects of slow build-up of soil residues. Fears on this score may not, however, be well-founded, since all heavy metals tend to get locked up in insoluble forms among the soil minerals. Many soils contain more unavailable native mercury than is applied as a pesticide. Nevertheless, in the present climate of public concern about possible environmental effects of pesticides, even if unproved, the use of mercury compounds is coming into disfavour.

Nickel Salts

Water-soluble nickel salts, such as the chloride, are far from non-toxic to higher plants if used in excess, but they are sufficiently more active against some fungal parasites, particularly wheat rusts, to permit their use for effective control. Some degree of curative action (i.e. killing of the established parasite) and systemic effect is claimed. The observation is of particular interest because wheat rust is very erratic in its incidence. In most years there is negligible damage. Occasionally, about one year in six, very serious loss of crop can result. To insure the crop, considerable stock-piling of fungicide is necessary and this is much more feasible when a heavy inorganic chemical with many other uses is the one which has to be called into action.

Organomercury Fungicides

Mercury, more easily than most other metals, forms stable organic compounds in which it is linked to carbon. Such compounds are all fungitoxic, many of them much more so, under practical conditions of application, than inorganic salts. The advantage is often associated with greater volatility, which permits a more uniform redistribution among dressed seeds during storage than can be obtained in initial mixing. It also permits some desirable movement in the soil air space.

Excessive volatility can, however, give rise to user-hazard. Casualties have occurred in factories handling ethyl mercury chlorides (EMC). These probably arose from disproportionation

$$2C_2H_5HgCl \rightleftarrows (C_2H_5)_2Hg + HgCl_2$$

yielding the more volatile mercury diethyl (b.p. 160°C). For this reason, and because the toxicity of EMC to men and animals is high, it is little used nowadays and the main mercury compounds which are still used in agriculture are phenylmercury acetate (PMA) and methoxyethylmercury silicate (MEMS) or acetate (MEMA). They are used exclusively as seed dressings or for fungal control in turf.

It should be noted that mercury and organomercury salts have much more covalent character than those of most metals and one cannot assume that these salts are effectively involatile. Mercuric chloride, for example, has a boiling point of 301°C and mercuric oxide is noticeably volatile in steam. The choice among the wholly covalent and partly electrovalent substituents on the mercury atom is largely determined by solubility, volatility and stability rather than by inherent fungitoxicity. There is a general tendency towards larger substituents than ethyl in the organic part. Phenylmercury and methoxyethyl mercury compounds are now more widely used for the reasons given above.

Synthesis of Organomercury Compounds

The compounds can almost always be prepared by the Grignard synthesis, but for the two main classes of organomercury compounds, cheaper industrial routes are available. The Kharasch method for diethylmercury chloride reacts lead tetraethyl with mercuric chloride. Lead tetraethyl is made in large tonnage as a motor fuel additive and is itself prepared by reaction of a sodium-lead alloy with ethyl chloride. Schematically,

$$Na_4Pb + 4C_2H_5Cl \rightarrow 4NaCl + Pb(C_2H_5)_4$$

(in fact the alloy must be much richer in lead and the excess reused), followed by

$$2HgCl_2 + Pb(C_2H_5)_4 \rightarrow 2C_2H_5.HgCl + (C_2H_5)_2PbCl_2.$$

The reactions are general for alkyl substituents, but the availability of lead tetraethyl directed chief attention to ethylmercury compounds. Other anions can be left attached to the ethylmercury cation by choice of the appropriate mercuric salt or a mixture of mercuric oxide and the appropriate acid. Thus, ethylmercury acetate (EMA) can be made by reaction of mercuric oxide, acetic acid and lead tetraethyl at 90-95°C.

In the preparation of ethylmercury compounds by this route, the reactants are usually ball-milled in the dry state with an inert mineral diluent. The lead compounds remain in the mixture, which is further diluted if necessary with inert powder and wetting agent to give the formulated product directly. If a pure product is required, soluble ethylmercury acetate is first produced which is extracted with water from the insoluble lead compounds. From this solution a less soluble ethylmercury compound can be precipitated.

Use of a higher proportion of tetraethyl lead enables mercury diethyl to be formed. Ethylmercury salts can be prepared by heating the appropriate acid and mercury diethyl, e.g.

$$Hg(C_2H_5)_2 + CH_3CO_2H \rightarrow C_2H_5HgOCOCH_3 + C_2H_6$$

By reaction of aniline with 4-toluene sulphonyl chloride at 80-90° to give 4-toluene sulphonanilide, followed by treatment with ethylmercury acetate at 20-30°C, ethylmercury 4-toluenesulphonanilide is obtained. This is a somewhat less toxic compound than ethylmercury chloride and has been used as a seed dressing.

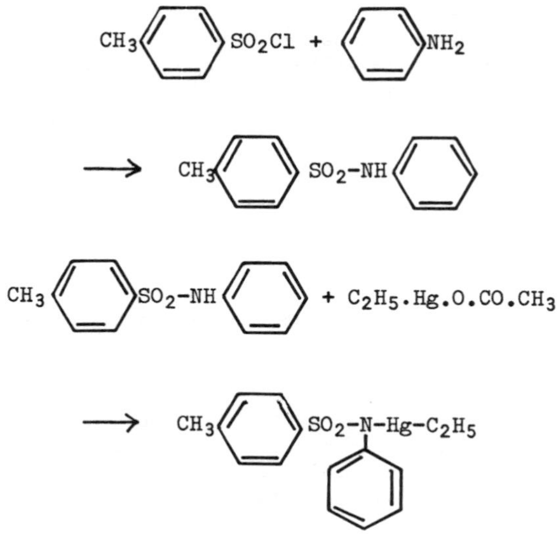

ethylmercury 4-toluenesulphonanilide

Some organomercury compounds can be obtained more directly by displacement of hydrogen in aromatic compounds or addition of ethylenic compounds. Thus the important phenylmercury acetate (PMA) can be obtained in 90% yield by heating benzene and mercuric acetate in glacial acetic acid under pressure at 110°C for 2 hr.

$$C_6H_6 + Hg(O.CO.CH_3)_2 \rightarrow C_6H_5.Hg.O.CO.CH_3 + CH_3CO_2H$$

The more reactive ring in 2-chlorophenol can react even with mercuric oxide to give hydroxymercury-2-chlorophenol which has also been used as an agricultural fungicide for seed dressing.

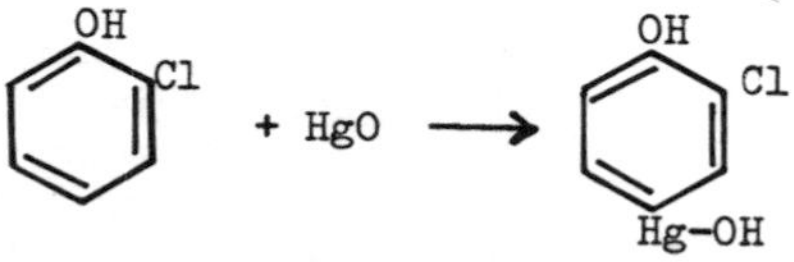

hydroxymercury-2-chlorophenol

A suspension of mercuric acetate in methanol absorbs 1 mol of ethylene to give methoxyethylmercury acetate (MEMA).

$$CH_3OH + CH_2 = CH_2 + Hg(O.CO.CH_3)_2 \rightarrow$$
$$CH_3OCH_2CH_2\text{-}Hg\text{-}O.CO.CH_3 + CH_3CO_2H.$$

Less soluble methoxyethyl mercury salts, such as the silicate (MEMS), are effective seed dressings.

Phenylmercury salts can add on to tertiary amines to give quaternary nitrogen compounds in which one of the four valences of the N atom binds it to Hg. Such a compound is phenylmercury triethanolammonium lactate, a water-soluble compound with a high affinity for wool which has found more application for moth-proofing than for control of fungi on crops.

Organotin Compounds as Fungicides

In 1950 a fruitful collaboration began between the International Tin Research Council in London and the Institute for Organic Chemistry in Utrecht. The research on fungitoxic properties of organotin compounds has been reviewed by van der Kerk. A review by Ascher and Nissim on the potential value of these compounds as insecticides can also usefully be consulted.

The organic compounds of tin have advantage over those of mercury in that tin is not an inherently toxic element. The toxicity arises from reaction of the molecular compound. It is much greater in those compounds of tetravalent tin in which three covalent Sn - C links are present than in compounds of the type R_2SnX_2 or SnR_4 where R as usual signifies a hydrocarbon radical and X an anionic group.

The biggest usage of an organotin compound is for stabilization of some transparent plastics against photochemical attack, for which purpose tributyltin oxide, $(C_4H_9)_3Sn\text{-}O\text{-}Sn(C_4H_9)_3$ is preferred. This compound has considerable biocidal properties and has found such use in rot-proofing of fabrics and as a constituent of anti-fouling paints which prevent the growth of barnacles and algae on ships' bottoms. Although expensive for timber preservation, it is to date the most effective compound for protecting wood piles in marine harbours against the destructive attack of the toredo worm.

Despite the non-toxicity of elementary tin, the development of organotin compounds suffered a serious set-back through one of the worst incidents involving death by poisoning in recent history. It happened in France in 1954 when a number of innocent users of a new skin disinfectant were fatally poisoned. The compound was nominally the relatively harmless diethyltin diiodide but was found to contain a high proportion of the very toxic triethyltin iodide. It is not rational for an error of ignorance or carelessness to bring into disrepute all related compounds - but it was inevitable. Only now have safer compounds, with proper knowledge of their risks, become accepted. In most countries agricultural use is restricted to ornamentals and root crops.

The compounds now of most importance are triphenyltin salts. These are less toxic to mammals than the trialkyl, particularly triethyl, tin salts and it is perhaps an advantage that the desired compounds are more toxic than the associated impurities. If appropriate precautions are taken in handling triphenyltin acetate, the presence of diphenyl and tetraphenyltin will create no new problem.

The trialkyltin compounds are too phytotoxic for safe use on growing crops. Success for tin compounds in this field came with the introduction of triphenyltin acetate (fentin acetate) and triphenyltin hydroxide (fentin hydroxide).

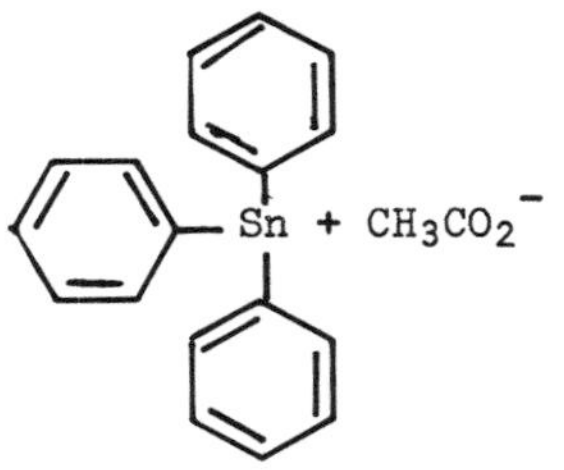

fentin acetate

This compound has been used for control of leaf spot diseases of sugarbeet and celery and blight of potatoes. There is use in other crops too, but on others again the compound is too damaging. Although more phytotoxic than "fixed coppers" at the same dosage, it is used, for successful fungus control, at only about one-tenth of the dosage. Higher yield increases of the crops mentioned have been found to follow control with triphenyltin than have followed control with copper. It has been suggested that triphenyltin has some growth stimulant effect at low dosage, but the view is more generally held that "copper shock" produces a growth depression, without visible damage, which partly off-sets the effect of fungus control.

Two related compounds which have found use as protective foliar fungicides are decyltriphenylphosphonium-bromochloro-triphenylstannate (decafentin) and di-|tri-(2-methyl-2-phenylpropyl)tin| oxide (fenbutatin oxide).

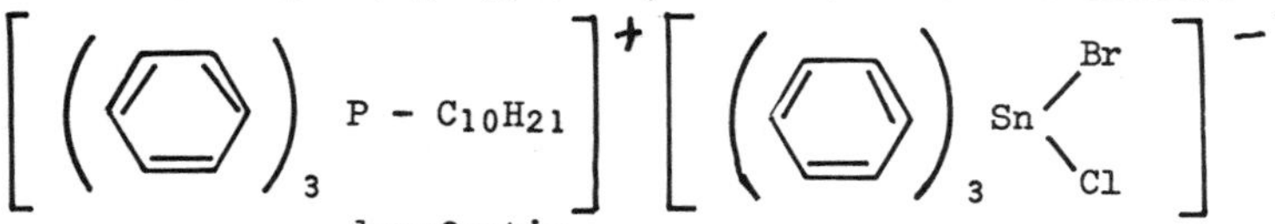

decafentin

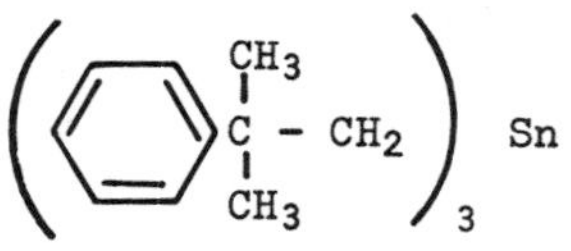

fenbutatin oxide

A tin compound which has found extensive use, not as a fungicide, but as an acaricide, is tricyclohexyltin hydroxide (cyhexatin).

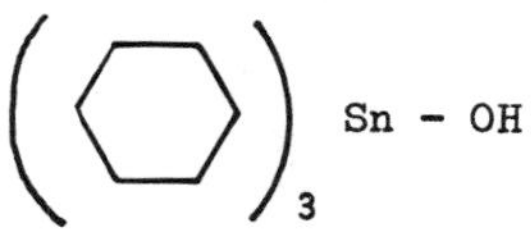

cyhexatin

A related compound, also used as an acaricide, is 1-tricyclohexylstannyl-1,2,4-triazole (tricyclotin).

Chemistry of Organotin Compounds

Trialkyl and triaryl tins are fairly stable cations forming salts with strong acids. Weak acids form salts which are extensively hydrolysed in water. Thus triphenyltin acetate, a white solid, soluble to about 20 ppm of water, is fairly rapidly converted, in contact with water, to the much less soluble oxide. There is some degree of covalent behaviour but not so extensive as with corresponding mercury compounds, and ionic replacement reactions are very rapid. All the compounds form sulphides of very low solubility and react rapidly with mercaptans. Affinity of the triorganotin radicals for - SH groups in biochemical systems is probably the basis of their toxic behaviour.

Tin can be effectively combined with hydrocarbon radicals only by the use of Grignard intermediates. The preferred industrial procedure is to use the alkyl or aryl magnesium halide in slight excess so as to obtain first the neutral non-polar tetraalkyl or tetraaryltin which can be removed by extraction with non-polar solvents or by distillation. The tetra compound is then reacted back with stannic chloride, a reaction which, in the case of tin, gives preponderantly the compound having the average composition, thus

$$3\ SnR_4 + SnCl_4 \rightarrow 4\ R_3SnCl.$$

The desired salt is then obtained by metathesis.

The organotin compounds are inevitably expensive, because of the scarcity of available elementary tin and the rather expensive reaction process. They are however very active. The big demands in the plastics field for organotins has stimulated work on conditions for maximum production efficiency.

Further Reading

Copper Development Association, Copper Compounds in Agriculture (London, 1957)

Evans, E., Fungal Diseases and Control by Chemicals, (Blackwell, Oxford, 1968)

Large, E.C., The Advance of the Fungi, (Jonathan Cape, London, 1958)

Torgeson, D.C., Fungicides, Volumes I and II, (Academic Press, London & New York, 1967)

Chapter 13

SULPHUR, ORGANOSULPHUR AND OTHER ORGANIC FUNGICIDES

Sulphur

Elementary sulphur - in the form of flowers of sulphur, produced by sublimation and applied as a dust - was the first successful protective fungicide. It was used against the powdery mildew of vines. It is also toxic to spider-mites and a single treatment often had both effects, since mite and powdery mildew attacks are at their worst during hot, dry weather. Sulphur is not now used to any significant extent solely against spider mites except for the species producing the big-bud deformity in black currants.

Elementary sulphur is now usually ground to a finer powder than "flowers", with the addition of a mineral diluent (see Formulation chapter), particularly when the product is made wettable and intended for application by spraying after suspension in water. Wettable powders, pastes and flowable formulations contain particles of 1-10 microns; micronized sulphurs are 0.5-0.8 microns; colloidal sulphurs are 0.2-0.5 microns. Sulphur is not highly active as a fungicide but it is cheap and, therefore, "cost effective" even at high application rates. It is also an extremely safe chemical to use.

Sulphur is one of the substances which can be brought into a "classical" colloidal suspension and has particular interest in this state since remarkably monodisperse material can be made, usually by the slow reaction of sodium thiosulphate in dilute solution with mineral acid. A much cheaper way of producing a concentrated suspension is by the interaction of hydrogen sulphide and sulphur dioxide simultaneously introduced into water containing stabilizing agents. Colloidal sulphur concentrates prepared in this way are still obtainable, but they have the usual storage and transport difficulties of this type of formulation.

A previously much used liquid sulphur product is the so-called "lime-sulphur". This is a clear reddish-yellow liquid, stable in the absence of air and light, obtained by adding elementary sulphur to a boiling water slurry of slaked lime. Both lime and sulphur pass into solution, concentrates with as high a density as 1:3 being easily obtained. The sulphur is mainly present as polysulphide ions, $-S-(S)_x-S-$, where x represents a scatter of numbers in the neighbourhood of 2. The oxygen of the lime appears mainly as thiosulphate. A very approximate representation of a typical reaction, in which 50 kg of calcium oxide and 100 kg of sulphur are finally brought into solution in 1000 litres is

$$6\ CaO + 21\ S \rightarrow 6\ Ca^{++} + 2\ S_2O_3^{=} + 3\ S_4^{=} + S_5^{=}.$$

There is a slight residual alkalinity and the polysulphide ions are stable in absence of light and air. The corresponding free acids deposit sulphur and liberate hydrogen sulphide. This decomposition occurs when the diluted spray takes up atmospheric carbon dioxide on exposure, the initially clear pale yellow spray liquid becoming milky over a period of a few minutes after spraying.

The alkalinity and salinity of lime-sulphur sprays and the evolution of hydrogen sulphide make them more aggressive and more phytotoxic than sprays of suspended sulphur. Phytotoxicity is increased by the presence of oil, and lime-sulphur should not be tank-mixed with emulsified pesticides. On chemical grounds it is also a rather incompatible product. Its alkalinity contra-indicates its admixture with easily hydrolysed organic pesticides, such as parathion. The hydrogen sulphide evolved inactivates heavy metals present so that it cannot be used with lead arsenate or with sprays intended to contribute deficient trace-metals, such as manganese or iron. Lime-sulphur is therefore usually employed by itself. It is effective for the control of powdery mildews on vines, hops and some other fruit, apple scab and some leaf-spot diseases. Used at sufficient dilution, damage to the crop can be avoided except on particular varieties known to the grower as "sulphur-shy". More concentrated lime-sulphur sprays are used against the overwintering spores of some diseases, such as leaf-curl of peach, during the dormant period of the host. However, because it is caustic and disagreeable to use, lime-sulphur is mostly being replaced by products more easy and pleasant to handle.

Lime-sulphur is, of course, chemically incompatible with copper sprays because of the precipitation of cuprous sulphide. There is also something complementary, and perhaps antagonistic, in the biological action of these fungicides. The spread of some diseases can be arrested by either copper or sulphur treatment, but, against most diseases, one or the other is clearly advantageous. Thus powdery mildews are much more susceptible to sulphur than to copper, while the reverse is true of the downy mildew diseases. When choice is made between them, sulphur is always used against the Oidium disease of rubber trees but copper always against potato blight and blister blight of tea.

Copper is one of the minor elements essential for all forms of life. Either excess or deficiency can be lethal, but, for most forms of life, including man, the right internal adjustment is made over a wide range of availability in the environment. The fungi appear to be more critically dependent on the external availability of copper than most organisms. A rather obvious theory of the fungicidal action of sulphur products held that they produced critical copper deficiency by insolubilizing copper as the sulphide or a mercaptide. No clear evidence in support of this theory has, however, been brought forward. Another theory drew a parallel between the action of elementary sulphur in cross-linking ("vulcanizing") raw rubber and its fungicidal effect. This theory was cruder and even more improbable, since the conditions of reactions are widely different. Nevertheless it was very productive in causing tests to be made of the various sulphur compounds developed in the rubber industry to accelerate and control the vulcanization reaction. This led to the discovery of what have been to date the most widely used of organic fungicides, produced on a scale comparable with that of the herbicide 2,4-D and the insecticide DDT at its peak. These are the dithiocarbamates.

Dithiocarbamates and Related Compounds

Chemistry

Carbon disulphide reacts with primary or secondary alkylamines to form alkylamides of dithiocarbonic acid. The half amides are formed very readily. They are acidic and usually called dithiocarbamic acids. The free acids are not stable but the salts are much more stable and so the reaction is normally carried out in the presence of aqueous alkali to give a solution of the alkali metal salt:

$$RR'NH + CS_2 + NaOH \rightarrow RR'N.CS.SNa + H_2O$$

If the reaction is carried out with excess amine in place of alkali and the resulting alkylammonium salt of the dithiocarbamic acid is heated, a thiourea is formed, H_2S being liberated. If the alkali salt is reacted with an alkyl halide, a rather stable ester is formed. Many such compounds have been prepared but have no importance as fungicides. Their formulae are given to illustrate the structures.

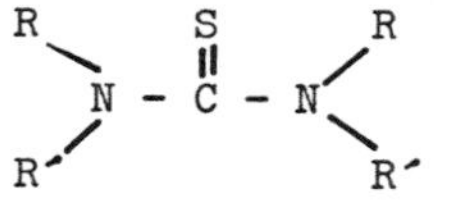

tetraalkylthiourea

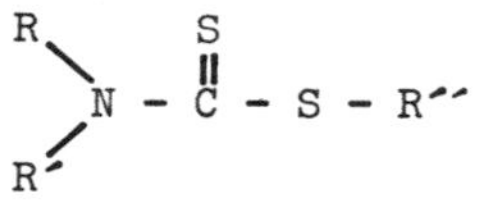

alkyl ester of dialkyl dithiocarbamic acid

The dithiocarbamates are very reactive compounds and it is impossible here to deal with all the reactions which they can undergo involving oxidation and loss of sulphur. The book by Thorn and Ludwig should be consulted for further information on chemistry, biochemistry and fungicidal action. Two reactions of great importance must, however, be mentioned.

When dithiocarbamate salts are oxidized under mild conditions, a reaction occurs analogous to the formation of disulphides from mercaptans. The products are called thiuram disulphides.

$$2\ RR'N.CS.SNa + I_2 \rightarrow RR'N.CS.S - S.CS.NRR' + 2NaI$$

In the laboratory preparation of these compounds, iodine is the preferred oxidizing agent since there is no danger of the oxidation proceeding too far, but, for industrial preparation, chlorine is preferred for economic reasons. It must be introduced slowly under conditions of good agitation.

Oxidation of dithiocarbamates to thiuram disulphides can occur on exposure to air as can the formation of monosulphides, with liberation of sulphur. Most thiuram disulphides are crystalline compounds. They are neutral in reaction, soluble in organic solvents and of low solubility in water. The tetramethyl compound (thiram) was the first compound in the whole class to find use as a fungicide and is still valuable as a seed dressing against soil fungi.

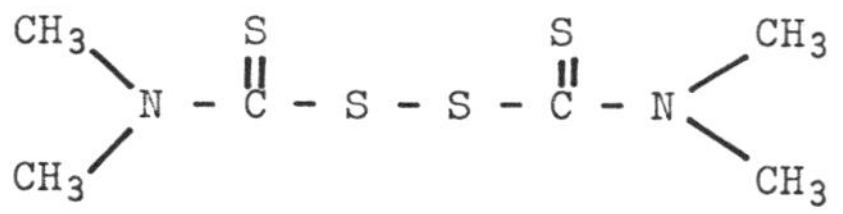

thiram

The other reaction which must be mentioned is confined to the compounds where only one alkyl group is attached to an N atom. This is decomposition into hydrogen sulphide and the alkyl isothiocyanate

$$RNH.CS.SH \rightarrow RNCS + H_2S.$$

This reaction occurs, along with more complex side reactions, whenever a solution of dithiocarbamate is acidified, although, as a laboratory route to isothiocyanates, it is preferable first to react the dithiocarbamate with a chloroformate. It is generally considered that the excellent soil fungus control which can be achieved by solutions of sodium monomethyldithiocarbamate (metham-sodium) is due to the action of the methylisothiocyanate slowly liberated. The evidence for this is that the effect is more widely distributed through the soil than could be explained by slow diffusion in soil water and that it is similar to the effect of injected methylisothiocyanate itself or of other compounds which can generate this volatile substance.

Metham-sodium is used at high dosage, in the range of 50-200 kg/ha for sterilization of soil to be used for very high value crops, such as glasshouse beds and land favourably situated for very early cropping. The liberated isothiocyanate is effective against nematodes, many soil insect pests and weed seeds as well as fungi. The more stable and less phytotoxic zinc salt (ziram) and ferric salt (ferbam) have been used as agricultural fungicides and the nickel salt has been used to control bacterial diseases in rice. Some use has also been made of zinc propylene bisdithiocarbamate (propineb).

However, by far the most important dithiocarbamates for use as protectant fungicides on growing crops are those derived from ethylene diamine, known as the ethylene bisdithiocarbamates

$$H_2NCH_2CH_2NH_2 + 2CS_2 + 2NaOH \rightarrow$$
$$NaS.CS.NHCH_2CH_2NH.CS.SNa.$$

nabam

The sodium salt (nabam) is fungicidal but tends to damage the host leaves and has no useful persistence as a fungicide under wet conditions. The persistence was found to be greatly increased by addition of zinc sulphate and lime in the spray tank. The insoluble zinc salt (zineb) and the manganous salt (maneb) were then prepared in the factory by the obvious methods and marketed in wettable powder form.

Nabam is usually manufactured by running a 20-60% solution of ethylene diamine and sodium hydroxide in water down a packed column up which carbon disulphide vapour is ascending. The reaction is highly exothermic and use of equipment with no moving parts obviates fire hazard from the extremely inflammable carbon disulphide. A solution of nabam, ready for use, is withdrawn from the bottom of the column and excess carbon disulphide is condensed from the top of the column and returned to the base.

Maneb is made by adding manganous chloride to the solution of nabam and filtering off the precipitated solid and drying it at 50°C in a forced draught.

Zineb is usually made by reacting the ammonium salt of ethylene bisdithiocarbamic acid with zinc oxide at 20-30°C. This gives crystals with an average size of one micron. If zineb is prepared from nabam and a zinc salt the product is colloidal and amorphous and not as stable as the crystalline material.

The heavy metal salts of dithiocarbamates are all of low solubility in water. Chemically they are not simple salts. They have greater solubility in some organic solvents than in water and the zinc salt of dimethyldithiocarbamate has been distilled in short-path vacuum apparatus. These facts, together with absorption spectra differing from those of the separate ions indicate a considerable degree of covalent or chelate structure.

Since a divalent metal can therefore link two dithiocarbamate groups which can also be linked by oxidation to the disulphide structure, it is evident that, in the products formed from ethylene bisdithiocarbamates, there are possibilities of polymeric structure. In some commercial products, such polymerization has been claimed to confer greater stability of deposits.

Zineb and maneb now rank in tonnage alongside elementary sulphur and fixed coppers. They are used as spray or dusts against a wide variety of fungus diseases of foliage. They have partly displaced fixed coppers for potato blight control, although they are more advantageously used in admixture with copper compounds. They have a wide spectrum of activity but are, curiously, least active against the diseases susceptible to elementary sulphur. They are extremely safe to use and have very low toxicities to men and animals.

Two variants which have found widespread use are the mixed manganous and zinc ethylene bisdithiocarbamates (mancozeb) and a mixture of the ammoniates of zineb and ethylene bisdithiocarbamic acid with bi- and tri-molecular cyclic anhydrosulphides and disulphides (metiram). A vast number of mixed formulations and combinations of maneb and zineb with other fungicides have been introduced as propietary products.

Substituted Phenols

Most phenols, particularly chlorinated phenols, are toxic to microorganisms. Cresols and higher homologues contribute to the fungicidal action of creosote oils in timber and solutions of pentachlorophenol (PCP) have replaced creosote impregnation when the colour and odour of the latter are objectionable. Some alkylchlorophenols are widely used constituents of household disinfectants. All these compounds are, however, too phytotoxic to permit their use for fungus control in agriculture.

A substituted phenol which has long been used to prevent mould growth on stored cotton goods is the anilide of salicylic acid.

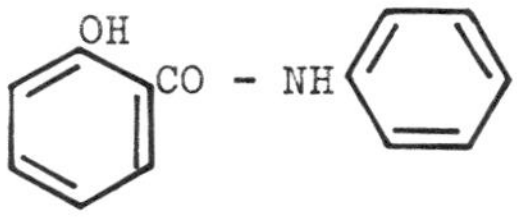

salicylanilide

Another phenolic compound of value in rot-proofing of cellulose textiles is dichlorophen prepared by condensation of 4-chlorophenol with formaldehyde with sulphuric acid as catalyst.

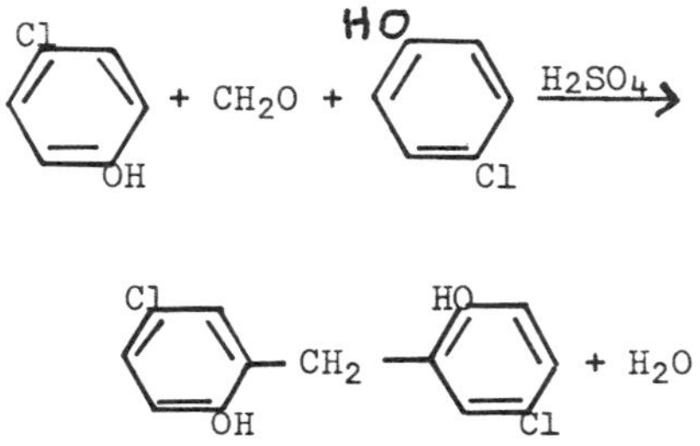

dichlorophen

Relatively simple phenols, particularly 3-cresol, have found a specialized fungicidal use in esterified form. The 3-cresyl acetate is not among the most active compounds when assessed by concentration effective in culture medium, but it is self-distributing through the vapour phase and non-corrosive and has been used for prevention of mould development in electrical and optical apparatus in tropical use. The total amount of mould growth which accidental nutrients can support inside a pair of binoculars may be extremely small, but a very thin spread of mycelium over lens surfaces can have a serious effect.

A phenol of interesting properties and having minor use on crops is 8-hydroxyquinoline (oxine).

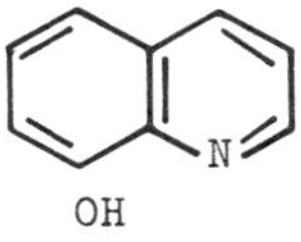

oxine

This compound was the first to show any significant systemic effect, in the elm tree against the Dutch elm disease, but it does not give adequate control. Oxine has considerable chelating powder for heavy metals, particularly copper, and the cupric oxine complex has shown greater fungicidal powder in some situations.

2-Phenylphenol is used in the preservation of packed fruit against moulds, but for this use the unsubstituted diphenyl is used more extensively although of lower activity in culture medium tests. The explanation is that the preservative is desirably applied to the wrapping medium only and the more volatile compound is more rapidly redistributed. This usage has of course also to meet a very critical standard with regard to off-flavour and residues.

Dinitrophenol Derivatives

Substituted 2,4-dinitrophenols, especially 6-methyl-2,4-dinitrophenol (DNOC), were earlier used as rapid-acting selective herbicides. These compounds are very toxic to men and animals and so their use has been largely discontinued. They were also found to be very toxic to spider mites, which infest fruit trees, but were too phytotoxic to be used for control except in winter washes in the dormant season to kill the spider mite eggs (see chapter on Oils).

However, by increase in size of the alkyl group adjacent to the OH, and by esterification of the OH, useful selectivity against mites with adequately reduced phytotoxicity was obtained. Similar substitutions also appeared to promote activity against some fungi, particularly the powdery mildews.

It is probable that the esters are not active as such, their activity being consequent on release of the free phenol by hydrolysis. In addition to changes in absolute solubility and in partition ratio between oily and aqueous biophases the rate of hydrolysis introduces another factor which can modify the toxicity. No simple correlation is observed with rates of hydrolysis in simple solutions, and it is probable that enzyme-catalysed hydrolysis is of major importance.

The most successful compounds for mite control have been the substituted acrylic esters. In particular, 6-sec.butyl-2,4-dinitrophenyl 3,3-dimethylacrylate (binapacryl) has been shown to be an effective acaricide of very low phytotoxicity, having some activity also against apple mildew. Some carbonates have also been used such as isopropyl 6-sec.butyl-2,4-dinitrophenyl carbonate (dinobuton).

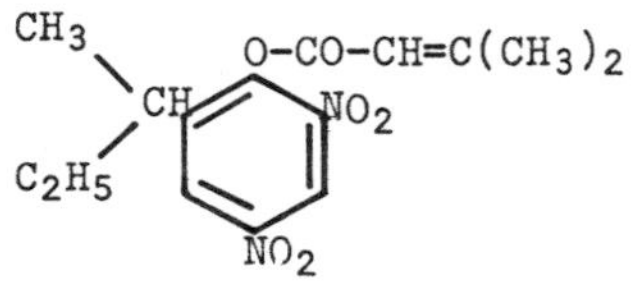

binapacryl

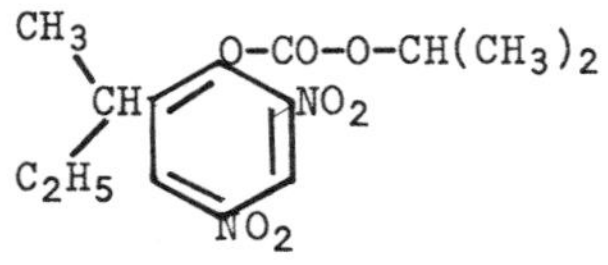

dinobuton

The product in this class which has proved to be most useful as a fungicide and which has been used extensively for many years for the control of powdery mildews is a mixture of about 65-70% of 2,6-dinitro-4-octylphenyl crotonate (dinocap-4) and 30-35% of 2,4-dinitro-6-octylphenyl crotonate (dinocap-6).

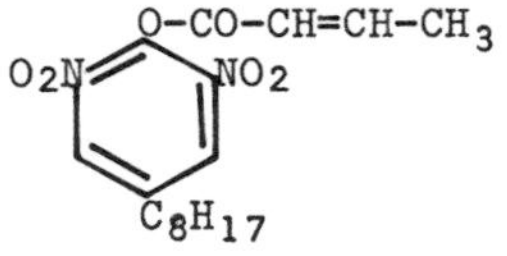

dinocap-4

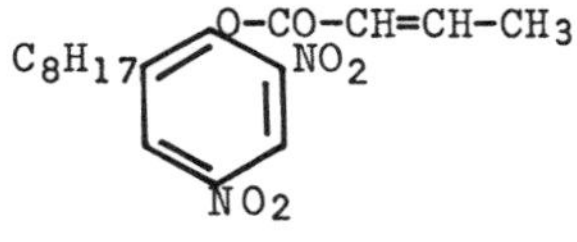

dinocap-6

A related product is a mixture of methyl 2,6-dinitro-4-octylphenyl carbonate (dinocton-4) and methyl 2,4-dinitro-6-octylphenyl carbonate (dinocton-6).

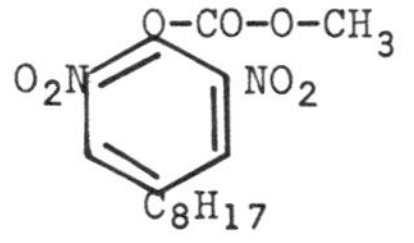

dinocton-4

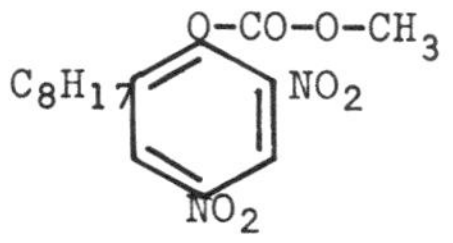

dinocton-6

It has been found that the fungicidal activity of dinocap, especially towards powdery mildews, is mainly due to the dinocap-4 whereas the acaricidal activity is mainly due to the dinocap-6. This appears also to be true for dinocton-4 and dinocton-6. In general, 2,6-dinitro-4-alkylphenol derivatives tend to be fungicidal whilst 2,4-dinitro-6-alkylphenol derivatives tend to be acaricidal.

The situation is made more complex because the octyl side chain in dinocap and dinocton is not a single species but a mixture of the isomeric 1-methylheptyl, 1-ethylhexyl and 1-propylpentyl groups. The products are made by condensing commercial secondary octanol, which is a mixture of species, with phenol which gives a mixture of 2-octylphenols and 4-octylphenols. This is then dinitrated to give a mixture of 2,4-dinitro-6-octylphenols and 2,6-dinitro-4-octylphenols.

In biological studies, it was shown that, in the 2,6-dinitro series the compounds having more compact alkyl substituents were more fungicidally active than those with more nearly straight chains whereas, in the 2,4-dinitro series, they were less fungicidally active.

The story now revealed is an interesting object lesson in the desirability of establishing the biological performance of chemicals in the first place with isolated pure materials. It is not, of course, in most cases practicable to supply the agricultural market with purified chemicals but it is desirable first to find out what compound to aim at.

The general method used for the preparation of alkylphenols is addition of an olefine and a phenol under rather drastic reaction conditions using sulphuric acid, an activated clay or a Friedel-Craft catalyst. These catalysts can also function to produce the olefine from an alcohol by elimination of water so that, if an alcohol and the phenol are used as starting products, the overall reaction appears to be a condensation. That direct condensation is not a true representation is shown by the non-formation of an n-alkylphenol from an n-alcohol and by the fact that cresols cannot be made by this route. From butanol, sulphuric acid and phenol a mixture of 2- and 4-s-butylphenols is obtained.

$$CH_3CH_2CH_2CH_2OH \xrightarrow{H_2SO_4} CH_3CH_2CH{=}CH_2 + H_2O$$

$$CH_3CH_2CH{=}CH_2 + C_6H_5OH \xrightarrow{H_2SO_4} \begin{array}{l} CH_3CH_2\searrow \\ \qquad\qquad CH\text{-}C_6H_4OH \\ CH_3\nearrow \end{array}$$

It will be evident that the same product would be obtained if secondary had been used in place of n-butanol. From higher secondary alcohols there are of course more possible isomers. From isobutanol the tertiary-butylphenols are obtained.

It is possible to obtain some control of the 2:4 ratio by choice of catalyst and reaction conditions. In particular, if a large excess of strong sulphuric acid is used, the phenol is sulphonated, mainly in the 4-position, before addition of the olefin. Further heating, after dilution with about 20-30% of water, causes desulphonation and this procedure gives a high yield of the 2-alkylphenol. The 2- and 4-alkylphenols are, however, easily separated by distillation, the 2- having significantly lower b.p. In the case of the sec-butylphenols the 4-compound has sufficient use as a plasticizer to make this separation of mixed products economical.

Chloronitrobenzenes and Related Compounds

Nitration of chlorobenzenes increases the reactivity of the C-Cl groups and compounds of this type are often lachrymatory and irritant. Many show activity on vegetative growth, of which scorch of leaves and arrested development of buds are the most frequent symptoms. They are toxic to most fungi in culture tests.

Pentachloronitrobenzene (quintozene) has the longest established usage of all these compounds. It is applied as a seed dressing. In massive dosage, up to several hundred kg per hectare, it has been incorporated by rotovation as a soil sterilant particularly against virus-carrying nematodes.

Another seed-protecting fungicide of this type is 1,2,4,5-tetrachloro-3-nitrobenzene (tecnazene). When used to control dry rot (*Fusarium caerulum*) in stored potatoes, it was found also to inhibit sprouting. It must not, of course, be used on seed potatoes and is suspected by some authorities of producing off-flavour. A mixture of 1,2,3-trichloro-4,6-dinitro and 1,2,4-trichloro-3,5-dinitrobenzenes has been used against soil fungi.

A related compound, which has cyano groups instead of nitro groups, is tetrachloroisophthalonitrile (chlorthalonil) which is manufactured by chlorination of isophthalonitrile in the vapour phase at 400-500°C. It has been used as a foliage spray on some crops and also as a soil fungicide.

A compound which may also be included in this group is 2,6-dichloro-4-nitroaniline (dicloran) which is mainly used against the grey mould of lettuce.

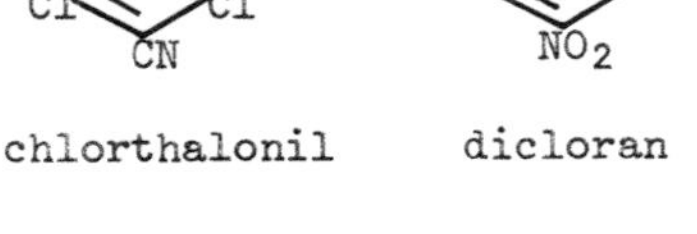

chlorthalonil   dicloran

Quinones

Exploration of quinones as fungicides was initiated by the identification of the natural product juglone from rot-resistant walnut heartwood. Two simple chloro-substituted quinones found considerable use as seed-dressings and are still in demand. They are 2,3-dichloronaphthoquinone (dichlone) and tetrachlorobenzoquinone (chloranil).

Dichlone is manufactured by reaction of 1-naphthol with chlorine at 80-120°C. The reaction is exothermic and better yield and purity are obtained if the first exothermic part of the reaction is carried out at less than 40°C followed by completion of the reaction at 80-120°C than if the whole reaction is carried out at 80-120°C. Reaction is carried out in a solvent mixture of 25 parts 96% sulphuric acid, 71 parts glacial acetic acid and 4 parts water.

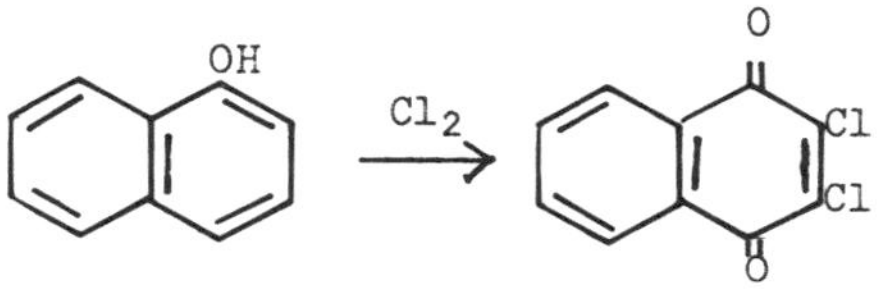

dichlone

Chloranil can be manufactured by a number of methods but the only economical process is reaction of cyclohexane with hydrochloric acid and oxygen. The reactants are mixed and passed at 220-240°C over a catalyst of copper chloride, cobalt chloride and ferric chloride deposited on alumina.

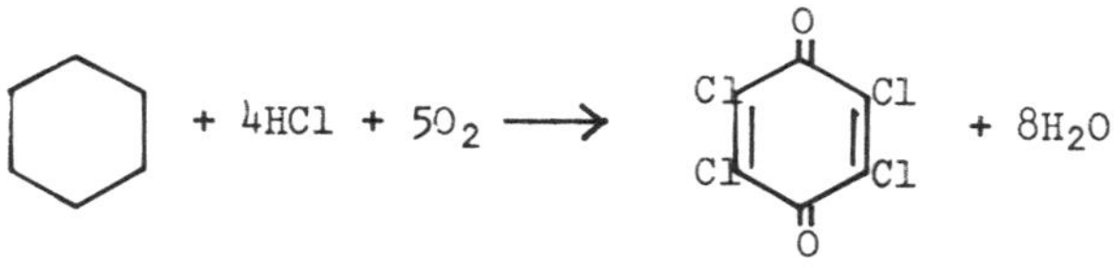

chloranil

A fungicide of more complex quinone structure is 2,3-dicyano-1,4-dihydro-1,4-dithia-anthraquinone (dithianon) which is active against apple scab and other pome fruit diseases, except apple mildew.

The dioxime, benquinox, is also a quinone derivative and, like the simpler quinone compounds, is mainly useful as a seed dressing.

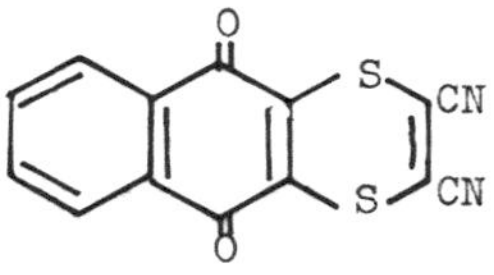

dithianon

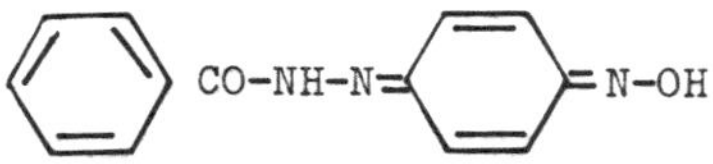

benquinox

Trichloromethylmercapto Compounds

A very widely used fungicide is N-trichloromethylthio-4-cyclohexene-1,2-dicarboximide (captan). The very reactive perchloromethyl mercaptan can be condensed with various imines to give quite stable compounds having the $-N-S-CCl_3$ group, all of which show fungicidal properties. In captan, the imine is that of tetrahydrophthalic acid, the anhydride of which is first prepared by condensation of maleic hydride with butadiene.

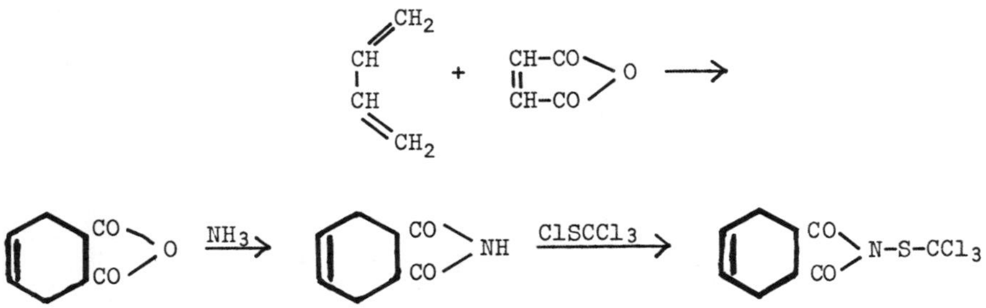

captan

Captan is a high-melting crystalline solid, formulated as a dust or wettable powder. It is used in seed dressings and also by application to foliage for protection against many leaf-spot diseases. It is widely used against apple scab and is considered to improve the appearance of apple skin as well as reduce the incidence of this disease.

The corresponding derivative of phthalimide itself, folpet, has also been used for similar purposes, but less successfully. There appears to be more risk of damage to fruit and, although the product is more effective than captan against potato blight, it is not competitive with fixed coppers, dithiocarbamates or, more recently, triphenyltin salts. Its main use has been on ornamentals.

A later related compound is captafol, in which the trichloromethyl group of captan is replaced by 1,1,2,2-tetrachloroethyl. In some quarters this is considered the most effective compound to date against potato blight, but is at an economic disadvantage with the more widely used products mentioned above.

Two related compounds which contain the $-SCFCl_2$ group instead of the $-SCCl_3$ group, have found some use. These are dichlofluanid and tolylfluanid.

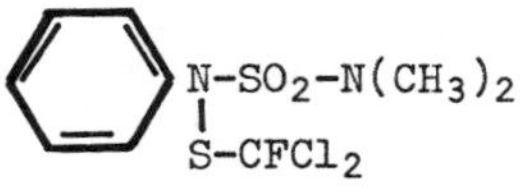

dichlofluanid

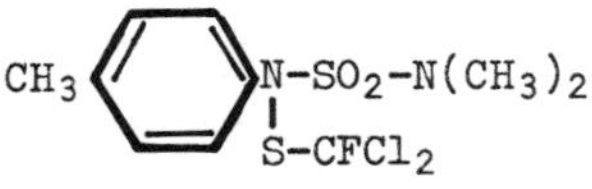

tolylfluanid

Cationic Surfactants and Relatives

Surfactants are compounds in which a large hydrophobic group (usually a paraffin chain) is attached to a strongly hydrophilic group which may be anionic, neutral or cationic. In the latter class it may be a quaternary nitrogen group, cationic at all pH, or an amine group requiring at least slight acidity to make it functional.

During investigations of cationic surfactants as formulation constituents (wetting, emulsifying, etc, agents) it was soon evident that they had, as a class, rather powerful bactericidal action. This is widely exploited in their use, officially approved in many countries, as alternatives to steam for the sterilization of dairy equipment. The most active of easily prepared compounds for this purpose is dodecylbenzyldimethyl ammonium chloride, prepared by methylation of dodecylamine and addition of benzyl chloride.

Compounds of this type are less effective against fungi than against bacteria and there is considerable risk of leaf damage if they are used against crop diseases. Compounds in which a strongly basic, but not quaternary, group terminates the paraffin chain have, however, shown useful activities against some fruit diseases, notably apple scab.

The most widely used and effective is dodecyl guanidine acetate (dodine), usually formulated as a 65% wettable powder. It is used to combat scab on apples and pears and leaf spot on cherries and strawberries.

The compound is manufactured by reaction of dodecylamine with cyanamide in the presence of acetic acid at 140-160°C in an autoclave.

$$C_{12}H_{25}NH_2 + H_2NCN + CH_3COOH$$
$$\longrightarrow C_{12}H_{25}NH\text{-}\underset{\underset{NH}{\|}}{C}\text{-}NH_2.CH_3COOH$$

dodine

Another fungicidal product of this type is 2-heptadecyl-2-imidazoline acetate (glyodin) which is used for control of foliage disease on a wide variety of fruits. Ethanol and stearic acid are reacted to give ethyl stearate which is heated with ethylene diamine to give N-(2-aminoethyl) stearamide which is then heated at 200°C under reduced pressure and the 2-heptadecyl-2-imidazoline distilled out. This is then reacted with acetic acid in isopropanol to give glyodin.

$$C_{17}H_{35}COOH + C_2H_5OH \longrightarrow C_{17}H_{35}COOC_2H_5 + H_2O$$

$$C_{17}H_{35}COOC_2H_5 + NH_2CH_2CH_2NH_2 \longrightarrow C_{17}H_{35}CONHCH_2CH_2NH_2$$

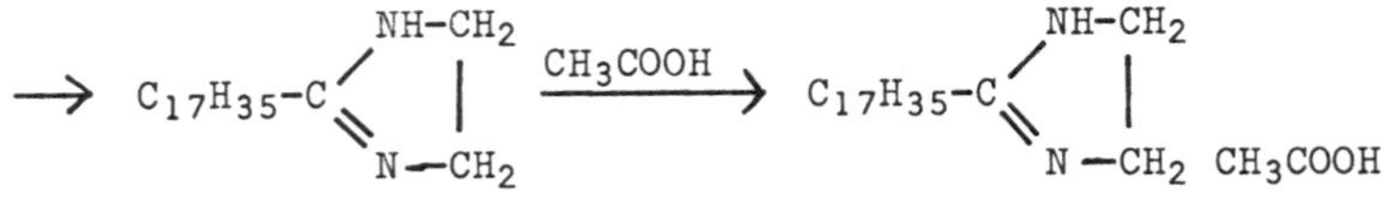

glyodin

Imidazolines and oxazolines formed from fatty acids of a wide range of chain lengths have been made and tested but the optimum chain length has always been found to be in the range of C_{14} to C_{17}. For good fungicidal effect a high lipoid solubility may be necessary or perhaps a high degree of surface adsorption, either of which requires a long chain. That the effect does not increase indefinitely with increasing chain length may well be due to decrease of absolute solubility.

A commercial fungicide related to dodine but which has two amidine groups and a chain containing nitrogen atoms is guazatine.

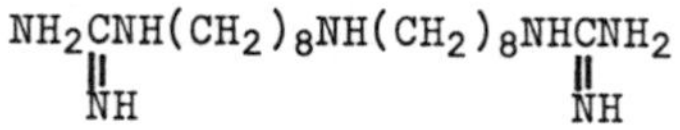

guazatine

Other Organic Compounds

It is impossible in a short general book to list all the compounds which have held at least a minor place or been subject to extensive field evaluation as protective fungicides. They include a diverse range of unrelated chemical types and some of the more successful are illustrated below.

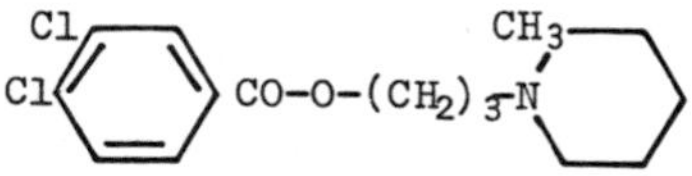

piperalin

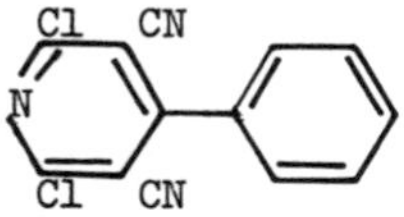

pyridinitril

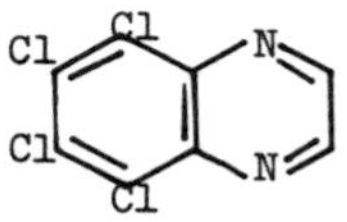

chlorquinox

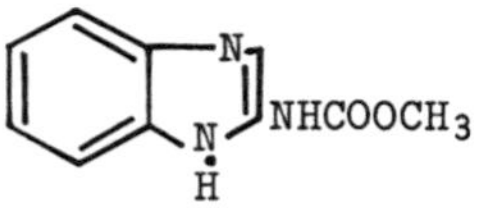

carbendazol

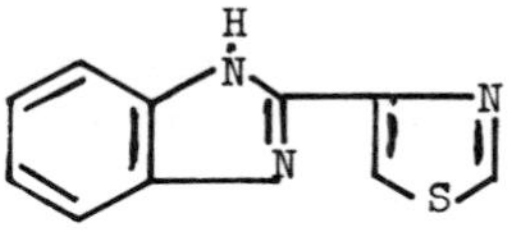

thiabendazole

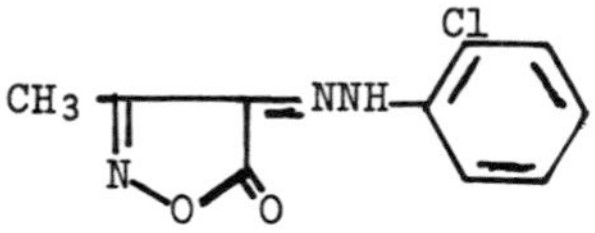

drazoxolon

However, the coppers, dithiocarbamates and captan provide adequate protective fungicidal effect against most fungi and they are cheap because they are manufactured on a very large scale from low-priced raw materials. The more complex organic compounds, which require several stages in their synthesis, cannot compete on the basis of "cost-effectiveness" even if they are more active and have, consequently, lower application rates. During the last decade the pesticides industry has therefore concentrated on discovery and development of fungicides which would be not just protective but would have eradicant or curative action on fungal diseases which had already become established in the plant. To do this the chemicals must be able to pass into the plant tissues and attack the fungus from within without injury to the host, that is, they must be systemic or, at least, partly systemic. A wide range of compounds of this type have now been produced and are dealt with in the next chapter.

Further Reading

Baker, K.F. & Snyder, W.C., Ecology of Soil-Borne Plant Pathogens, (John Murray, London, 1965)

Evans, E., Plant Diseases and their Chemical Control, (Blackwell, Oxford, 1968)

Sharvelle, E.G., Chemical Control of Plant Disease, (University Publishers, Texas, 1969)

Thorn, G.D. & Ludwig, R.A., The Dithiocarbamates and Related Compounds, (Elsevier, London, 1962)

Chapter 14
SYSTEMIC FUNGICIDES

Introduction

The idea that it might be possible actually to cure a fungal disease by applying some chemical to the foliage or roots of an infected plant had occurred to agricultural chemists as early as the 19th century when the nature of fungal diseases was first recognized and the first inorganic protective fungicides such as Bordeaux mixture were introduced. Various attempts had some modest success - scab on potato tubers was controlled with mercuric chloride, anthracnose on vines with copper sulphate, and fireblight on pears with zinc chloride. Application of lithium salts to the roots of wheat and barley prevented development of powdery mildew. When organic protective fungicides such as the dithiocarbamates and captan were developed from 1940 onwards a vast number of organic compounds were tested to see if they could control established infections. In general, the compounds were either not sufficiently effective or, if they did kill the fungus, they also caused considerable damage to the host plant. This was not surprising because of the similarity of the biochemistry of the fungi and of higher plants, and the problem of selectively inhibiting a fungal infection within the living tissues of a plant without harming the plant itself was thought to be insoluble.

Nevertheless, two facts sustained the hopes of chemists in the post - World War 2 period. The first was that most plants are resistant to most fungi, and this resistance was shown in many cases to be due either to the presence within the plant of natural antifungal substances or to the production by the plant of antifungal chemicals - the phytoalexins - as a response to the invading organism. Some of these naturally occurring compounds were isolated but they did not give effective control when topically applied to the foliage because they did not penetrate to the appropriate sites of action - the plant produced them exactly where they were needed. Nevertheless, if plants could produce selective systemic fungicides which were not phytotoxic, there seemed no reason why chemists should not also eventually do so.

The second fact was the remarkable success of the medicinal chemists in this era in discovering synthetic chemicals which would control bacterial diseases in humans without harming the host. If systemic bactericides were possible, why not systemic fungicides? It was, in fact, from the field of chemotherapy that the first breakthrough came.

The isolation of penicillin in 1940 started the discovery of a whole range of chemotherapeutic antibiotics. The effects of many of these in plants were studied and it was found that streptomycin is readily translocated and would control some fungal diseases in plants such as the downy mildew of hops. The most interesting antifungal antibiotic was griseofulvin which was shown to be readily translocated, to have low phytotoxicity and to be effective systemically against a considerable number of fungal diseases of plants.

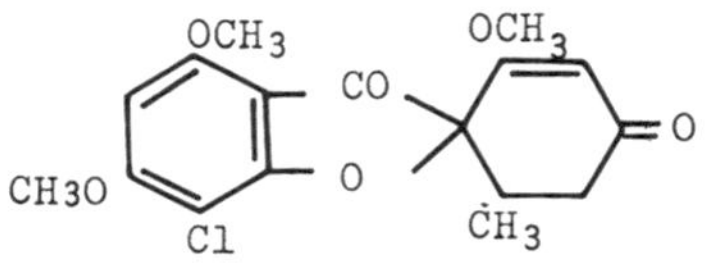

griseofulvin

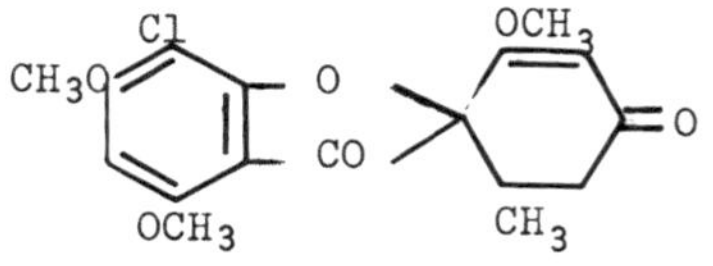

epigriseofulvin

Griseofulvin has asymmetric centres at positions 2 and 6' and antifungal activity is confined to the (+)- isomer. The diastereoisomer, epigriseofulvin, is inactive. Griseofulvin does not inhibit germination of fungal spores but, at very low concentrations, causes stunting and distortion of the fungal hyphae, probably by affecting nucleic acids. It is broken down in plants and in soil so its use in practical agriculture is limited.

Other antibiotics have since been developed as systemic fungicides, particularly in Japan. Cycloheximide, obtained as a by-product in manufacture of streptomycin, is very active against a wide range of fungi but its use in agriculture is limited by its considerable phytotoxicity. Blasticidin has been used for control of paddy blast on rice, at concentrations as low as 5-10 $\mu g/cm^3$, and so is 100 times as active as organomercury fungicides. The polyoxins have been shown to be effective against a number of fungi. Kasugamycin, like blasticidin, controls paddy blast but has the advantage of being less phytotoxic and less toxic to men and animals. Cellocidin will control rice leaf blight and is interesting in having a very simple chemical structure for an antibiotic. It can be easily manufactured from 1,4-dihydroxybutyne or from fumaric acid.

$$\begin{array}{l} CH{\equiv}CH \\ + \; 2CH_2O \end{array} \longrightarrow \begin{array}{l} C - CH_2OH \\ ||| \\ C - CH_2OH \end{array} \longrightarrow \begin{array}{l} C - CONH_2 \\ ||| \\ C - CONH_2 \end{array}$$

cellocidin

Cycloheximide, blasticidin and kasugamycin all act by inhibiting protein synthesis in the fungus. Antibiotics vary considerably in their toxicities to men and animals. Streptomycin is very safe (LD50 9000 mg/kg); griseofulvin (LD50 400 mg/kg); kasugamycin (LD50 2000 mg/kg) and polyoxin (LD50 1500 mg/kg) are relatively safe but cycloheximid (LD50 133 mg/kg), cellocidin (LD50 89 mg/kg) and blasticidin (LD50 39 mg/kg) are rather toxic.

The search for relatively simple organic systemic fungicides which could be manufactured cheaply was continued, but it was not until the mid 1960's that success was achieved. Since then, a considerable number of synthetic products have come into commercial use and are having a profound effect on practical control of fungal diseases. This tendency will continue as new and more effective compounds are discovered, as they certainly will be.

Organophosphorus Compounds

Organophosphorus compounds had been shown to be outstandingly successful systemic insecticides; they are readily taken up by and translocated in plants and they are, in general, not phytotoxic. These are two essential requirements for a systemic fungicide so it was logical for chemists to search in this group for a compound with the required fungicidal activity. This was eventually found in 5-amino-2-|bis-(dimethylamido) phosphoryl|-1,-2,4-triazole, which was effective against the powdery mildew of barley. Many analogues were made and tested and this led to the first commercially successful systemic fungicide, 5-amino-2-|bis-(dimethylamido) phosphoryl|-3-phenyl-1,2,4-triazole (triamiphos).

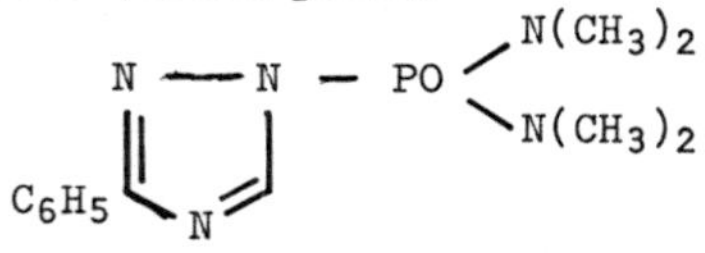

triamiphos

Triamiphos is very toxic to men and animals so its use has mainly been limited to control of powdery mildew on ornamentals such as roses. Its dermal toxicity is much less than its oral toxicity (LD50 oral 20 mg/kg, dermal 1500 mg/kg). It acts by inhibiting synthesis of DNA in the fungus.

Another group of organophosphorus compounds which have been shown to have systemic fungicidal activity are the O,O-dialkyl-phosphorothiolates. Of these, two compounds have been developed commercially and are used to control paddy blast in rice, O,O-diethyl-S-benzylphosphorothiolate (kitazin) and O,O-diisopropyl-S-benzylphosphorothiolate (kitazin P).

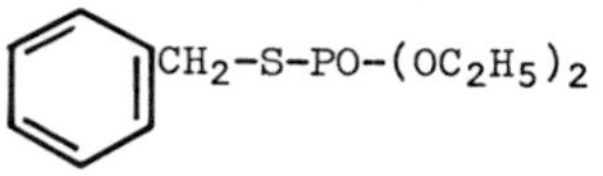

kitazin

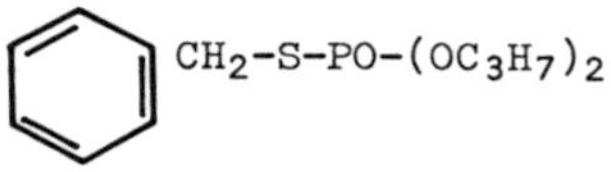

kitazin P

Kitazin is fairly water-soluble so can be applied as a granular formulation to the water in the paddy fields from which it is rapidly absorbed through the roots and translocated in the plant. A single application lasts for three weeks and produces a remarkable increase in yield. The compound acts by inhibiting esterases in the cell wall synthesis system of the fungus. Both kitazin and kitazin P are very much less toxic to men and animals than is triamiphos so they can be used safely (LD50 oral 238 mg/kg).

In general, very few organophosphorus compounds are suitable for practical use as systemic fungicides because of their very high toxicities to men and animals. Amongst those which have been developed are O,O-diethyl-(1,3-dihydro-1,3-dioxo-2H-isoindol-2-yl) phosphonothioate (ditalimfos) and O,O-diethyl-O-(6-ethoxycarbonyl-5-methylpyrazolo-|2,3,a| pyrimidin-2-yl) phosphorothioate (pyrazophos). Ditalimfos has particularly low toxicity (LD50 oral 5660 mg/kg).

Benzene Derivatives

1,4-Dichloro-2,5-dimethoxybenzene (chloroneb), obtained by chlorination of hydroquinone dimethyl ether, is used as a seed treatment on sugar beet and as a soil treatment for beans and cotton. It is taken up by the roots and concentrated in them and in the lower parts of the stems, so it is not fully systemic. It is very specific and controls only the Rhizoctonia fungi. It acts by inhibition of DNA synthesis at the nucleotide polymerization stage. It is a very safe compound to use as its oral and dermal toxicities to men and animals are very low indeed (LD50 oral 11,000 mg/kg, dermal 5000 mg/kg).

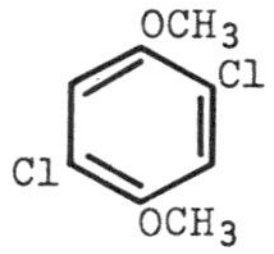

chloroneb

Salicylanilide has been mentioned in the previous chapter as a protective fungicide for textiles. Two related anilides have been found to have some systemic action but specifically against the brown rust of barley and the yellow rust of wheat. They are 2-toluanilide (mebenil) and 2-iodobenzanilide (benodanil). Of these, benodanil has more activity and a greater margin of crop safety. Only the 2-substituted benzanilides are active. They are easily made by reaction of aniline with the appropriate benzoyl chloride or ester.

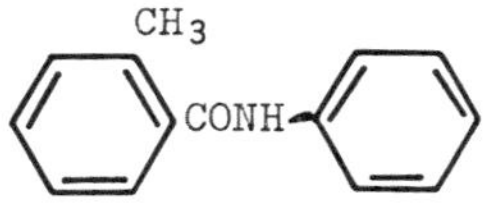

mebenil

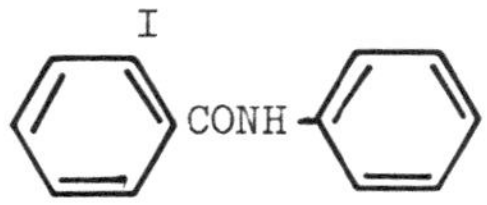

benodanil

It has recently been shown that the benzoyl group in mebenil can be replaced by crotonyl. Cis-crotonanilide is active, but too phytotoxic for practical use, whereas trans-crotonanilide is inactive. It will be seen that the crotonanilide structure is essentially part of the mebenil molecule. Butyranilide is inactive.

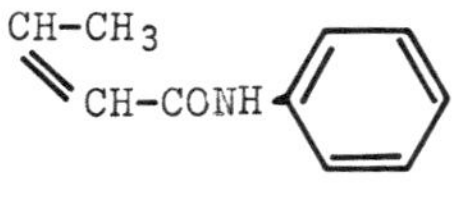

cis-crotonanilide

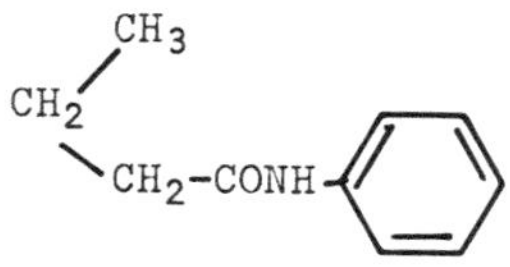

n-butyranilide

Mebenil and benodanil both act by interfering with the electron transport system in the fungus. They are very safe to use as they both have very low oral and dermal toxicities to men and animals (LD50 oral 6000 mg/kg, dermal 10,000 mg/kg).

Two commercially important systemic fungicides are 1,2-bis-(3-ethoxycarbonyl-2-thioureido) benzene (thiophanate) and 1,2-bis-(3-methoxycarbonyl-2-thioureido) benzene (thiophanate-methyl). They are manufactured by reaction of phenylene diamine with isothiocyanoformic ester (obtained from methyl chloroformate and potassium thiocyanate) at room temperature in a non-hydroxylic solvent.

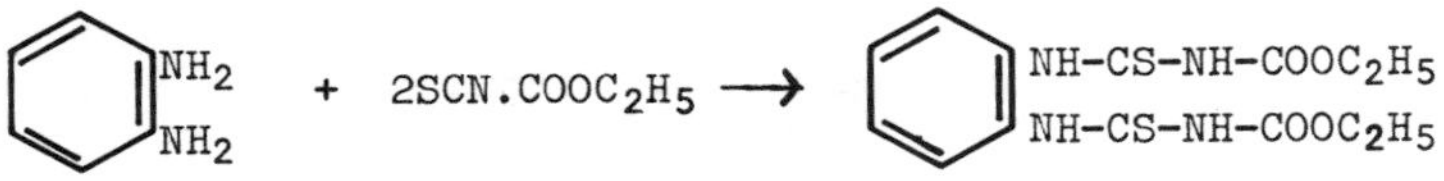

thiophanate

Thiophanate and thiophanate-methyl are highly active systemically against barley and cucumber mildews but not against apple mildew. In the plant they are rapidly converted to ethyl benzimidazol-2-yl carbamate and methyl benzimidazol-2-yl carbamate respectively and these are, in fact the active fungicidal principles.

These benzimidazole compounds interfere with DNA synthesis and inhibit mycelial growth. Because cyclisation is necessary to produce an active compound the 1,3- and 1,4- analogues of thiophanate are not active because they cannot undergo ring-closure. It is interesting to note some structural similarity of thiophanate with the ethylene bis-dithiocarbamates.

1-(3,4-Dichloroanilino)-1-formylamino-2,2,2-trichloroethane (chloraniformethan) is a systemic compound which is effective against mildew on spring barley. It has achieved considerable commercial success. It has a low toxicity to men and animals (LD50 oral 2500 mg/kg, dermal 500 mg/kg). It is manufactured by reaction of 3,4-dichloroaniline with $CCl_3.CHCl.NHCHO$, which is made from chloral and formamide.

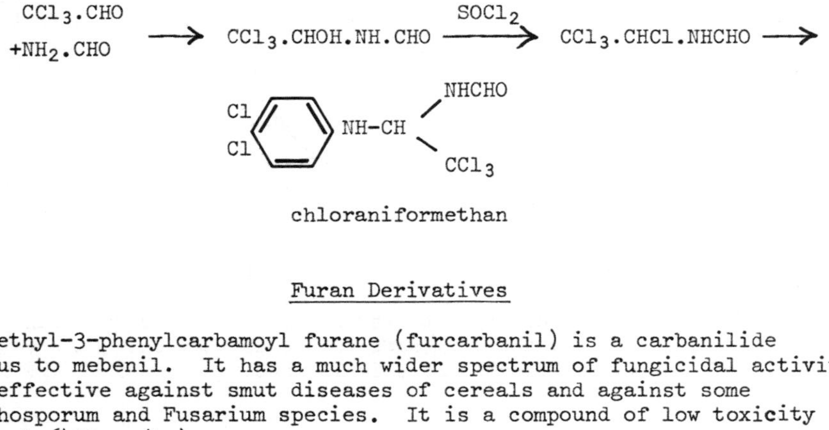

chloraniformethan

Furan Derivatives

2,5-Dimethyl-3-phenylcarbamoyl furane (furcarbanil) is a carbanilide analagous to mebenil. It has a much wider spectrum of fungicidal activity and is effective against smut diseases of cereals and against some Helminthosporum and Fusarium species. It is a compound of low toxicity (LD50 oral 6400 mg/kg).

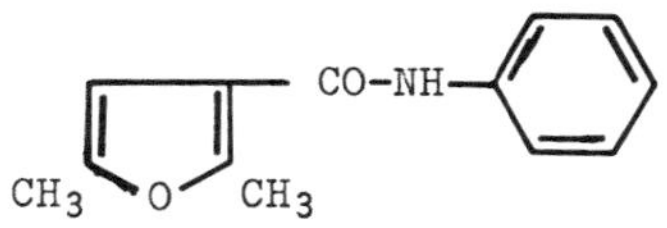

furcarbanil

Pyran Derivatives

A carbanilide similar to furcarbanil but derived from pyran instead of furane is 2,3-dihydro-6-methyl-5-phenylcarbamoyl-4H-pyran (pyracarbolid). It is active against rusts, smuts and Rhizoctonia species. It is a compound of very low toxicity indeed (LD50 oral 15,000 mg/kg).

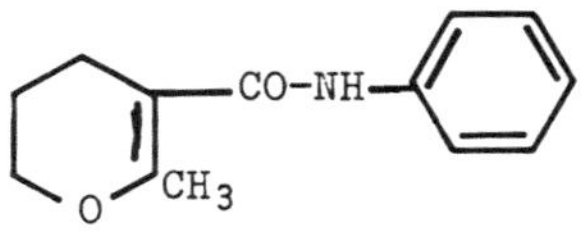

pyracarbolid

Pyridine Derivatives

Bis-(4-chlorophenyl)-3-pyridine methanol (parinol) is very effective by foliar application in very low concentrations against powdery mildews on cucumbers, apples, roses and vines. It is not fully systemic but has what is called "translaminar" movement, that is, it protects the underside of the leaf when applied to the upper surface. A problem with purely protective fungicides is that they tend to be deposited only on the upper surfaces of the leaves so that mildew can grow unchecked on the undersides.

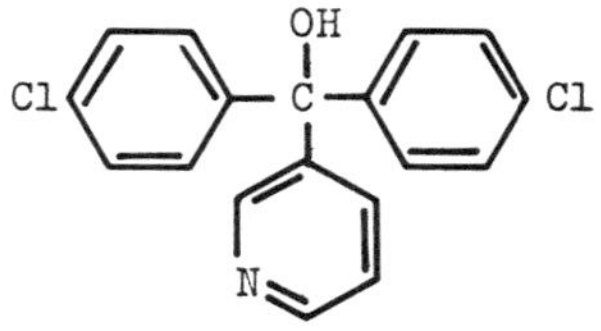

parinol

The corresponding 2- and 4- substituted pyridines are not active.

Pyrimidine Derivatives

A pyrimidine analogue of parinol has very similar activity, and is particularly effective also for control of scab on apples. Like parinol; it is not fully systemic but enters the leaf and stops development of the fungus within it. It is 2,4-dichloro-α-pyrimidin-5-yl-benzhydrol (triarimol). However, the toxicity of triarimol, to men and animals is somewhat suspect. Related compounds are α-(2-chlorophenyl)-α-(4-fluorophenyl)-5-pyrimidine-methanol (nuarimol) and α-(2-chlorophenyl)-α-(4-chlorophenyl)-5-pyrimidine methanol (fenarimol). Fenarimol is safe to use (LD50 2500 mg/kg) and is active at low concentrations (40 ppm) against apple mildew. It also controls a wide range of other mildews.

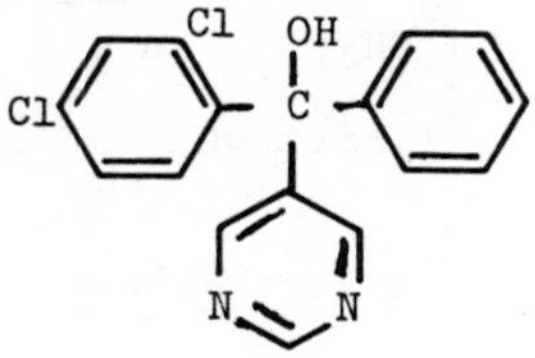

triarimol

The most important pyrimidine derivatives, which have had widespread commercial impact, are the 2-amino-4-hydroxypyrimidines. Two compounds are of particular importance, 5-n-butyl-2-dimethylamino-4-hydroxy-6-methylpyrimidine (dimethirimol) and 5-n-butyl-2-ethylamino-4-hydroxy-6-methylpyrimidine (ethirimol).

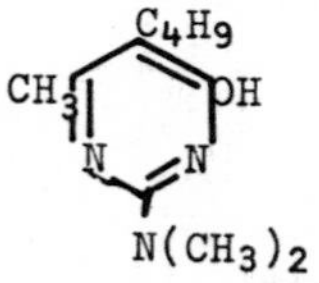

dimethirimol

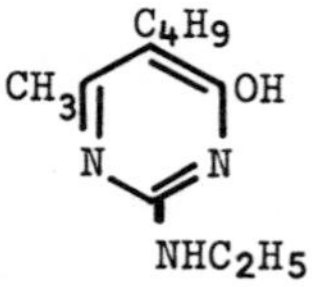

ethirimol

Dimethirimol is particularly effective against the powdery mildew of cucurbits such as cucumbers while ethirimol is more active against the powdery mildew of cereals. They are fully systemic and move freely throughout the plants when applied to the roots. They act by inhibiting spore germination, probably by interfering with tetrahydrofolic acid metabolism. They are very safe compounds to use (LD50 oral 4000 mg/kg, no effect dermally).

Fungicidal activity is affected by the size of the alkyl group in the 5-position and reaches a maximum at C_4H_9. Straight chain alkyl groups are better than branched chain. In general the nature of the substituents on the amino group in the 2-position is not very significant.

Ethirimol is manufactured by condensation of ethyl butylacetoacetate (obtained from ethyl acetoacetate and butyraldehyde) with ethylguanidine (obtained from cyanamide and ethylamine). Dimethirimol is manufactured likewise from dimethylguanidine.

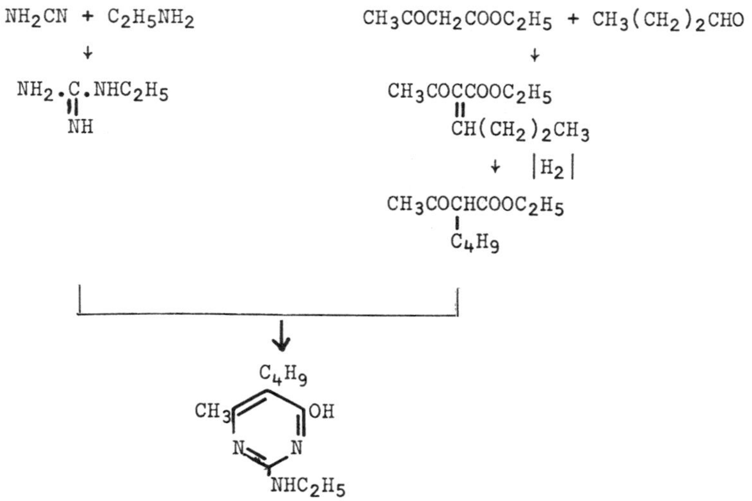

ethirimol

The dimethylsulphamoyl ester of ethirimol (bupirimate) is active against apple mildew and against a number of mildews of soft fruits, whereas ethirimol is mainly active against cereal mildew. This widening of the spectrum of activity is probably due to effects on penetration and transport to the site of action. It is, in fact, broken down to ethirimol in the plant.

Piperazine Derivatives

N,N'-bis-(1-formamido-2,2,2-trichloroethyl)-piperazine (triforine) is a systemic fungicide effective by foliar application against powdery mildew on cereals, apples and cucumbers and against apple scab. It is an interesting compound in that it appears not only to move upwards to the leaves when applied to the roots but also downwards to the roots when applied to the leaves. This is an unusual effect because it is against the main flow of sap in the plant. It is an effect which would be required in any systemic fungicide which was to be active against root fungi by foliar application.

Triforine is chemically related to chloraniformethan and is manufactured by condensation of piperazine with $CCl_3.CHCl.NH.CHO$. It is safe to use as it has low toxicity (LD50 oral 6000 mg/kg).

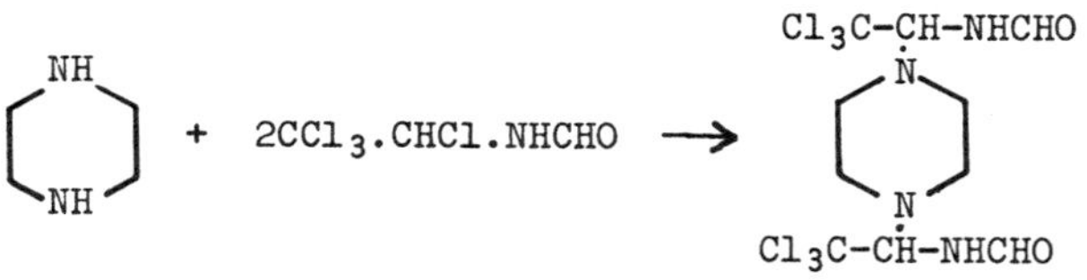

triforine

5,6,7,8-Tetrachloroquinoxoline (chlorquinox) appears to have limited systemic activity against mildew on spring barley.

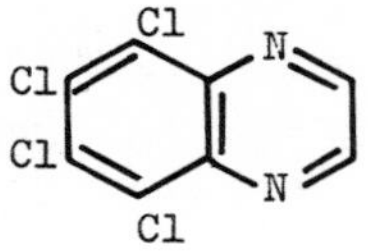

chlorquinox

Triazole Derivatives

1-(4-chlorophenoxy)-3,3-dimethyl-1-(1,2,4-triazol-1-yl)-2-butanone (triadimefon) is active at low concentrations (40 ppm) against apple mildew, and a number of other diseases. It probably acts by inhibiting biosynthesis of steroids. It is safe to use (LD50 oral 500 mg/kg).

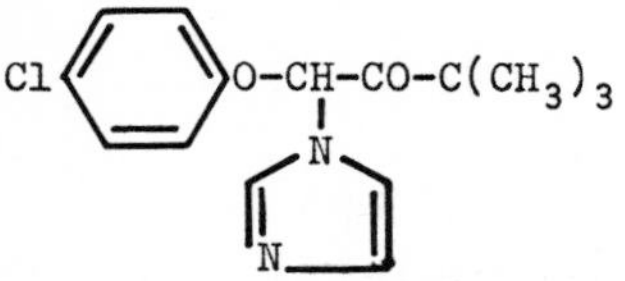

triadimefon

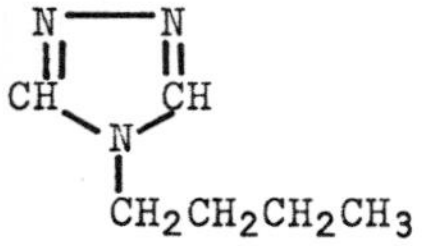

butrizol

Another triazole which is claimed as a systemic fungicide is 4-butyl-4H-1,2,4-triazole (butrizol).

Triazine Derivatives

3,5-Dioxo-2,3,4,5-tetrahydro-1,2,4-triazine (azauracil) is systemically active against powdery mildews by root or foliar application, at concentrations as small as 0.3-0.6 ppm. It was the only compound out of 82 purines and pyrimidines tested which completely inhibited development of mildews. It appears not to prevent germination but stops the growth of the fungus during formation of the first haustorium, by inhibition of uridinemonophosphate.

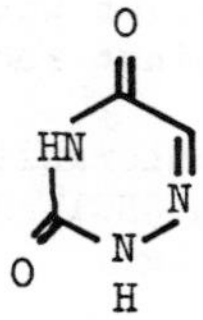

azauracil

Oxazine Derivatives

4-Tridecyl-2,6-dimethylmorpholine (tridemorph) is active against barley mildew by root uptake but less so by foliar application. As the size of the alkyl group in 4-alkyl-2,6-dimethylmorpholines is increased both fungicidal activity and phytotoxicity increased to a maximum at C_{13}, so tridemorph has a low safety margin for crop damage. The 4-cycloalkyl-2,6-dimethyl-morpholines retain fungicidal activity but are much less phytotoxic. The best compound, which is used commercially, is 4-cyclododecyl-2,6-dimethyl-morpholine (dodemorph). They are made by reacting tridecyl chloride or cyclododecyl chloride with 2,6-dimethylmorpholine obtained from catalytic vapour phase ammoxidation of dipropylene glycol.

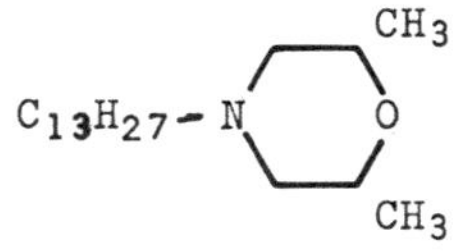

tridemorph

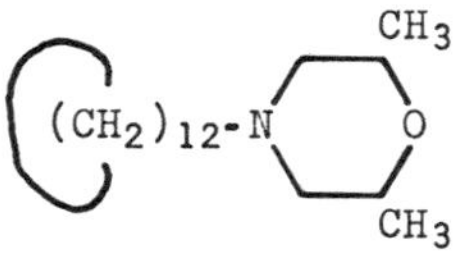

dodemorph

Other alkyl groups in the 2,6 positions are less effective than methyl. Tridemorph and dodemorph are less active against wheat mildew than against barley mildew. They act by affecting the permeability of cell membranes to the fungus. They have low oral toxicities (LD50 oral tridemorph 1270 mg/kg, dodemorph 4800 mg/kg) but they can cause fairly severe skin irritation.

Oxathiin Derivatives

Two cyclic compounds closely related to salicylanilide which have good systemic activity against the rusts, bunts and smuts of cereals and some soil fungi of cotton are 2,3-dihydro-6-methyl-5-phenylcarbamoyl-1,4-oxathiin (carboxin) and the corresponding sulphone (oxycarboxin).

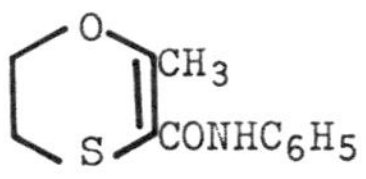

carboxin

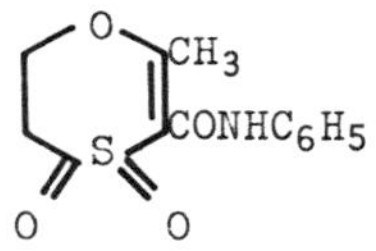

oxycarboxin

Substitution in the phenyl group or its replacement by an alkyl group reduces activity.

Carboxin and oxycarboxin probably act by interfering with the electron transport system in the tricarboxylic acid cycle. They have low oral and dermal toxicities (LD50 oral 3200 mg/kg LD50 dermal 8000 mg/kg) and so are very safe compounds to use.

Benzimidazole Derivatives

2-Substituted derivatives of benzimidazole are a class of systemic fung-icides some members of which have achieved considerable commercial success. These include methyl (1-butylcarbamoyl)-2-benzimidazole carbamate (benomyl), 2-(4'-thiazolyl)-benzimidazole (thiabendazole) and 2-(2'-furyl)-benzimid-azole (fuberidazole).

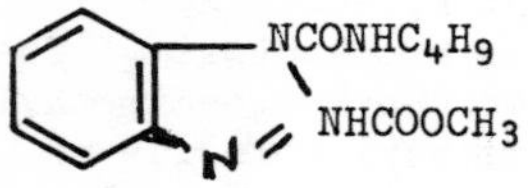

benomyl

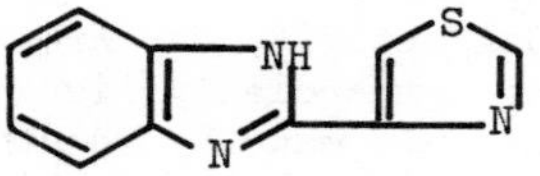

thiabendazole

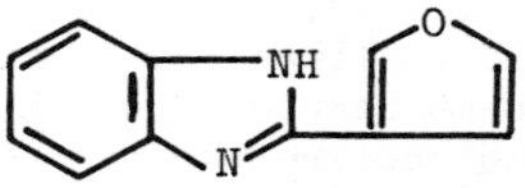

fuberidazole

These three compounds are all very active systemically against a wide range of fungal diseases of fruit and vegetables. They all inhibit mycelial growth rather than spore germination and probably act by interfering with synthesis of DNA. They all have very low toxicities (LD50 oral benomyl 9590 mg/kg, thiabendazole 3320 mg/kg, fuberidazole 1110 mg/kg) so are very safe to use.

Benomyl is an important commercial product and is manufactured from phenylene diamine according to the following process.

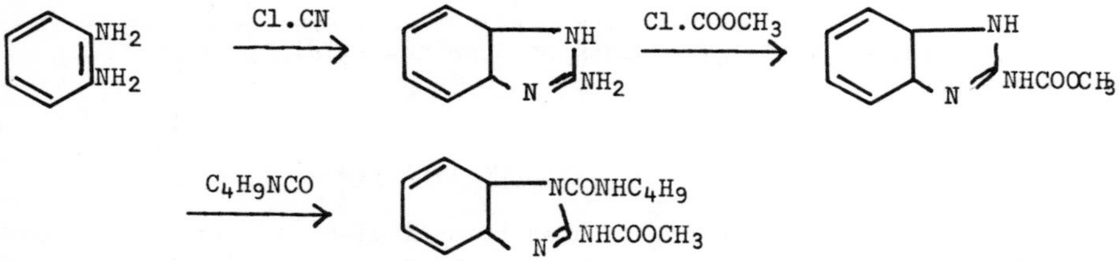

General Comment

The oral and dermal toxicities quoted show that systemic fungicides are amongst the safest pesticides which are commercially used, and they appear not to have any adverse environmental effects.

It is interesting to note that, whereas most protective fungicides act by interfering with energy production and transport processes in the fungus, all known systemic fungicides, with the exceptions of carboxin and oxycarboxin, act by interfering with biochemical synthetic processes of synthesis in the fungus. This is also true for systemic insecticides and for systemic antibacterials used in human medicine. This is probably because energy processes are basically similar for all forms of life and so it is not possible to achieve selective action in a systemic compound.

The biochemical mode of action of systemic fungicides may be a factor in the biggest problem with regard to this type of pesticide, namely, development of resistant strains of the pathogen. If only one specific biosynthetic reaction is involved then only a small genetic mutation would be required to counteract the activity.

Although protective fungicides have been used for many years, very few examples of induced resistance have been reported and, where they have, resistance has always disappeared rapidly when use of the fungicide was discontinued. The resistance induced by systemic fungicides appears, on the other hand, to persist. There seems good reason, therefore, to use systemic fungicides judiciously, as with antibiotics in medicine, and not to use any single compound continuously for long periods but to "ring the changes" on various systemic fungicides and, possibly, alternate with protective fungicides. Apart from using systemic fungicides only where economically necessary, great attention should be paid to methods of application to confine the fungicide strictly to the crop plants and to avoid "drift".

Systemic fungicides, because of their biochemical action, tend to be highly specific to certain species of fungus. The compounds developed so far tend to be more effective against the surface fungi such as mildews than against deep-seated fungal diseases. However, the wide diversity of chemical stomatures which have been shown in the last decade to have systemic fungicidal activity makes it likely that specific cures for most fungal diseases will eventually be discovered.

Further Reading

Marsh, R.W., (Ed)., Systemic Fungicides, (Longman, London, 1972)

Chapter 15
CHEMICALS FOR WEED CONTROL

Historical

The object of agriculture is to replace the flora natural to the land by vegetation better suited to the needs of man. Wild species, not necessarily those which would eventually dominate if man withdrew from the scene, are ever-pressing competitors of his cultivated varieties. The age-old mechanical methods of cultivation are largely designed to suppress these competitors. The plough, the hoe and the hands of the labourer are still the most widely-used weedkillers over most of the cultivated world. Chemicals, however, are now making the major contribution in the nations with advanced agriculture and high labour costs and this has led to important modifications of cultural practices.

Most evident are modifications to traditional rotation of crops. A large proportion of the land in eastern England, where the climate is very suitable for cereal production, now carries this crop much more often than the 2 years out of 4 which was usual up to 20 years ago. Barley is frequently grown for 4 or 5 years successively. The consequent disappearance of the land-hoed root crop and the grazed leys has led to enormously increased pressure from two very competitive grass weeds - wild oat and blackgrass - which are unaffected by the weedkiller - MCPA - most widely used over this period. Their rise was not due directly to reduced competition from other species, since competition which would suppress wild oat would suppress the barley first. It was due to the reduction in hoeing and grazing which were the previous means of control.

It is not the unique prerogative of the chemist to create other problems when he solves the first. The history of mechanical civilization is full of such events. New products have been developed to control these grass weeds. The natural flora of the world is more versatile than man himself. The battle with weeds has had a very long history and will probably have a long future.

For 100 years or more chemicals have been used for "total" weed control - the elimination of all vegetation on railway tracks, timber yards, unmetalled roads, etc. The earliest were crude products used in massive doses. They included crushed arsenical ores, oil wastes, thiocyanates from coal-gas washing and creosote. They met objections on grounds of toxicity, messiness and competitive demand within industry for the products which could be refined from them. Their place is now largely taken by more complex products of chemical synthesis, but chlorates, very rapid in action, and borates, very non-selective, are still used to some extent despite the fire-risk with the former. About 15000 t/year of inorganic herbicides are manufactured in the U.S.A. but are mostly used in non-crop situations.

The earliest selective chemical treatments, at the beginning of the century, were made on cereals with sulphuric acid and with soluble copper salts.

Sulphuric acid destroys the integrity of the plant tissue with which it comes in contact and opens the way to rapid advance of saprophytic (feeding on dead tissue) micro-organisms. Copper salts are toxic to all forms of plant life although copper at the trace level is an essential element. Both give rise to corrosion problems on exposed steel in machinery. Copper salts are no longer used. Sulphuric acid still finds limited use in the destruction of potato haulm at the end of the season to facilitate mechanical harvesting, and to a yet more limited extent for the control of weeds in onions. These essentially non-selective toxicants can be used selectively in cereals mainly by a physical mechanism. All the grasses, including cereals, have nearly vertical leaves, which are also difficult to wet. The base of the leaf is well designed to throw off water drops which descend the blade, and the growing point is well sheathed. Most broad-leaved weeds, on the other hand, have growing points exposed in the leaf axils where water drops tend to collect. The cereal therefore receives a much smaller dose than the weeds and what is received lies in a less vital site.

Foliage and Soil Applications

Modern herbicides depend for their selectivity on biochemical factors. Many are now used by application to the soil. Some inhibit root growth: others enter the tissues of the plant and kill systemically. The behaviour of the latter class is far less dependent on soil type and rainfall and most of the successful soil-applied herbicides have no direct action on roots.

Application to the soil is often loosely called "pre-emergent" because it is carried out before the weeds are accessible to spray, but it is better to distinguish between "presowing" applications carried out before the crop seeds are put in and "pre-emergent" application which is carried out afterwards but before the crop seedlings emerge. In the latter case a compound having no action by the soil may be used, the seedbed being prepared as far as practical in advance of sowing which is then carried out with the minimum of disturbance. In this way selectivity is obtained by having the weed seedlings accessible while the crop seedlings are still below the soil. This is often referred to as the "stale seedbed" method.

Pre-emergent spraying of a soil-acting herbicide is sometimes effective when presowing treatment would damage the crop, selectivity being assisted by the crop being a large-seeded one sown deep. Its roots grow away from the herbicide while the weed seeds are germinating in a shallower layer and take up a lethal dose of surface-applied herbicide. The moderate biochemical selectivity of simazine in favour of the field bean crop is made fully safe by this method of application.

✓ Dinitrophenols and Dinitroanilines ✓

The modern period may be considered to begin with the use in France in 1933 of the sodium salt of 2-methyl-4,6-dinitrophenol (DNOC) as a selective herbicide in cereals. It was more efficient than acid or copper but suffered from two disadvantages. The dry salt is extremely inflammable and the compound is toxic and easily adsorbed through the skin. The first was overcome by changing to a suspension of the free acid, usually with added ammonium sulphate, which was found to be more effective as well as less inflammable.

The second caused trouble. The safety record of spray operators in Britain has been on the whole very good, but DNOC has been responsible for most of the few fatalities.

DNOC is not, in fact, as toxic as many insecticides which have appeared since. Its bad record is due to its place in the history of pesticide use. It came in as a weedkiller, on a much larger scale than insecticides. It was used by operators much less accustomed to toxic sprays than were the orchardists who were the main users of insecticides. Its use spread widely during the war when safety was given least attention. It came before the growth of present government control measures, which it did much to bring about. Had it today been newly introduced, rather than nearly discarded, its record would be much better.

Some would give DNOC a third bad mark - its intense yellow colour, difficult to remove from wool and skin. This is not very logical. A strong colour is a good safeguard, since it enables contamination to be easily noticed and therefore avoided. It is, in fact, now established practice in this country for agricultural products containing scheduled poisons to contain also a red or violet warning colour.

DNOC is hardly used at all nowadays as a herbicide but the related 2-sec. butyl-4,6-dinitrophenol (dinoseb) and 2-tert.butyl-4,6-dinitrophenol (dinoterb), and their acetates, are more toxic to most weeds and less toxic to mammals than DNOC and are used on a small scale in the U.K. as post-emergent selective herbicides in peas, and in the U.S.A. as pre- and post-emergent herbicides to control small annual broad-leaved weeds in a variety of crops. Dinoterb is more selective in cereals than dinoseb, less dependent on climatic conditions, particularly temperature, and has a longer period of use during the growing season. Dinoseb and dinoterb are manufactured by nitration of the appropriate alkyl-substituted phenols which are themselves made by vapour-phase reaction of phenol with an olefine over a suitable catalyst. About 4000 t/year are manufactured in the U.S.A.

A number of nitrophenyl ethers have been developed as herbicides, the most commercially important of which are nitrofen and fluorodifen.

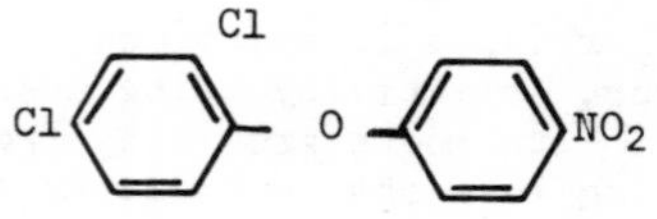

nitrofen

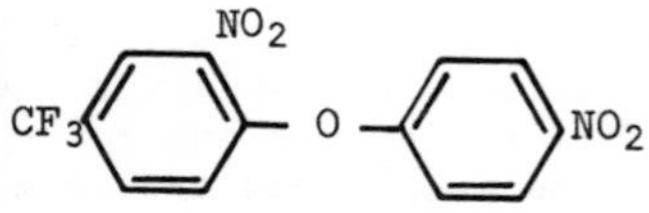

fluorodifen

They are manufactured by reacting the sodium salt of the appropriate phenol with 1-chloro-4-nitrobenzene. The aromatic chlorine is made reactive by the electron-withdrawing effect of the nitro group. Nitrofen is a pre-emergence herbicide, toxic to various broad-leaved and grassy weeds when left as a thin layer on the soil surface but it rapidly loses its activity if incorporated into the soil. It is used for weed control in cereals and various vegetables. Fluorodifen is used mainly on soya beans but, like nitrofen, loses its activity if incorporated into soil. About 600 t/year are used by U.S.A. farmers.

Two recently introduced products of this type are nitrofluorfen and oxy-fluorfen.

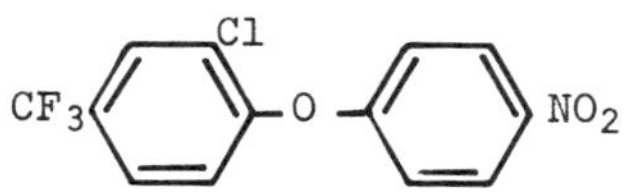

nitrofluorfen

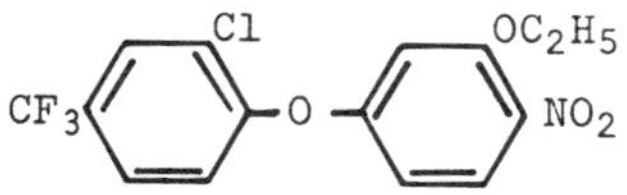

oxyfluorfen

The herbicidal activity of the dinitrophenols encouraged chemists to explore the dinitroanilines because these have long been commercially available as dyestuffs intermediates. The initial observation that 2,6-dinitroaniline had significant general herbicidal activity, whereas 2,3- and 2,4-dinitro-anilines did not, focussed attention on this isomer and stimulated attempts to achieve selectivity by introducing substituents into the benzene ring and onto the amino group. This led to one of the most commercially successful herbicides of the past decade, **trifluralin.** Toluene is chlorinated to give 1-chloro-4-trichloromethylbenzene which is reacted with hydrogen fluoride in the vapour phase over a catalyst to give 1-chloro-4-trifluoromethylbenzene which is then dinitrated and the product reacted with dipropylamine to give trifluralin.

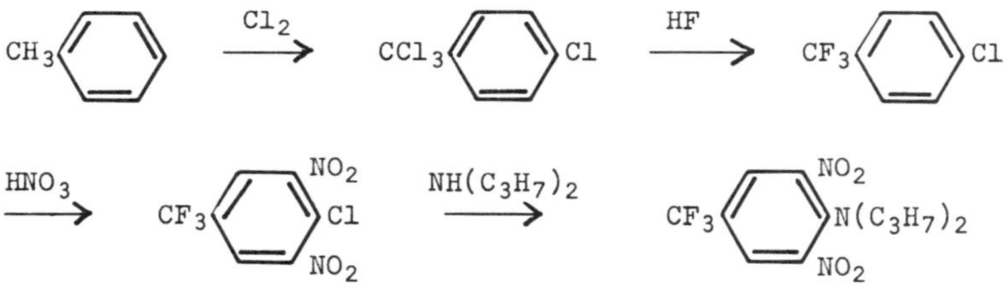

trifluralin

About 12000 t/year of trifluralin are manufactured in the U.S.A. where it is mainly used by pre-sowing incorporation into soil for weed control in many important crops, particularly soya beans and cotton. It is effective against many of the most troublesome grassy weeds as well as against broad-leaved weeds. It kills weed seedlings as they are produced by the germinating seeds by inhibiting both root and shoot growth and development. The mode of action is not certain but it is probable that inhibition of mitosis may be the main lethal effect since effects on cell division in the roots of germinating seedlings have been observed, but there are also indications of interference with photosynthesis and respiration. There is evidence that the lipid contents of various plants may influence the selectivity of trifluralin. Also, it is clear that, in established plants, little trifluralin is translocated from roots to aerial portions.

Changes in substituents in the 3- and 4-positions of the benzene ring and on the amino group do not essentially change the type of herbicidal activity, provided that the 2,6-dinitroaniline structure is retained, but do produce variations in patterns of selectivity and degrees of activity. A wide range of molecular structure is therefore possible and a number of products which have advantages in certain situations have been introduced commercially. Thus dipropalin has more foliar contact activity than trifluralin, isopropalin controls grasses and broad-leaved weeds in tomatoes and tobacco crops, oryzalin is a selective, surface-applied herbicide for soya beans, ethalfluralin is particularly useful for weed control in cotton and prosulfalin for weed control in turf. The most widely-used compound after trifluralin is nitralin, of which about 4000 t/year are manufactured in the U.S.A.

2,6-Dinitroanilines

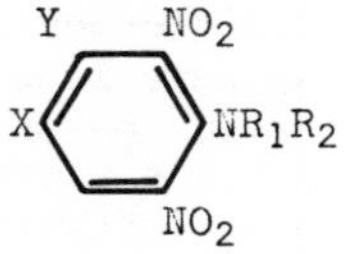

Name	X	Y	R_1	R_2
trifluralin	CF_3	H	C_3H_7	C_3H_7
benfluralin	CF_3	H	C_2H_5	C_4H_9
nitralin	CH_3SO_2	H	C_3H_7	C_3H_7
dipropalin	CH_3	H	C_3H_7	C_3H_7
isopropalin	$(CH_3)_2CH$	H	C_3H_7	C_3H_7
oryzalin	NH_2SO_2	H	C_3H_7	C_3H_7
chlorethalin	CF_3	H	C_3H_7	C_2H_4Cl
profluralin	CF_3	H	C_3H_7	$CH_2CH(CH_3)_2$
butralin	$(CH_3)_2CH$	H	H	$C_2H_5CHCH_3$
penoxalin	CH_3	CH_3	H	$(C_2H_5)_2CH$
ethalfluralin	CF_3	H	C_2H_5	$CH_2{=}C(CH_3)CH_2$
prosulfalin	$(CH_3)_2S{=}NSO_2$	H	C_3H_7	C_3H_7
dinitramine	CF_3	NH_2	C_2H_5	C_2H_5
prodiamine	CF_3	NH_2	C_3H_7	C_3H_7
fluchloralin	CF_3	H	C_3H_7	CH_2CH_2Cl

All these products are manufactured by methods similar to that used for trifluralin. An aromatic compound with a chlorine group in the 1 position is dinitrated, which directs the nitration into the 2,6 positions and the nitro groups thus introduced then activate the chloro group sufficiently for it to be reacted with an amine.

Phenoxyacid Herbicides

These were developed during World War 2 as products of war-directed research. They were the key compounds for the very rapid expansion of chemical weed control in the last 30 years. Without their development at the right time, farmers would have been much slower to take up the new techniques. These compounds had four advantages which were vital at this stage. They were cheap to produce. They had so wide a margin of selectivity in favour of the cereal crop that little skill was needed in their use. They were virtually non-toxic to man and stock. They came forward at a time when maximum home production of food was essential and labour on farms very scarce.

The phenoxy acids were undoubtedly suggested by the structure and activity of the then newly discovered endogenous plant hormone - or auxin - indolyl-3-acetic acid (IAA).

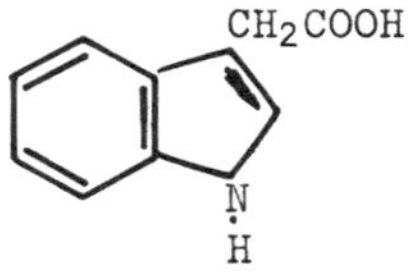

IAA

This compound is produced in the growing shoot tips and controls cell elongation and root initiation. It was thought that synthetic imitations might interfere with these essential actions. Then came the once-in-a-century stroke of luck. No sensible chemist could really expect 2,4-dichlorophenoxyacetic acid (2,4-D) to be a plausible imitiation of indolylacetic acid in an

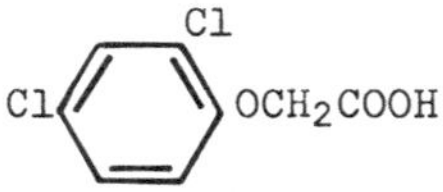

2,4-D

activity so biochemically specific and delicate as hormonal control. Yet it was found that this compound had a growth-modifying activity even greater than that of the natural auxin. Unlike the natural auxin it was not subject to internal regulation of concentration, so that it produced lethally abnormal growth. In low concentration it could induce root growth from stem cuttings, but the margin of safety between induction of healthy root growth and the induction of excessive root thickening was far too small. It was the lethal effect of excessive growth in the wrong tissues which was exploitable, especially as this effect was far less in grasses than in most other species.

With hindsight, now that the biochemistry of the modes of action of both natural auxins and of phenoxyacetic acids are understood, it is apparent that substances of different chemical structures can have the same common mode of action provided they meet the structural requirements for the stereospecific interactions with cell protein which result in stimulation of RNA and DNA polymerase and other enzyme activities and consequent production of the characteristic auxin effects.

So, the "hormone" weedkillers were born. By one of those rare accidents a compound very simple for the industrial chemist to produce turned out to possess very good selective properties as a weedkiller in cereals, without any significant effect on animals and at a time when such properties were very much needed.

Phenoxyacetics

The manufacture of 2,4-D and its close relation MCPA (4-chloro-2-methylphenoxyacetic acid) is very simple. Phenol is first directly dichlorinated or 2-methylphenol directly monochlorinated, in each case the desired positions of chlorination being those naturally favoured. The substituted phenol, as its sodium salt, is then reacted with sodium monochloroacetate, a Williamson ether synthesis of the most simple type. The reaction is carried out very efficiently, without the need for anhydrous conditions, by heating in concentrated solution in water. The sequence is simply, for MPCA,

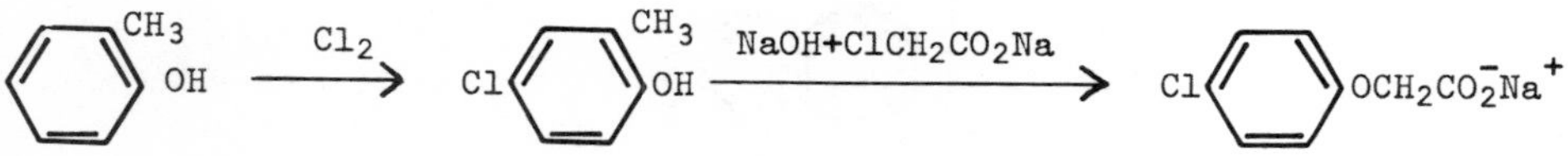

MCPA

with a yield of 80% or more, there being some by-product 6-chloro-2-methylphenoxyacetate and 4,6-dichloro compounds.

In early production the by-products were not removed. Even the sodium chloride was retained in solution, limiting the concentration of the sodium 4-chloro-2-methylphenoxyacetate to about 10%, but with the advantage that the microflora of the vast areas of the farmers' fields disposed, without any cost, of the organic by-products. There can have been few chemical processes where the directly obtained products of reaction in solution could be sold direct to the user.

Economics of packing and transport later dictated the production of more concentrated products from which the sodium chloride must be removed. This is carried out by hot separation of the free phenoxyacetic acid after acidification, into an organic solvent from which it is extracted back into a more concentrated alkali solution. Other refinements have been made in the process. Good distribution of chlorine at the chlorination stage is essential to reduce loss in an unwanted mixture of unchlorinated and dichlorinated products. In manufacture of MCPA the Williamson condensation can be carried out before chlorination, with the advantage that the chlorination has even greater preference for the 4-position as opposed to 6(the 6-chloro compound is virtually inactive). The chlorination of 2-methylphenoxyacetic acid is best carried out in 1,2-dichloropropane at 80°.

MCPA has an important advantage over 2,4-D in that the sodium salt is much more soluble. Most concentrates contain about 25% of the active acid in the saline form, some containing mixed sodium and potassium to decrease the temperature at which the salt may freeze out. The sodium salt of 2,4-D, on the other hand, is only about 2% soluble: the potassium not much more. In order to make an economically acceptable concentrate the acid must be neutralized with an amine, dimethylamine and mixed ethanolamines being most common.

From the beginning, 2,4-D was almost exclusively preferred to MCPA in the U.S.A. The reverse was, and still is, true in the U.K. Over the continent of Europe the two are more nearly equal in usage. The difference between British and American choice was dictated by two factors. 2-Methylphenol was relatively abundant in Britain as a product of coal-tar distillation. The American coal-fields were less rich in this product but, on the other hand, the production of synthetic phenol was much more advanced. The second factor was agronomic. Both products can do some damage to the cereal crops if overdosing occurs or if sprayed too early or after the critical "jointing" stage. In this respect MCPA is the safer of the two. This was of less consequence in the U.S.A. since the agronomic practice was to keep weeds at a low competitive level rather than produce almost complete kill. Lower dosage was therefore acceptable. These compounds are not direct killers: rather they disorganize the weeds which then succumb to drought and competition. A better effect is therefore produced when the surface soil is too dry to enable plants with a deficient root system to survive. Even the lower dose is often lethal in the drier U.S.A.

2,4-D and MCPA have been described at some length because they were key compounds in the history of the subject. Without the fortunate combination of cheap production, safety to the operator and the need for no great skill, chemical weed control would not have become so quickly a very important tool in modern farming. Without these pioneers, many later compounds, requiring more skill and judgment in use, would probably have never become commercial. One must, however, be careful, in giving them an important place in history, not to emphasize the history to the exclusion of the compounds. They are still, in terms of world tonnage, the most important of all selective weed-killers. The U.S.A. manufactures about 20,000 t/year of 2,4-D.

The only other phenoxyacetic acid of importance is the 2,4,5-trichloro (2,4,5-T). It has advantages over the others, particularly in an ester formulation, against woody species. In this case, the biological preference is not for the 2,4,6-compound, obtained easily by direct chlorination. This is not active. Fortunately an easy alternative process is available. The 2,4,5-trichlorophenol is prepared by alkaline high-pressure hydrolysis of the symmetrical tetrachlorobenzene, just as phenol itself can be made from monochlorobenzene. Because 2,4,5-trichlorophenol is less reactive than 2,4-dichlorophenol the condensation with sodium monochloroacetate is usually carried out at about 150° in a higher-boiling organic solvent such as amyl alcohol, rather than in water.

About 4000 t/year of 2,4,5-T are manufactured in the U.S.A. and are mainly used for brush control in non-crop situations. Considerable public concern was aroused when it was discovered after 25 years use that many commercial samples of 2,4,5-T contained up to 30 ppm of a very highly teratogenic and embryo-toxic compound, 2,3,7,8-tetrachlorodibenzo-p-dioxin (TCDD) which is formed if hydrolysis of tetrachlorobenzene to give 2,4,5-trichlorophenol is carried out at too high a temperature. Manufactory processes have now been modified so that the dioxin content is well below the 0.5 ppm level which is regarded as safe.

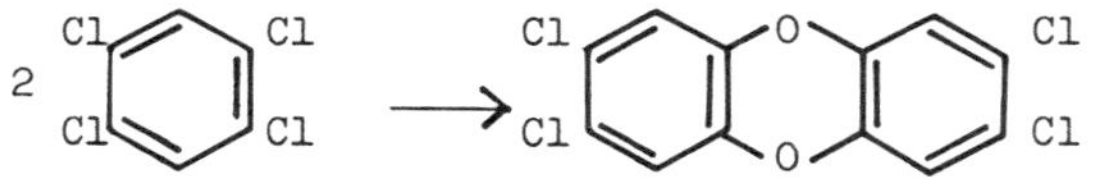

TCDD

Although some adverse effects have been reported in animals with doses of purified phenoxyacetic acids far beyond those likely to be encountered under normal conditions of use, there is no reason to believe that their agricultural use presents any danger to man or animals and, as they are rapidly broken down in vegetation or soil, they present no lasting hazard to the environment.

Phenoxybutyrics

An extension of the phenoxyacetic herbicides is due to Wain in the U.K., following earlier observations of Zimmerman. The ω-phenoxy derivatives of fatty acids higher than acetic are not themselves active as herbicides. They can, however, be degraded biochemically by the β oxidation process to derivatives of the fatty acid with two fewer carbon atoms (similar to the process which occurs in the metabolism of the acids of true fats, which always occur in nature with an even number of C atoms).

$$RCH_2CH_2CH_2CO_2H \xrightarrow{O_2} RCH_2COCH_2CO_2H \xrightarrow{H_2O} RCH_2CO_2H + CH_3CO_2H$$

When the 3-propionic derivative is degraded it probably goes through the unstable phenoxyformic acid to the phenol, but 4-(2,4-dichloro or 4-chloro-2-methylphenoxy)-butyric acid (2,4-DB or MCPB) gives the herbicidal derivative of acetic acid. Wain found that there are considerable differences between species in their ability to oxidize the higher acids. A new mechanism of selectivity is thereby introduced. It was discovered that leguminous species generally, but by no means invariably, are resistant to the phenoxybutyric acids. These compounds are therefore applied to cereals undersown with leguminous crops. While clover species and lucerne escape damage, "trefoil" is sensitive. The biochemistry is not as simple as at first supposed. In lucerne, 2,4-dichlorophenoxybutyric acid is decomposed rapidly but the phenoxyacetic acid is only a minor product. Sensitivity is probably mainly determined by the relative rates of β oxidation and attack on the benzene ring. The phenoxybutyrics are prepared by condensation of the substituted phenol with butyrolactone, preferably in a mixture of nonane and butanol at 160°.

MCPA and 2,4-D were effective against most of the weeds which were common in cereal fields 20 or 30 years ago. They are still widely used, although these weeds have become, in consequence, far less important. The phenoxybutyrics extended the range of tolerant crops rather than the range of susceptible weeds. Against certain weeds, one or other of these compounds proved superior, but against some none was effective. To the farmer these "hormone resistant" weeds took on greater importance. This was partly because, becoming more weed-conscious, he noticed them more, partly because changes in rotation and cultivation practice were giving them a better chance.

Phenoxypropionics

Some resistant weeds were grasses - wild oat, blackgrass, couch - and required different compounds. It was discovered that the phenoxy-2-propionic acids were very much more effective on chickweed than the corresponding acetic acids. The 4-chloro-2-methyl compound (mecoprop) has been extensively developed and found to deal also with cleavers, a weed almost unaffected by the phenoxyacetics and causing great trouble in the harvesting of grain. To a lesser extent it had an improved effect on mayweeds. The 2,4-dichloro compound (dichlorprop) is also used and the 2,4,5-trichloro compound (fenoprop) has found wide commercial application for control of woody plants. The phenoxypropionic compounds are prepared in an analogous way to the acetics, 2-chloropropionic acid replacing chloroacetic acid.

An important new development is the discovery that 4-phenoxyphenoxypropionic acids will control grasses and other graminaceous weeds in broad-leaved crops such as soya bean, cotton, oil-seed rape, sugar beet and peanuts. The margin of selectivity of these compounds is so great that they can be safely used post-emergent. They are the first compounds with sufficient safety margin to permit post-emergent use, and control of grassy weeds in broad-leaved crops has hitherto had to rely entirely on use of pre-emergent herbicides. The most promising compounds at the moment are diclofop-methyl and trifop-methyl.

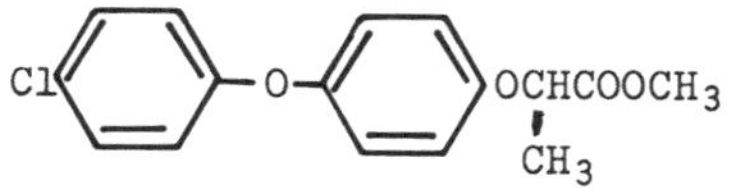

diclofop-methyl

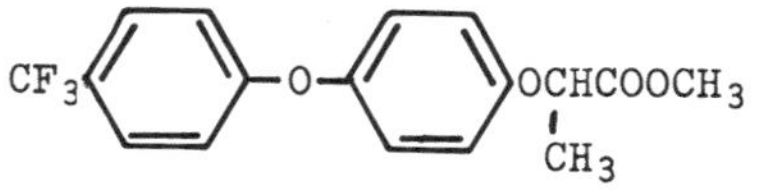

trifop-methyl

The 2,4-dichlorobenzyl analogue of diclofop-methyl, clobenprop-methyl is also effective.

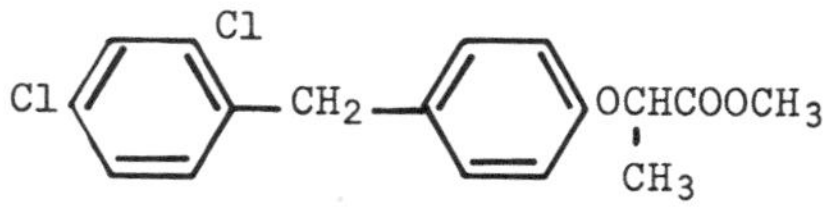

clobenprop-methyl

Benzoic Acid Herbicides

2,3,6-Trichlorobenzoic acid (TBA) was found to be very effective against cleavers at a dose so low that the crop was not damaged. In conjunction with MCPA it was effective against mayweeds and chickweed.

TBA is a difficult compound to make in high yield because the biologically desirable place for the chlorine substituents is not the place where they most naturally go by the obvious chemical process. The yield of the right isomer by direct chlorination of benzoic acid or benzoyl chloride is negligible. The successful route starts by monochlorination of toluene. The 2-chlorotoluene is separated from other isomers by distillation and then further chlorinated, yielding first mainly 2,5-dichloro and then a mixture of 2,5,6- and 2,4,5-trichloro. With careful choice of conditions, a mixture containing about 60% of the desired isomer can be obtained. This is then oxidized with nitric acid to the corresponding mixture of benzoic acids.

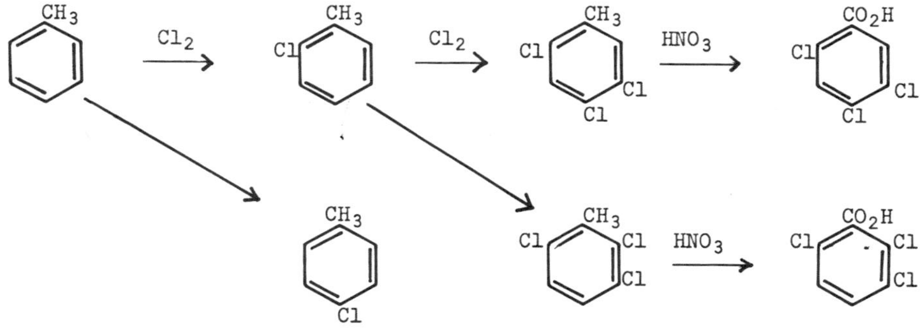

TBA

The scheme is oversimplified, since other isomers, in smaller quantity are produced, together with some di- and tetra-chloro compounds. The 4 substituted compounds are virtually inactive. The 2,6-di and 2,3,5,6-tetrachloro compounds have a similar, but less powerful, activity than that of the 2,3,6.

Most of the 2,4,5-trichloro acid can be removed from the final product by partial acidification and extraction, the 2,6-disubstituted acids being considerably stronger than those having a hydrogen atom remaining in the 2- or 6-position.

3,6-Dichloro-2-methoxybenzoic acid (dicamba) is used in combination with MCPA for selective weed control in cereals and is particularly effective against weeds of the genus Polygonum against which 2,4-D and MCPA are only moderately effective.

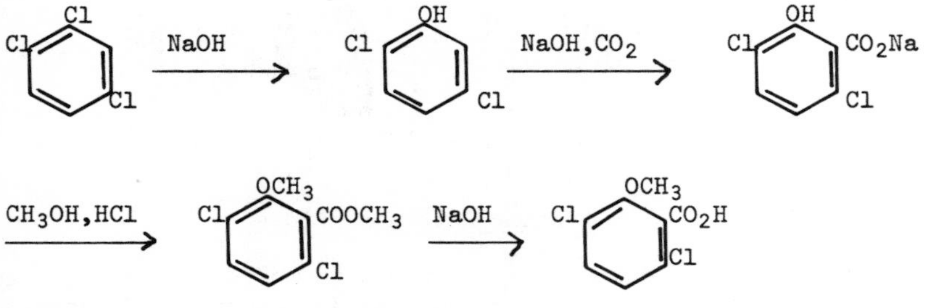

dicamba

TBA and dicamba, like the phenoxy aliphatic acids, are plant disorganizers rather than direct killers, but they produce very different symptoms. The phenoxy compounds produce contortions of the stems and leaf stalks and proliferation of stumpy roots, sometimes on the aerial parts. The benzoics produce very narrowed leaves and buds, but with much less twisting and without abnormal root growth. All hormones can produce gross swelling of the stems in some species and the benzoics particularly make the stem very brittle so that treated plants are often found in the field with the appearance of having been nipped off by rodents.

Tha analogue of dicamba with an extra chlorine atom, 3,5,6-trichloro-2-methoxy-benzoic acid (tricamba) has also found some commercial application.

The most commercially useful of the benzoic acids is 3-amino-2,5-dichlorobenzoic acid (chloramben). It is manufactured by chlorination of benzoic acid to 2,5-dichlorobenzoic acid, which is then nitrated and reduced.

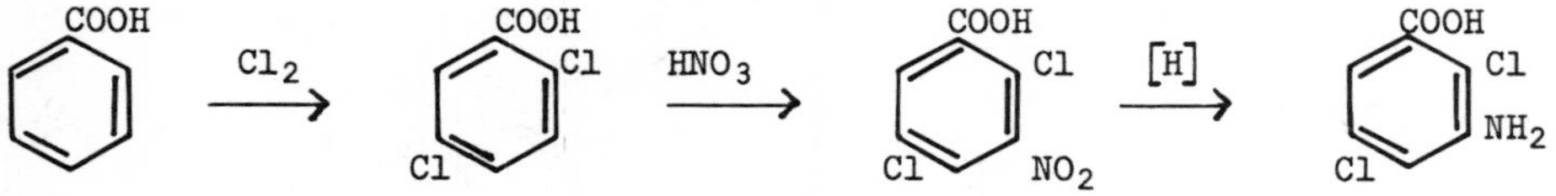

chloramben

Chloramben is a selective pre-emergent herbicide which is widely used on soya beans, carrots and cucurbits. It has a very low mammalian toxicity

(LD50, 5600 mg/kg), and is effective against both grassy and broad-leaved weeds. About 10,000 t/year are manufactured in the U.S.A., whereas manufac ture of TBA and dicamba together in the U.S.A. total only 500 t/year.

An interesting development of the benzoic acids is N-1-naphthylphthalamic acid (naptalam) which inhibits seed germination and is widely-used as a pre-emergent herbicide for soya beans, potatoes, peanuts and cucurbits. It appears to act mainly by root absorption and has little contact activity. It has a very low toxicity to mammals (LD50, 8200 mg/kg). About 2000 t/year are manufactured in the U.S.A., by reaction of phthalic anhydride and 1-naphthylamine in benzene or xylene at room temperature.

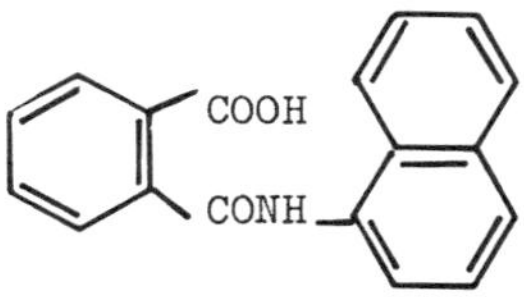

naptalam

Pyridine Acid Herbicides

Since the original plant hormone - indole-3-acetic acid - which was isolated is a heterocyclic compound it is not surprising that chemists looked extensively at acid derivatives of heterocyclic compounds of all kinds. Somewhat surprisingly, when one considers how fruitful the phenoxy acids and benzoic acids have been, these investigations have produced few active compounds. The only one which has achieved any degree of commercial success is 4-amino-3,5,6-trichloropicolinic acid (picloram). It is manufactured by chlorination of 2-methylpyridine, reaction of the product with ammonia and then hydrolysis.

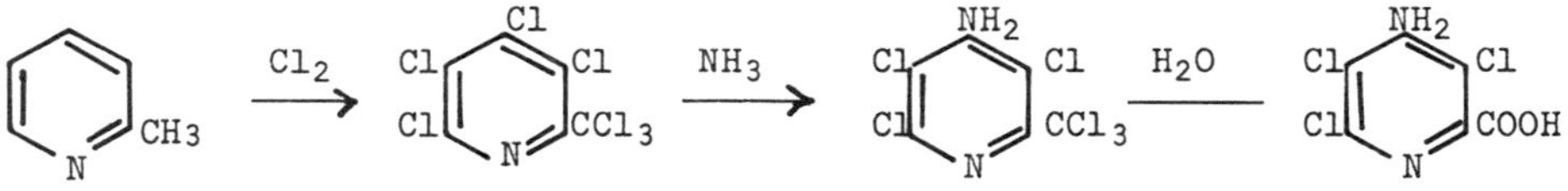

picloram

It is a systemic herbicide which produces epinasty and leaf-curling, and is rapidly absorbed by leaves and roots and translocated to the growing shoots. Most broad-leaved weeds are sensitive to it, but most grasses are resistant. It is one of the most active herbicides known and has been used against annual weeds at rates as low as 15 g/ha, although perennials need higher dose rates. It is also a very persistent compound in the soil so has fallen somewhat into disfavour.

A pyridine acid similar to the phenoxyacetic acids has recently been introduced, namely, (3,5,6-trichloro-2-pyridyl)-oxyacetic acid (triclopyr).

Chlorinated Aliphatic Acids

There is now available a wide range of herbicides which can deal with the non-grass weeds of cereals. Many are used in combinations to give "broad spectrum" control or to meet particular weed situations. Grass weeds are not dealt with by these compounds or mixtures and have in recent years very greatly increased in importance. Wild oat and blackgrass can, in bad cases, suppress the crop and always have a depressing effect on crop yield.

The first compounds to show the opposite selectivity were trichloroacetic acid (TCA) and, later 2,2-dichloropropionic acid (dalapon). They are toxic to grass species generally and have only minor and transient effects on most other plants. Higher dosage rates are needed to control grasses by these compounds than are necessary with the "hormones" for control of most non-grass species. Both are usually applied as solutions of the sodium salts, but both salts are supplied to the consumer in the solid state to avoid hydrolysis in storage. The hydrolysis of dalapon is a normal Cl by OH replacement, giving rise to pyruvic acid, but in the case of TCA there is C-C fission, giving chloroform and carbonate

$$CH_2.CCl_2.CO_2^- + 2OH^- \longrightarrow CH_3.CO.CO_2^- + H_2O + 2Cl$$

$$CCl_3.CO_2^- + H_2O \longrightarrow HCCl_3 + HCO_3^-$$

Dalapon is generally the more effective, particularly by foliage application, but the lower cost of TCA makes it more nearly equivalent in price per hectare for soil application. The rhizomatous perennial, couch grass, is sprayed with dalapon in cereal stubble, but its treatment with TCA must be carried out after cultivation, preferably in two applications with a second cultivation in between. About 2500 t/year of dalapon are manufactured in the U.S.A.

Three other compounds in this group which have found commercial application are TFP (CHF_2CF_2COOH), erbon and alorac.

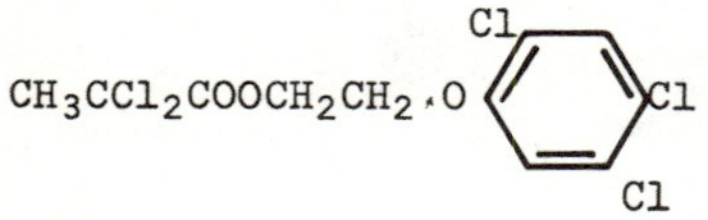

erbon

$$CCl_3COCCl{=}CClCOOH$$

alorac

None of these compounds can be used in cereals and it is not safe to sow cereals within 6 months of their application. They can only contribute to the control of wild oat and blackgrass by application in some other crop or preferably to the soil prior to sowing. TCA has been used in this way to prevent the establishment of wild oat seedlings in sugar beet, peas and Brassica crops. The treatment produces a remarkable modification of the leaves of peas and Brassicae in that the "bloom" consisting of micro-crystalline waxes (mainly paraffins and saturated alcohols and ketones of about 30 C atoms) does not form. The early leaves are more glossy and less blue in appearance and are wetted by rain or sprays. The crop grows out of this effect but it is necessary, when peas are sprayed with dinoseb for control of other weeds, to reduce the dose if they are growing in soil which has been treated with TCA. The dose of TCA used in this treatment is 7 kg/ha (compare 0.25-1.5 kg/ha of "hormones" are used post-emergent), but for treatment of couch grass in autumn as much as 30-40 kg/ha is necessary. The use of TCA against seedling wild oat and blackgrass is now only a minor one as it is becoming possible to control these weeds in cereals with other compounds referred to below.

Carbamate Herbicides

Another war-time product, at first thought to be highly selective against grasses, was isopropyl phenylcarbamate (propham). It was most effective via the soil to which it was applied as a fine suspension of the crystalline solid (soluble only to about 70 ppm in water). Although effective at lower doses than TCA or dalapon, its selectivity is not so clear. It is now largely replaced by the 3-chloro compound (chlorpropham) in limited use in some vegetable crops and sugarbeet, and by the methyl 3,4-dichlorophenyl carbamate (swep) which is used for control in rice.

chlorpropham

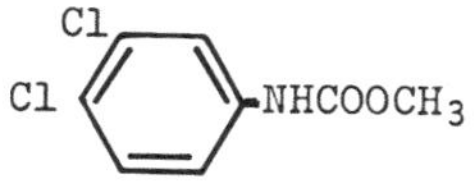

swep

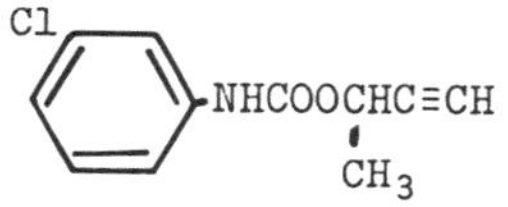

chlorbufam

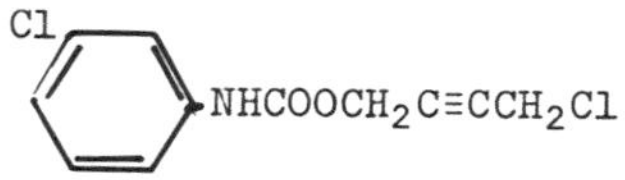

barban

Chlorbufam and barban are related compounds with acetylenic substituents. Chlorbufam is used for weed control in sugar beet and some vegetables. Barban is highly active against oat species, including the wild oat, and can be used selectively for control of this weed in wheat and barley by post-emergent spraying of an emulsion type. This is the "narrowest" selectivity yet exploited commercially, and much more care is necessary with this compound than when a grass-inactive hormone is used for control of non-grass weeds. Barban must be sprayed at the right stage of development of the weed (between one leaf only expanded and two expanded, one visible) otherwise control is not good. Some varieties (including Proctor) of barley are sensitive and its use on these varieties must be avoided.

A particularly interesting double carbamate, phenmedipham, successfully kills most important annual weeds when sprayed on the young sugarbeet crop.

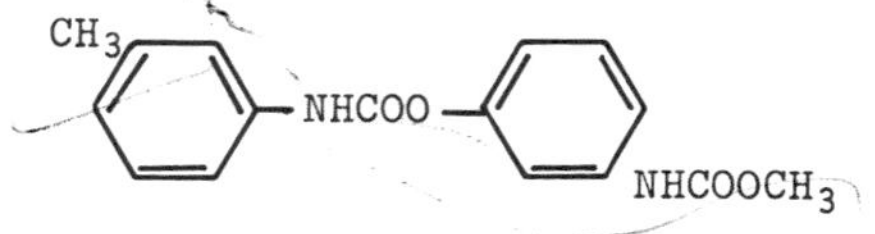

phenmedipham

A particular interesting set of compounds are the carbamates of 4-aminobenzenesulphonamide, namely, asulam, carbasulam and nisulam.

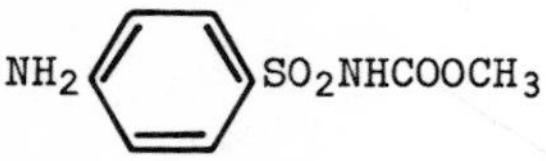

asulam

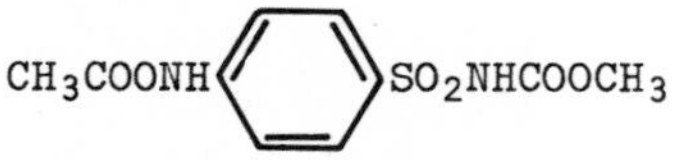

carbasulam

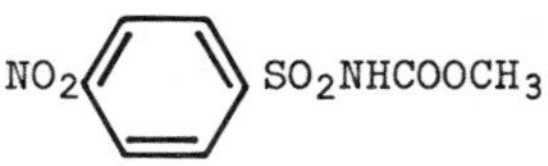

nisulam

Asulam was originally introduced for the comparatively minor use of control of dock but it was found later to be very effective against bracken which is a troublesome and ubiquitous weed in pastures and rangeland. Although it is slow-acting it is very effective and it does not reduce the quality of the grass nor present any hazard to grazing animals as its mammalian toxicity is very low (LD50, 5000 mg/kg). It is manufactured by reaction of 4-aminobenzene sulphonamide with dimethyl carbonate.

A number of N-alkylcarbamates and N-alkylthiocarbamates are of interest as herbicides. These include karbutilate, terbucarb, dichlormate, EPTC, vernolate, diallate and triallate, thiobencarb, ethiolate, methiobencarb and thiocarbenil. In total, about 10,000 t/year of carbamates are manufactured in the U.S.A.

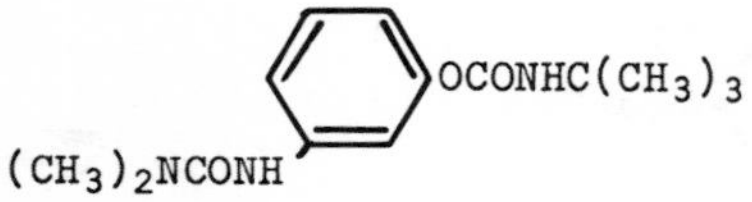

karbutilate

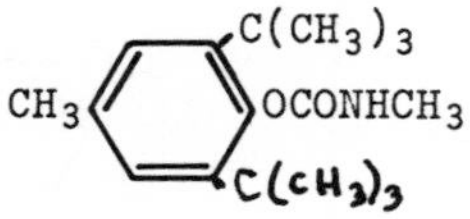

terbucarb

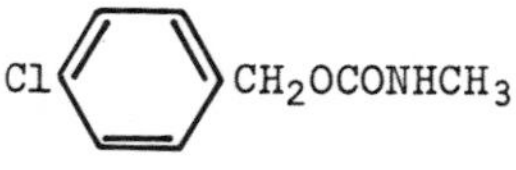

dichlormate

$C_2H_5SCON(CH_2CH_2CH_3)_2$

EPTC

$CH_3CH_2CH_2SCON(CH_2CH_2CH_3)_2$

vernolate

$CHCl{=}CClCH_2SCON[CH(CH_3)_2]_2$

diallate

$CCl_2{=}CClCH_2SCON[CH(CH_3)_2]_2$

triallate

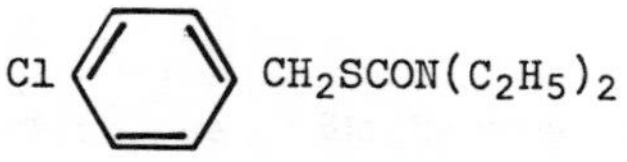

thiobencarb

$(C_2H_5)_2NCOSC_2H_5$

ethiolate

methiobencarb

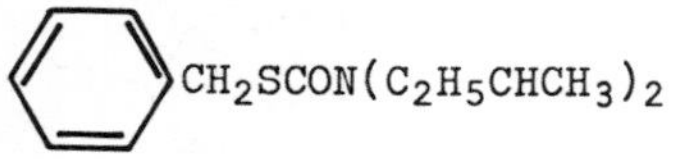

thiocarbenil

The carbamates are prepared by the action of phosgene on the amine and reaction of the product with the alcohol. The thiocarbamates are prepared by reaction of phosgene with the amine and then with a mercaptan. Schematically:

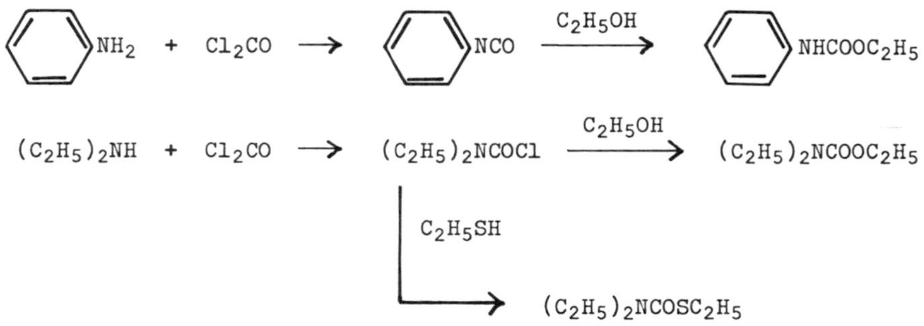

They have very low solubility in water and are used either as suspensions or emulsions. They have moderate persistence in the soil. In true solution they are rapidly hydrolysed under alkaline conditions but are largely protected by their low solubility. The thiocarbamates have significant volatility and their entry into the underground shoot tissue of the plant after soil application is considered to take place largely through the vapour phase.

Amide Herbicides

Herbicidal activity was discovered in the chloroanilides. The most commercially successful compound has been 3',4'-dichloropropionanilide (propanil) which is a contact herbicide of particular value for post-emergent control of grass weeds in rice, especially of barnyard grass, which is one of the most troublesome weeds of this crop. It has no pre-emergent activity. About 3000 t/year are manufactured in the U.S.A. by reaction of 3,4-dichloroaniline with propionic acid in the presence of thionyl chloride.

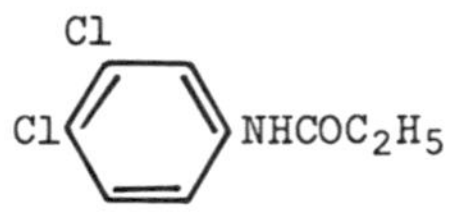

propanil

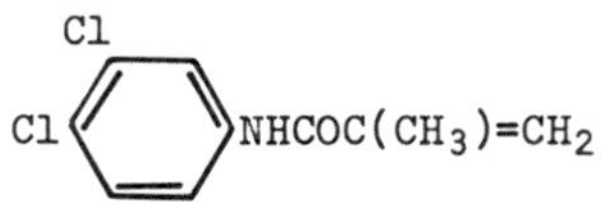

chloranocryl

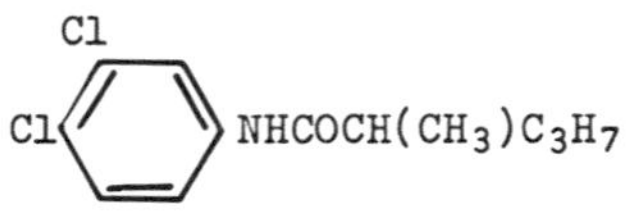

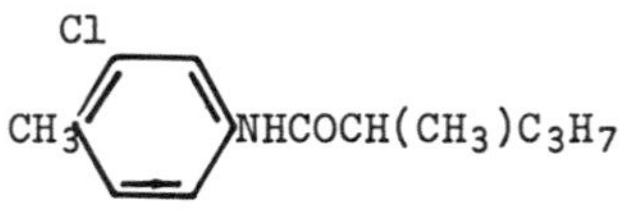

pentanochlor

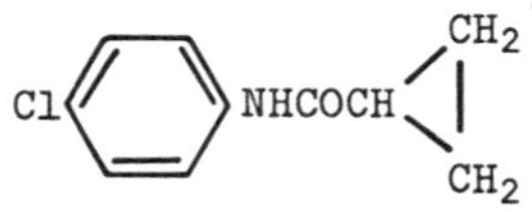

cypromid

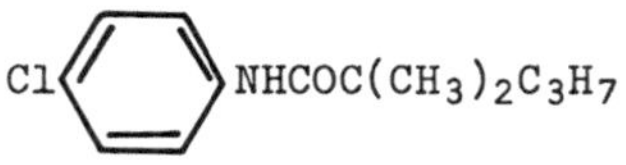

monalide

A number of other anilides, shown above, have found some commercial use. Pentanochlor can be used for selective weed control in tomatoes which are not susceptible to it.

It has long been known that some chloroacetamides are herbicidal and a systematic study of their phytotoxic properties led to the most commercially successful compound, N,N-diallylchloroacetamide (allidochlor), which is made by reaction of diallylamine with chloroacetyl chloride. It is a selective pre-emergent herbicide which is active against certain grasses, and has a wider range of activity than TCA or dalapon. It is of interest that the dialkylamides of the herbicidal TCA have negligible activity.

$$CH_2ClCON(CH_2CH{=}CH_2)_2$$

allidochlor

The herbicidal activity of chloroacetamides on the one hand and of chloroanilides on the other, led to investigation of N-aryl substituted chloroacetamides. This study yielded a number of herbicides of very great commercial utility, in particular, propachlor and alachlor.

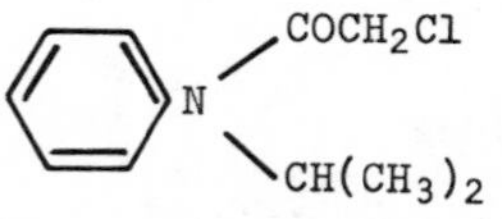

propachlor

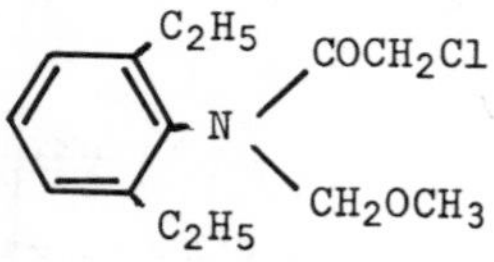

alachlor

Propachlor is a pre-emergent herbicide which is widely-used against annual grasses and certain broad-leaved weeds in maize, cotton, soya beans, sorghum, sugar cane and peanuts. It is one of the few herbicides which retain their effectiveness in peat and muck soils. In general, best weed control is obtained when rain falls within 10 days of application, but the product is still effective even in dry conditions. About 12,000 t/year are manufactured in the U.S.A., by reaction of chloroacetyl chloride with N-isopropylaniline in an inert solvent.

Alachlor is used for much the same applications as propachlor but it requires more moisture for activation than does propachlor. It is particularly safe for use on soya beans.

About 10,000 t/year are manufactured in the U.S.A. by reacting 2,6-diethylaniline with formaldehyde to give the N-formyl derivative, adding chloroacetyl chloride to this and reacting the product with methanol.

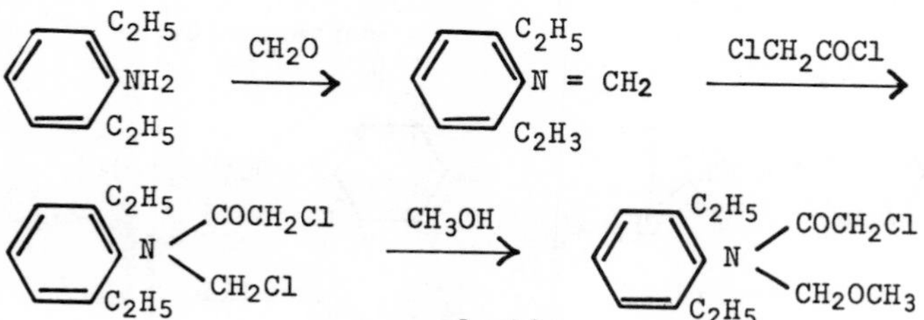

alachlor

Butachlor, which is alachlor with $-OC_4H_9$ instead of $-OCH_3$, is particularly useful for weed control in rice. Recent introductions are terbuchlor, which is butachlor with a tert.butyl group and a methyl group in the aromatic ring instead of the two ethyl groups, and diethacine-ethyl which is alachlor with $-COOC_2H_5$ instead of $-OCH_3$.

It appears that crop species which are resistant to these herbicides can rapidly detoxify them either by reaction of the aliphatic chlorine atom with exogenous substrates or by cleavage of the amide linkage. Their mode of action is not completely known but it is probable that they interfere with protein synthesis but that they react by a variety of mechanisms at a number of different sites.

A class of amides which has become of interest more recently is exemplified by benzoylprop-ethyl and flamprop-isopropyl.

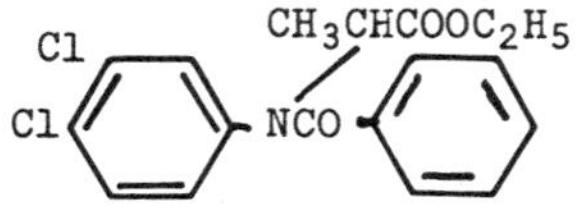

benzoylprop-ethyl

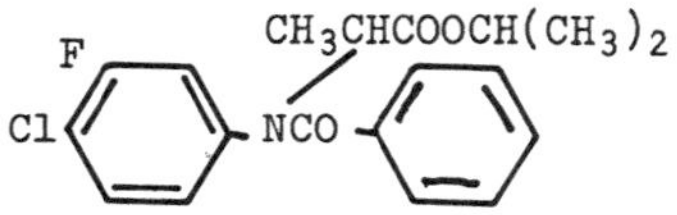

flamprop-isopropyl

These compounds have proved particularly successful for post-emergent control of wild oat in wheat in the U.K. and continental countries. They inhibit cell elongation in the wild oat and cause it to remain stunted. Benzoylprop-ethyl is manufactured by reaction of 3,4-dichloroaniline with ethyl 2-chloropropionate followed by benzoyl chloride.

A structurally related compound which has recently been introduced is benzipram.

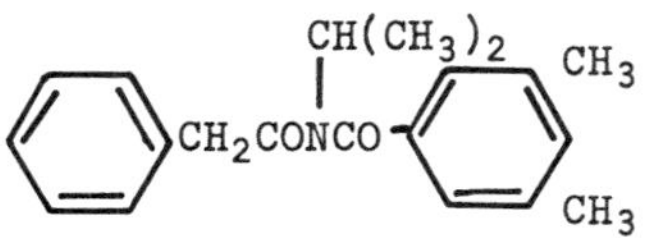

benzipram

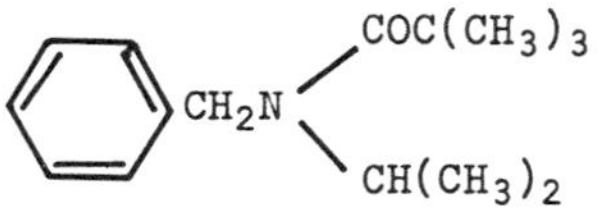

butam

Another amide herbicide which has been recently announced is butam, which has interestingly no chlorine substituents in the aromatic ring and no chloromethyl grouping.

A further new development is napropamide which is 2-(1-naphthoxy)-N,N-diethylpropionamide.

Nitrile Herbicides

Two interesting herbicides, related in some ways to the benzoic acids, are dichlobenil and chlorthiamid.

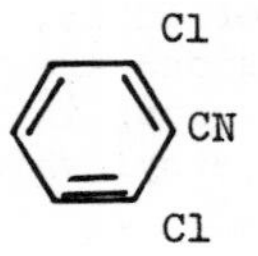

dichlobenil

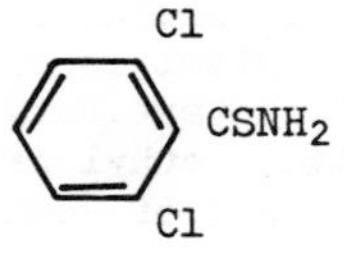

chlorthiamid

Dichlobenil is used as a pre-emergent selective herbicide mainly for weed control in orchards and amongst fruit bushes, and also to some extent for aquatic weed control. Chlorthiamid has the same applications as it is broken down to dichlobenil in the soil. They both inhibit actively dividing meristems and thus control annual and perennial weeds in the seedling stages. Dichlobenil is manufactured from 2,6-dichlorotoluene via the aldehyde and oxime.

Another interesting group of nitrile herbicides is exemplified by bromoxynil and ioxynil.

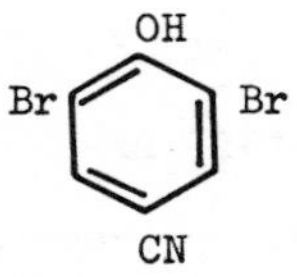

bromoxynil

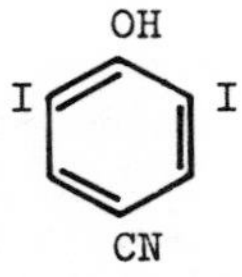

ioxynil

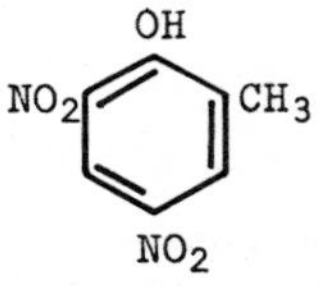

DNOC

These are both contact herbicides with some translocated activity and are used mainly for control of broad-leaved weeds in cereals. Their structure is reminiscent of DNOC and that biological activity is somewhat similar.

They are made by halogenation of 4-hydroxybenzaldehyde and conversion of the product to the oxime which is then dehydrated to give the nitrile.

Urea and Triazine Herbicides

These, together, comprise one of the most important and widely-used group of modern herbicides. A typical urea is monuron and a typical triazine is simazine.

monuron

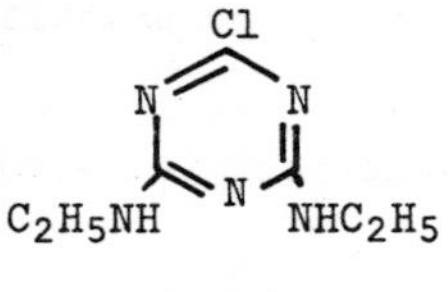

simazine

Numerous other members of the series are available. Other ureas have different substituents in the benzene ring and, in a few cases, one methyl on the second N atom is replaced by methoxy. Other "triazines" have different alkyl substituents on the side chain N atoms and the substituent on the third ring C atom may be methoxy or methylmercapto (CH_3S-) in place of Cl.

Ureas

Ureas of the general formula $RNHCON(CH_3)_2$ include fenuron (R = phenyl), monuron (R = 4-chloroophenyl), diuron (R = 3,4-dichlorophenyl), chlortoluron (R = 3-chloro-4-methylphenyl), fluometuron (R = 3-trifluoromethylphenyl), metoxuron (R = 3-chloro-4-methylphenyl), difenoxuron (R = 4|4-methoxyphenoxy|-phenyl), chloroxuron (R = 4-|4-chlorophenoxy|phenyl), isoproturon (R = 4-isopropylphenyl), tetrafluron (R = 3-tetrafluoroethoxyphenyl), chloreturon (R = 3-chloro-4-ethoxyphenyl) and fluthiuron (R = 3-chloro-4-chlorodifluoromethylthiophenyl).

Ureas of the general formula $RNHCON(CH_3)OCH_3$ include monolinuron (R = 4-chlorophenyl), metobromuron (R = 4-bromophenyl), linuron (R = 3,4-dichlorophenyl) and chlorbromuron (R = 4-bromo-3-chlorophenyl).

Other ureas include 3-benzoyl-3-(3,4-dichlorophenyl)-1,1-dimethylurea (fenobenzuron), 3-|3-(N-tert.butylcarbamoyloxy)phenyl|-1,1-dimethylurea (karbutilate), 3-cyclooctyl-1,1-dimethylurea (cycluron), 3-(hexahydro-4,7-methanoindan-5-yl)-1,1-dimethylurea (noruron), 1-butyl-3-(3,4-dichlorophenyl)-1-methylurea (neburon), 3-(4-chlorophenyl)-1-methyl-1-(1-methyl-2-propynyl) urea (buturon), 1-(2-methylcyclohexyl)-3-phenylurea (siduron), 1-methyl-3-(2-benzothiazolyl)urea (benzthiazuron), 3-(2-benzothiazolyl)-1,3-dimethylurea (methabenzthiazuron), 1-(5-ethylsulphonyl-1,3,4-thiadiazo-2-yl)-1,3-dimethylurea (ethidimuron), 1,3-dimethyl-1-(5-trifluoromethyl-1,3,4-thiadiazolyl-2-) urea (thiazafluron), 2,5-dimethyl-N-phenyl-1-pyrrolidinecarboxamide (cisanilide) 1-(N-ethyl-N-propylcarbamoyl)-3-propylsulphonyl-1H-1,2,4-triazole (epronaz) and 1-(5-butylsulphonyl-1,3,4-thiadiazolyl)-1,3-dimethylurea (buthiuron).

Triazines

	X	Y	name
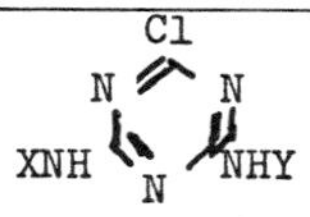	ethyl	ethyl	simazine
	ethyl	isopropyl	atrazine
	isopropyl	isopropyl	propazine
	isopropyl	cyclopropyl	cyprazine
	ethyl	sec.butyl	sebuthylazine
	ethyl	tert.butyl	tertbuthylazine
	ethyl	1-cyano-1-methylethyl	cyanazine
	ethyl	diethyl	trietazine
	isopropyl	diethyl	ipazine
	diethyl	diethyl	chlorazine
	diisopropyl	diisopropyl	siprazine
	cyclopropyl	1-cyano-1-methylethyl	procyazine
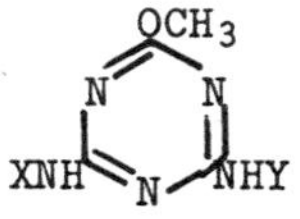	ethyl	ethyl	simeton
	ethyl	sec.butyl	secbumeton
	ethyl=	tert.butyl	terbumeton
	ethyl	isopropyl	atraton
	isopropyl	isopropyl	prometon
	methoxypropyl	methoxypropyl	methometon

<u>Triazines</u> (Continued)

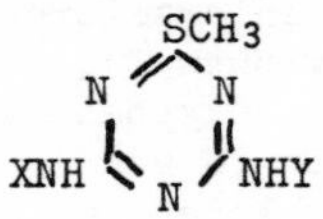

X	Y	name
methyl	isopropyl	desmetryne
ethyl	ethyl	simetryne
ethyl	isopropyl	ametryne
ethyl	tert-butyl	tertbutryne
isopropyl	isopropyl	prometryne
isopropyl	methoxypropyl	methoprotryne
isopropyl	azido	aziprotryne
ethyl	dimethylpropyl	dimethametryn
ethyl	1-cyano-1-methylethyl	cynatryn

Many of these compounds have only low solubility and are not effective by foliage application. They are very, but slowly, effective after application to soil. In massive dosage (5-10 kg/ha) they are used for "total" weedkilling on railway tracks and industrial sites, but have also very useful selective activity at lower dosages. Atrazine (=simazine with one ethyl group replaced by isopropyl) has a commanding position for weed control in the maize crop, although simazine is preferred in wet districts. The tolerance of maize to atrazine and simazine is an outstanding example of biochemical selectivity. It has been established that the maize plant contains a hydroxamic acid which is able to remove the Cl substituent very easily, probably by first adding oxygen to one of the adjacent ring N atoms. This reaction can take place <u>in vitro</u> in expressed maize sap.

Practice in Hungary provides an interesting example of adaptation of crop rotation to exploit fully the properties of available chemicals. Four or more successive maize crops are taken on suitable land, the first receiving a heavy dose of atrazine at sowing time. This gives an extremely clean crop. A rather lower dose is applied in the second year and none, or only a small "maintenance" dose, in the third year. By this method, there is time for the rather persistent herbicide to decompose in the soil before a sensitive crop is taken. Probably the later maize crops themselves contribute largely to the elimination of the herbicide by taking it up and decomposing it.

Some fruit crops, particularly raspberries and blackcurrants, and asparagus are rather tolerant of these herbicides. Top fruit orchards can be kept moderately clean with simazine, but are damaged by a dosage which will eliminate all weeds.

The partial resistance of many tree and bush species to these herbicides is probably helped by their greater size as compared with the annual weeds which are mainly controlled at the seedling stage. On a milligram/kilogram basis a mature tree requires a much larger dose and it has a greater proportion of its root system in lower levels of the soil than has a young seedling. Owing to adsorption of the herbicide on the colloidal components of the soil, chiefly the organic matter, the leaching of these herbicides downwards proceeds much more slowly than the movement of water itself, just as does an adsorbed solute in a chromatographic column. Since these herbicides were the first of the almost water-insoluble type to be widely used, their slow leaching is often put down to their low solubility, but this is of negligible importance compared with adsorption, except for compounds less soluble than simazine (5 ppm).

Among specialist uses of compounds of the phenylurea or aminotriazine type may be mentioned that of linuron, or prometryne for control of annual weeds in carrots or potatoes. The compounds are usually applied pre-emergence but have more action post-emergence than monuron or simazine, probably because they are more soluble. Linuron especially is fairly effective post-emergent and a second spray is often recommended. Desmetryne is useful for the control of many annual weeds in kale or other Brassica crops, for which purpose it is applied when the crop has three true leaves expanded. Selectivity is partly dependent on the ability of the crop leaves to reflect the spray droplets.

Other special uses are under investigation. Both series are almost embarrassingly rich in active compounds. Many of these show some potentially explitable selectivity. Owing to the confusion which could arise among farmers and the heavy cost of securing Government safety approval, it is not practical to produce commercially all the variants which could have some specific advantage for minor crops.

By far the most commercially useful ureas are diuron, of which about 3000 t/year are manufactured in the U.S.A., fluometuron (2000 t/year) and linuron (1000 t/year). The total amount of all other ureas manufactured in the U.S.A. is only 200 t/year.

Of the triazines, by far the most important is atrazine which is the largest tonnage herbicide of all, about 45,000 t/year being manufactured in the U.S.A. Manufacture of all other triazines in the U.S.A. amounts to about 3000 t/year.

The substituted ureas are made by a process analogous to the formation of carbamates, differing only in that both partners to the reaction with phosgene are amines. The substituted triazines are made from the trichlorotriazine (cyanuric chloride), itself made by trimerization of cyanogen chloride. Reaction of cyanuric chloride with excess amine readily replaces the chlorine atoms. As the successive chlorine atoms are replaced, the reactivity of those remaining decreases, a fact which greatly facilitates the production of compounds with different substituents. Replacement of the first chlorine atom occurs rapidly at below room temperature; replacement of the second requires moderate, and of the third, strong heating. Replacement of the third by methoxy or methylmercapto is more facile than by amino. The sequence is shown for the case of desmetryne.

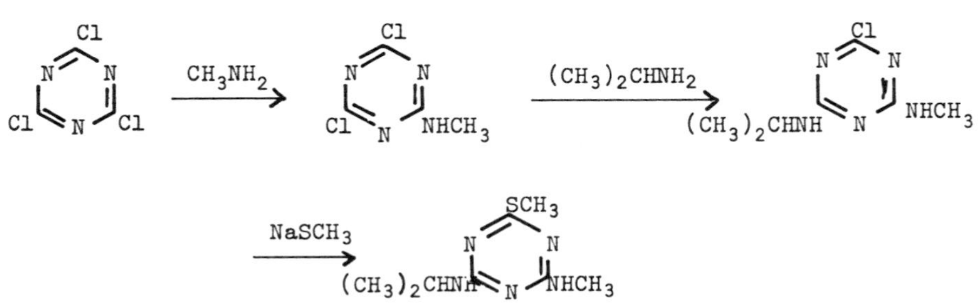

desmetryne

Uracil Herbicides

A useful group of herbicides is derived from the pyrimidine compound, uracil. The most commercially important compounds are isocil, bromacil and terbacil.

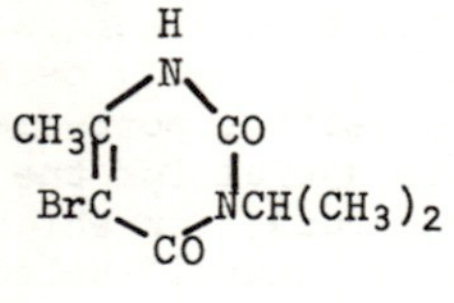

isocil

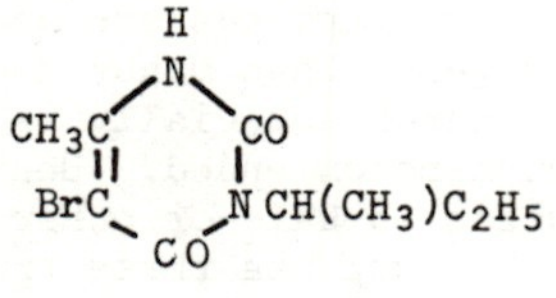

bromacil

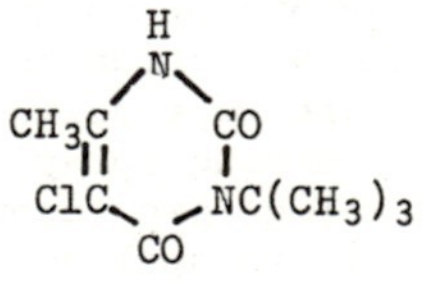

terbacil

These are powerful total herbicides which can be used for general weed control in non-crop situations. They are particularly effective against perennial grasses. Bromacil is also used selectively in citrus and sisal, and terbacil is used selectively in sugar cane, apples and peaches. These compounds appear to be general inhititors of photosynthesis and their selectivity arises from positional and physiological factors which prevent transport of the compound to an active biochemical site.

They are manufactured by reaction of ethyl acetoacetate with an N-substituted urea (prepared from phosgene and the appropriate amine) to give the uracil which is then halogenated, either with bromine or chlorine.

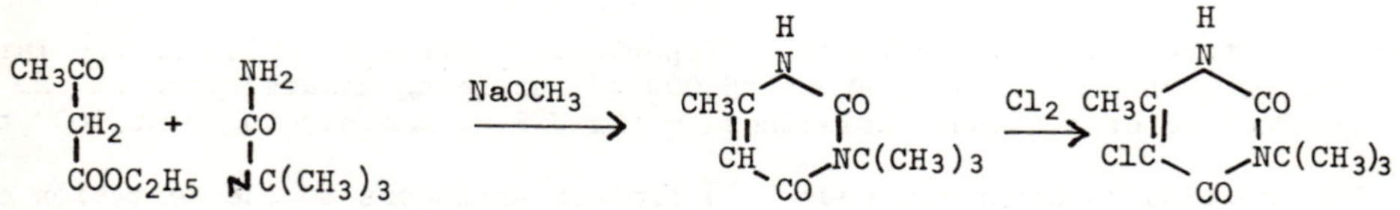

terbacil

Pyridazine Herbicides

Another useful group of herbicides is derived from pyridazinone and has the general formula.

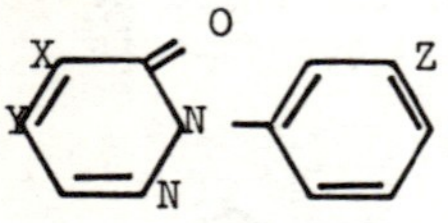

X	Y	Z	name
NH_2	Cl	H	pyrazone
NH_2	Br	H	brompyrazone
HOOC.CONH	Br	H	oxapyrazone
CH_3O	CH_3O	H	dimidazon
$(CH_3)_2N$	Cl	CF_3	merfluorazone
CH_3NH	Cl	CF_3	norflurazon
$CCl_3CHOHNH$	Cl	H	chlorazon

The most widely-used of these compounds has been pyrazone which is highly selective for pre-sowing, pre-emergent and post-emergent weed control in sugar beet. It is manufactured by reaction of phenylhydrazine with mucochloric acid followed by treatment of the product with ammonia. Norfluorazon is manufactured likewise from 3-hydrazinobenzotrifluoride, mucochloric acid and methylamine, and is particularly useful for selective weed control in cotton.

Bipyridylium Herbicides

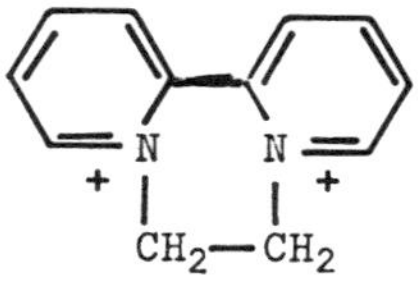

diquat cation

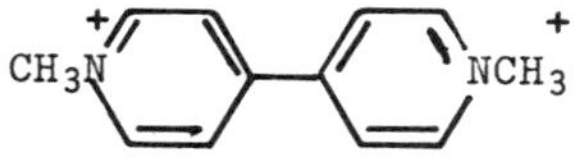

paraquat cation

The salts of the quaternised bipyridylium cations, diquat and paraquat, are freely water-soluble and kill top growth very quickly when applied to foliage. They are both general herbicides but are much better translocated than the old-fashioned contact herbicide DNOC. Both, particularly paraquat, are effective against grasses.

However, both paraquat and diquat are immediately absorbed very strongly onto the clay minerals which are present in every soil. They are so strongly bound to these minerals that they can be removed only by completely breaking down the structure of the soil, e.g. by heating with concentrated sulphuric acid. In practice, although physically present in the bound form, they are completely inert biologically.

They can be used, therefore, for cleaning up undergrowth in woodland, plantations, orchards and shrubberies and, with carefully directed application for inter-row weeding in field crops.

Paraquat has been widely used for destruction of existing growth prior to reseeding, thus offering the possibility of a chemical alternative to ploughing. Naturally this revolutionary use took some time to become accepted and there are doubtless situations where mechanical soil disturbance will still remain desirable. A problem of "chemical ploughing", particularly when applied to old grassland, was the mechanical problem of sowing the seed in the dense mat of fibre, but new and simple agricultural machinery has been developed for this purpose.

2,2'-Bipyridyl, the parent intermediate for diquat, is made by oxidative coupling of pyridine over heated Raney nickel

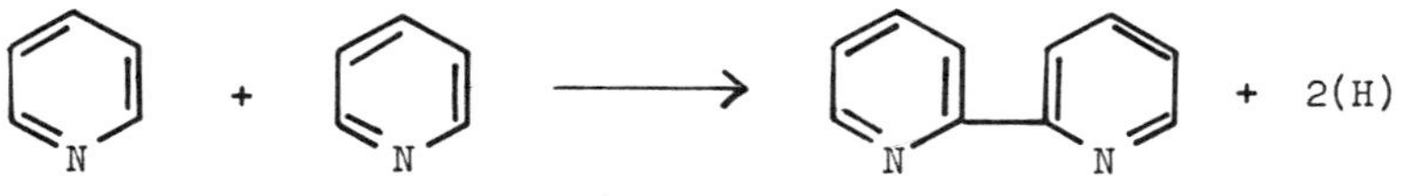

Direct coupling of pyridine in the 4,4' positions is not possible and all earlier methods gave mixtures of 2,2', 2,4' and 4,4'-bipyridyl. In the present process pyridine is reacted with sodium in liquid ammonia to give the transient pyridyl radical anion which immediately dimerizes to 4,4'-tetrahydrobipyridyl which is then oxidized by air to 4,4'-bipyridyl. This process gives only the required 4,4'-isomer and is an interesting example of application of a free radical reaction on an industrial scale.

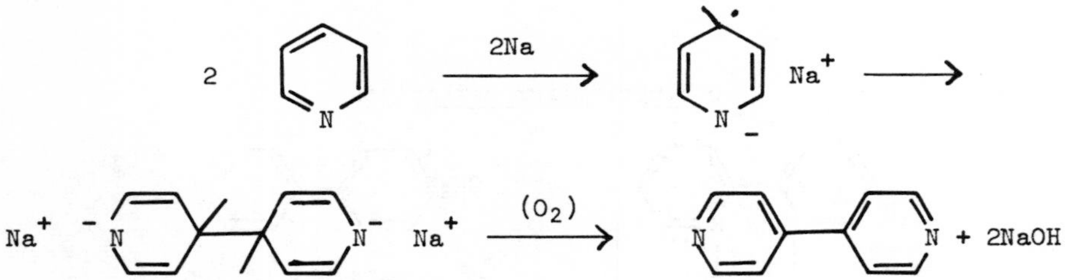

Other Heterocyclic Herbicides

A very large number of heterocyclic compounds, apparently quite unrelated to one another, have been reported to have herbicidal activity. It is impossible to make any logical classification of them. Very few have achieved any substantial commercial success.

Amongst the more important are the following.

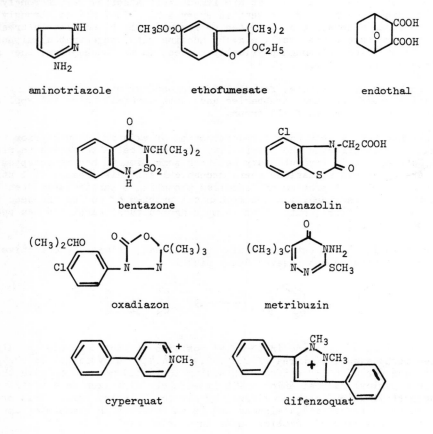

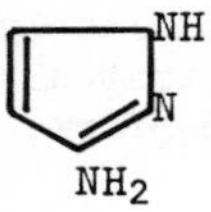

aminotriazole

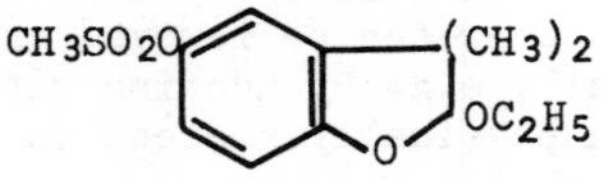

ethofumesate

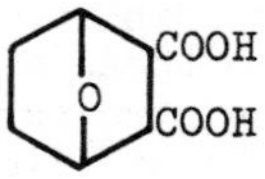

endothal

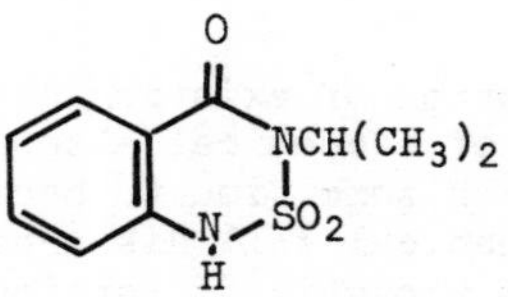

bentazone

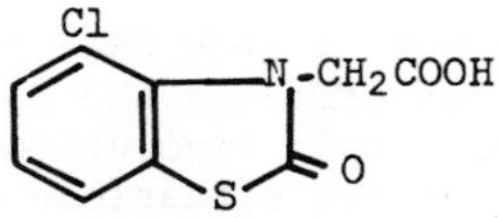

benazolin

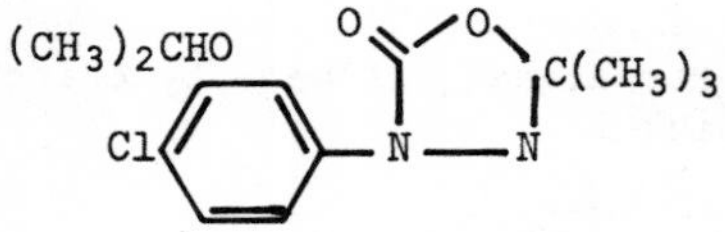

oxadiazon

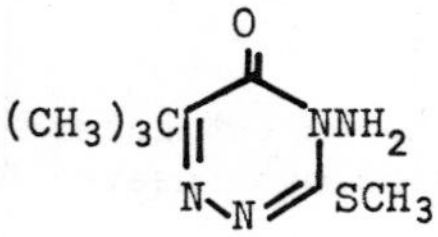

metribuzin

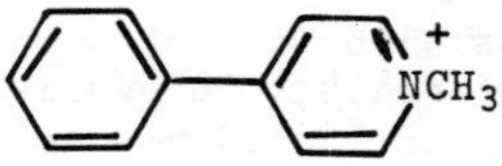

cyperquat

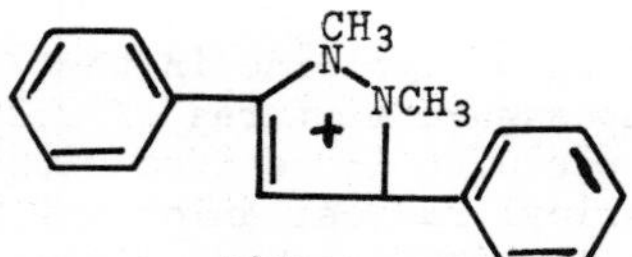

difenzoquat

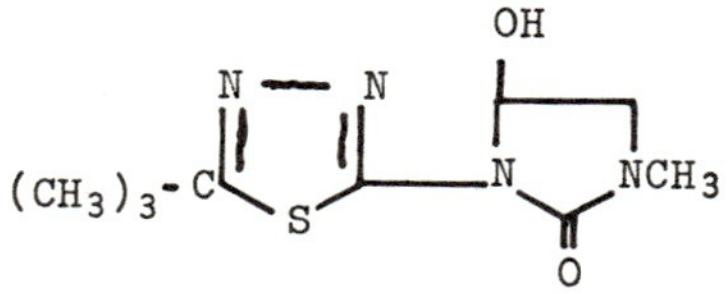

buthifazole

Aminotriazole is very water soluble and rapidly halts further growth of plants by inhibition of photosynthesis. It is very well translocated and, applied at the season of declining top growth, can penetrate to underground stems and rhizomes and so kill difficult weeds like couch-grass and horsetail.

Ethofumerate is a new herbicide which looks very promising for selective weed control in sugar beet.

Endothal is used for pre- and post-emergent weed control in sugar beet.

Bentazone is a contact herbicide which provides useful post-emergent control of some troublesome weeds in cereals.

Benazolin is a very highly specific post-emergent herbicide effective against cleavers and chickweed in cereals and in oil-seed rape.

Oxadiazon is used as a post-emergent herbicide in rice.

Metribuzin is a herbicide particularly for use in potatoes.

Cyperquat controls the ubiquitous purple nutsedge which is one of the most troublesome weeds in the world and is affected by very few herbicides.

Enough has been said to illustrate the wide range of compounds which exhibit herbicidal activity, and more are being produced every year. However, it must be remembered, that the total amount of "minor" organic herbicides manufactured in the U.S.A. is only about 3000 t/year whereas atrazine, 2,4-D, chloramben, trifluralin, the carbamates and the ureas each exceed 10,000 t/year.

Silicon Herbicides

The only commercial herbicide containing silicon is

2-chloroethyl-tri-(2-methoxy-ethoxy)-silane (etacelasil)

but others may be developed as the current interest in organosilicon chemistry progresses.

$$(CH_3OCH_2CH_2O)_3SiCH_2CH_2Cl$$

etacelasil

Phosphorus Herbicides

Although organophosphorus compounds have been developed mainly as insecticides, because of their anticholinesterase activity, a number of such compounds do show herbicidal activity. Their mode of action is not known but is not thought to depend on inhibition of cholinesterase although an enzyme similar to animal acetylcholinesterase has been discovered in plants and there is some evidence that acetylcholine does participate in phytochrome mediated processes.

Commercially used phosphorothioic acid derivatives include bensulide, a selective herbicide, and butifos, a cotton defoliant.

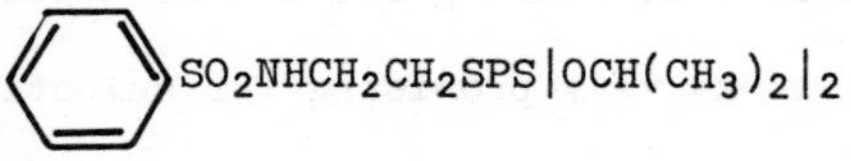

$(C_4H_9S)_3PO$

bensulide butifos

Some alkyl aryl N-alkylphosphoramidothionates have high herbicidal activity, and commercially used compounds include DMPA, amiprophos and metacrephos. Their main application is as pre-emergent contact herbicides for control of crabgrass in turf.

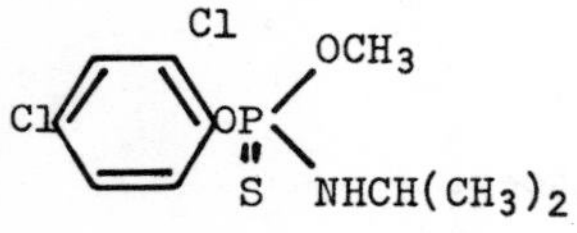

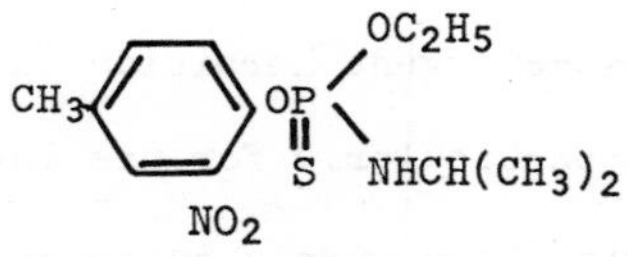

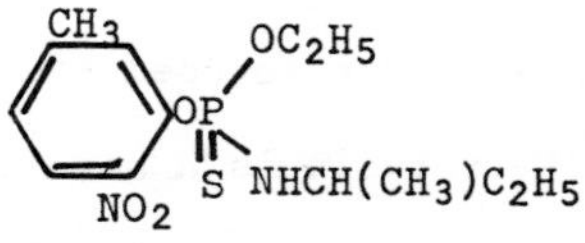

DMPA amiprophos metacrephos

It is interesting to note that phosphoramidates with a bulky alkyl group on the amido nitrogen atom have herbicidal activity whereas N-methyl or N-ethyl phosphoroamidates are generally insecticides.

By far the most important organophosphorus herbicide is a compound with a structure unlike that of any other herbicide, namely, N-(phosphonomethyl) glycine (glyphosate). It is manufactured from phosphorus trichloride, formaldehyde and glycine, and is generally used as the isopropylamine salt.

$$PCl_3 + CH_2O \longrightarrow Cl_2POCH_2Cl \xrightarrow{H_2O} (HO)_2POCH_2Cl$$

$$(HO)_2POCH_2Cl + NH_2CH_2COOH \longrightarrow (HO)_2POCH_2NHCH_2COOH$$

glyphosate

Glyphosate is a contact herbicide active only by foliar application. It is strongly absorbed onto soil and thus rendered biologically inert but, whereas the bipyridylium herbicides, paraquat and diquat, are inactivated by absorption onto the clay minerals in soil, glyphosate is believed to be inactivated by chelation with heavy metals, particularly iron. It is, therefore, much more strongly absorbed by soils with a high organic content, such as peat soils. Because of its inactivation by soil, it can, like the bipyridylium herbicides, be used for weed control in orchards, shrubberies and plantations, and also, using carefully directed application, for inter-row weeding in crops.

Glyphosate controls not only annual weeds but also the deep-rooted perennials in which it is translocated to the underground rhizomes and destroys them. It has been suggested that it acts by suppressing biosynthesis of phenylalanine by inhibiting chrorismate mutase or prephenate dehydratase.

Arsenic Herbicides

For many years sodium arsenite was used as a general contact herbicide but was discarded when the newer organic herbicides were introduced, because of the danger of toxic residues.

However, two organic arsenical compounds are still used on a large scale, namely, disodium methanearsonate (DSMA) and sodium dimethylarsenite (sodium cacodylate).

$CH_3AsO_3H_2(Na)_2$	$(CH_3)_2AsO_2H(Na)$
DSMA	sodium cacodylate

DSMA is used as a selective post-emergent contact herbicide for control of grass weeds in cotton. It is manufactured by reaction of sodium arsenite with methyl chloride. DSMA is much less toxic to men and animals (LD50, 1800 mg/kg) than sodium arsenite (LD50, 20 mg/kg).

Sodium cacodylate is used for weed control in non-crop areas and for killing unwanted trees. It is manufactured by reduction of DSMA with sulphur dioxide and methylation of the product with methyl chloride. It, also, has fairly low mammalian toxicity (LD50, 1350 mg/kg). Nevertheless, despite the comparatively low toxicities of DSMA and sodium cacodylate, it is probable that their use will be phased out because of the possibility of their conversion into more toxic products in the environment. At the moment, about 16,000 t/year are manufactured in the U.S.A.

Modes of Action of Herbicides

Much less is known about the biochemical action of herbicides than about that of organophosphorus insecticides. Many different types of action must be involved. The "hormone" herbicides, as has been mentioned, are essentially disorganizers of plant growth rather than direct killers and clearly interfere with, or compete with, the action of natural hormones. By contrast, the carbamates appear to be more truly toxic. They inhibit the development of juvenile tissue in shoot or root.

Some of the important compounds act mainly by drastic interference with the photosynthetic process. This is probably true of the anilide herbicides and certainly of the phenylureas and aminotriazines. These inhibit an early reaction in the photosynthetic chain in which water is oxidized with liberation of elementary oxygen. Some at least of the steps of *in vivo* photosynthesis can take place in a suspension of chloroplasts (the sub-cell units which contain the chlorophyll) in suitable solution containing a reducible model substance - usually ferricyanide ion. This is reduced, with liberation of oxygen, when the suspension is illuminated. The phenylureas and amino-triazines, at very low concentrations, inhibit this reaction.

Some of the later steps in photosynthesis are common to the basic processes of respiration, but the energy to drive them is obtained, via chlorophyll, from light rather than from oxidation of carbohydrate fuel. This reduction of pyridine nucleotide (the TPN → TPNH reaction) and the phosphorylation of adenosine diphosphate (ADP → ATP) are light-driven reactions in the chloroplasts and carry out important further reduction processes. These processes are inhibited by the bipyridilium herbicides which are reduced in the presence of light to free radicals and give rise to peroxides during the subsequent dark reoxidation to the quaternary ions. The presence of these compounds causes the reductive power of the whole photosynthetic process to be wasted in useless side reactions.

This is not, however, in its simplest form, the complete explanation of the action of diquat and paraquat, because although their action is much faster in the light than the dark, it is also, in the light, faster than death by starvation in the dark. Also the lethal effect of keeping plants in the dark is accelerated by their treatment with these compounds. The action of the phenylureas and triazines, however, seems to be fully linked with the effect of light. Seeds germinate normally in treated soil, but the seedlings die off when photosynthetic processes should be taking over. It is consistent with their interference with the early, unique, steps in photosynthesis, that the phenylureas and aminotriazines have very low toxicity to animals, in which photosynthesis plays no role.

The dinitrophenols and probably bromoxynil and ioxynil, act as respiratory accelerators by inhibition of oxidative phosphorylation. In the absence of this mechanism, the energy of sugar-oxidizing processes is largely diverted to useless side reactions.

Amitrole also acts on the photosynthetic process but in a third way. Chlorophyll is an inherently unstable substance in the presence of light but is normally returned to its initial state during the transfer of its light-derived extra energy to other reactions. New growth in plants treated with amitrol contains no chlorophyll. Either its synthesis, or its continued reformation, is inhibited.

Further Reading

Ashton, F.M. & Crafts, A.S., Mode of Action of Herbicides, (Wiley - Interscience, London & New York, 1973)

Audus, L.J., (Ed)., Physiology and Biochemistry of Herbicides, (Academic Press, London & New York, 1964)

Corbett, J.R., The Biochemical Mode of Action of Pesticides, (Academic Press, London & New York, 1974)

Fryer, J. & Makepeace, R., (Ed)., Weed Control Handbook, 6th Edition, Volumes I & II, (Blackwell, Oxford, 1970)

Kearney, P.C. & Kaufman, D.D., Herbicides, Volumes I & II, (Marcel Dekker, New York, 1975)

Street, J.C., (Ed)., Pesticide Selectivity, (Marcel Dekker, New York, 1975)

Urenovitch, J.V. & Dixon, D.D., Synthesis of Commercial Herbicides, Volumes I & II, (David Chemical, Pennsylvania, 1971)

Weed Science Society of America, FAO International Conference on Weed Control, (Illinois, 1970)

Chapter 16
PLANT GROWTH MODIFICATION

Man not only wants to populate his arable land with the species of plants best suited to his needs but he is also never fully satisfied with their natural manner of growth. He would, for example, prefer wheat to have shorter straw and cotton to shed its leaves as soon as the seed hairs have developed in the bolls. The possibility of using chemicals to produce such desirable modifications has long been considered and a number of noteworthy discoveries have been made in the past thirty years. Plant growth regulators (PGR), as they are generally called, are thought to have a great future potential in agriculture. Pesticide chemists have, up to now, concentrated mainly on discovering and developing chemicals which kill plants but which desirably are selective in their toxic effects so that weeds can be destroyed with minimal damage to crops. The more subtle targets of making plants grow the way we should like them to grow and of diverting the energy which they obtain by photosynthesis into those types of plant growth which are most desirable for the farmer, have only recently become the subjects of intensive research. It may be objected by the reader that such chemicals do not properly come under the heading of "Crop Protection and Pest Control" but they are likely to be related to herbicides, will be discovered, developed and manufactured by the pesticide companies, will need the same machinery for application as herbicides and will conveniently be marketed through the same commercial supply chains. For this reason, they are included in this book.

One reason why plant growth regulators have been the subject of less research effort than herbicides is that their screening and evaluation is much more difficult than that of herbicides. In herbicidal screening, selective toxic effects are sought by applying the candidate chemical to a range of major crop plants and major weeds. In many cases, the results of such tests are clear and unambiguous and the precise pattern of selective toxicity which is observed indicates where the commercial opportunities are likely to be. For plant growth regulators, however, the botanist is being asked to observe any unusual effect of the chemical on the habit of growth or metabolism of any crop plant and then to equate those observations to a field situation in which such an effect could be of economic value to the grower. The difficulty is essentially of defining a manageable number of targets and of setting up a screening system which would be comprehensive for those targets yet not too large, expensive and resource-demanding. A further problem is that, at the moment, it is very difficult to predict from effects observed in the glasshouse exactly what effects will be produced in the field on the total crop, so that candidate plant growth regulators have to be field tested at an early stage, and this makes great demands on money, time and effects of skilled staff. Furthermore, plant growth regulators may have to be tested over a number of seasons since a desirable effect produced in one season, e.g. increased yield of fruit, might possibly be offset by a swingback in the following seasons, or some long-term deleterious effects might be induced. Obviously, this is more of a problem with established perennial crops, such as orchards or vineyards, than with annual crops. Toxicity and residue studies have to be far-reaching since a useful plant growth regulator, in order to produce a lasting effect, might be required to persist for some time within the plant. The cost of discovery and development of a plant growth regulator is, therefore, very high and it is likely to be economical to develop and sell a particular compound only if there is a large potential international market in a major crop. Because of the possibility of long-term effects, profitable markets are more likely to be in annuals than in perennials.

It may be useful, at this stage, to try to define exactly what a plant growth regulator is. Possibly the best definition, even if it does not say very much specific, is "a chemical which modifies plant growth in a manner beneficial to the farmer". Such chemicals can be classified in three categories:

1 Chemicals which have a direct visible effect which is of value to the farmer by reducing labour costs, facilitating cultivation and harvesting or producing some other economically advantageous effect. Such effects might include the promotion or retardation of vegetative growth, rapid crop establishment and shortening of the growing season, control of fruit fixing, setting, abcission and quality, defoliation and prevention of flowering.

2 Chemicals which increase crop yields either by modifying the morphology or growth of the plant or by affecting the metabolism of the plant.

3 Chemicals which affect morphology, growth or metabolism in such a way as to provide greater protection to the plant against environmental effects such as drought, heat, cold, soil salinity, vulnerability to weather damage, or susceptibility to atmospheric pollutants.

Retardation of Vegetative Growth

In developed countries with intensive agriculture, such as those in Europe, yields per hectare of cereals have been steadily pushed up by applying greater amounts of nitrogenous fertilizers. In some weather and soil conditions this has caused harvesting difficulties and losses consequent on increased lodging. A compound which is extensively used, particularly in Holland and Germany, to reduce the length of the wheat straw and the risk of lodging is 2-chloroethyltrimethylammonium chloride, $ClCH_2CH_2N^+(CH_3)_3 Cl^-$, (chlormequat chloride). It is manufactured by reaction of ethylene dichloride with trimethylamine. Apart from acting as a growth retardant and producing compact plants with shortened internodes and petioles it may also affect the response of the plant to pests and pathogens, thus, it has been reported to decrease the susceptibility of tomatoes to verticillium wilt and of cabbages to aphid attack.

Runner type peanuts are difficult to harvest because of their spreading lateral shoots. A retardant spray which is widely used in the USA is N-dimethylaminosuccinamic acid, $HOOC(CH_2)_2CONHN(CH_3)_2$, (daminozide), which is manufactured by reaction of succinic anhydride with 1,1-dimethylhydrazine. Apart from its use on peanuts, it is also used to control the vegetative growth of fruit trees and to modify the stem length and shape of ornamental plants. Like chlormequat, it reduces internodal distances and is said to increase heat, drought and frost resistance.

As with cereals, higher planting densities and increased use of fertilizers have led to lodging problems in soya beans and the plants tend to grow tall and lanky. A compound which has been used to retard vegetative growth is 2,3,5-triiodobenzoic acid, but its performance has proved unreliable in practice.

Retardation of growth of grass, as an alternative to regular mowing, has long been a desirable target particularly for golf courses, cemeteries, airfields, roadside verges and amenity areas. A disadvantage for home lawns or regularly used grass areas is that grass rapidly becomes dirty and trampled, and mowing has the advantage that it continually presents a fresh, clean, upstanding grass surface. Nevertheless, there is a commercial market for such products and two of the most successful have been 6-hydroxy-3-(2H)-pyridazinone (maleic hydrazide) and 2-chloro-9-hydroxyfluorene-9-carboxylic acid (chlorflurecol-methyl).

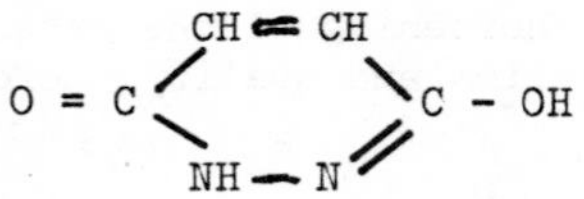

maleic hydrazide

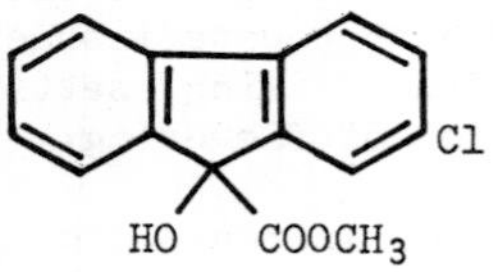

chlorflurecol-methyl

Tobacco plants, when topped to remove the flower buds, develop lateral shoots from the leaf axes. Maleic hydrazide is widely used to supress these shoots, but a cheap and effective material is also a mixture of the methyl esters of C_8 to C_{12} fatty acids $(C_8\text{-}C_{12}).COOCH_3$. Dimethyldodecylammonium acetate has also been used for this purpose.

Two recent compounds which appears to be particularly effective for reducing the growth of woody plants are ethyl hydrogen 1-propylphosphate and 1-propylphosphonic acid.

A large number of experimental compounds have been suggested for control of growth of ornamental plants but development of a new compound purely for this market is unlikely to be financially rewarding. Such compounds include α-cyclopropyl-4-methoxy-α(pyrimidin-5-yl) benzyl alcohol (ancymidol), 6-benzylamino-(9-tetrahydropyran-2-yl)-(9H)-purine; and tributyl-2,4-dichlorobenzylphosphonium chloride (chlorphonium chloride). The purine is the benzyl derivative of the naturally occurring growth regulator, kinetin, which plays an essential part in cell division. It has been suggested that it might be used to maintain vegetables in a fresh condition during transport to the customer.

An effect which may reasonably be included under suppression of vegetative growth is the prevention of sprouting in stored potatoes, and a compound which has been widely used for this purpose is 1,2,4,5-tetrachloro-3-nitrobenzene (tecnazene), which is made by nitration of tetrachlorobenzene. Use has also been made of isopropyl 3-chlorophenylcarbamate (chlorpropham).

Rapid Crop Establishment

One of the oldest types of plant growth regulators is the "hormone herbicide" which can be used to initiate root growth in cuttings and thus to facilitate rapid propagation. The margin of safety of the herbicides 2,4-D and MCPA is too small to allow them to be used for this purpose and the most generally utilised compounds are 1-naphthalene acetic acid and indolebutyric acid. These compounds have made possible the vegetative propagation of some choice varieties of ornamental plants which do not breed true from seed.

Control of Fruit

For most fruits the edible portion is not the true seed but some fleshy covering or receptacle, the function of which is to assist dispersal of the seed by providing a tempting food for mobile animals. In most cases, the plant breeder has selected strains which give very much bigger fruits, and has even succeeded in suppressing the true seed. Seedless raisins are well-known and a banana containing fertile seed is a rarity.

Usually, however, the fleshy fruit will not develop unless the flower is fertilized and viable seed is set. Nevertheless, some fruit can be produced parthenocarpically by spraying the flowers with 1-naphthylacetic acid or with 4-chlorophenoxyacetic acid. This technique is particularly widely-used on tomatoes.

Although many types of fruit trees thin out naturally by dropping a proportion of the first-formed fruit, there are some - for example, peaches - where it is necessary to thin in order to obtain reasonably-sized fruits. Chemical thinning is more economical than hand-thinning. Compounds which have been used for this purpose on apples are the insecticide 1-naphthylmethyl carbamate (carbaryl), 1-naphthylacetamide and 1-naphthylphthalamic acid (naptalam), and also the naturally-occurring plant growth regulator, abscisic acid.

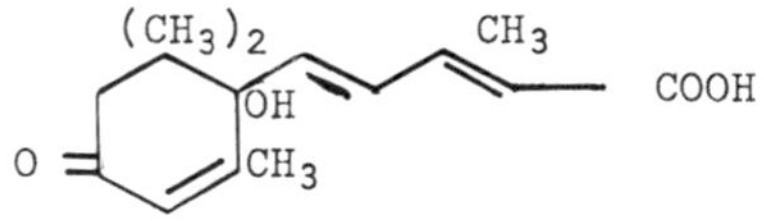

abscisic acid

On peaches, the main compounds used are 2-(3-chlorophenoxy) propionic acid and 2-(3-chlorophenoxy)-propionamide.

1-Naphthylacetic acid is also used to prevent developed apples from dropping before they are really ripe, which a number of varieties are very prone to do.

A number of compounds have been developed to prevent post-harvest "scald" of apples, which is a physiological effect. These include 1,2-dihydro-6-ethoxy-2,2,4-trimethylquinoline (ethoxyquinol) and diphenylamine.

Fresh oranges have to be harvested by hand because of the need to maintain a high standard of appearance. At present, pickers snip the stem of each orange but, if the fruit could be loosed sufficiently to allow pickers to twist it off the stem, then two-handed picking would be possible with consequent increase in harvesting speed and reduction in costs. Oranges which are used to make juice do not have to be treated so gently but they are generally harvested by tree shakers and catch nets, so loosening agents would be of utility here also to ensure total harvest.

Defoliation

Destruction of vegetative tissue in a mature crop can facilitate harvesting of roots or seeds. The action needed in this case, is, of course, directly herbicidal. There should be no systemic transfer to the organs which are required for storage and food, unless the herbicide were to improve storage while introducing no toxic hazard.

Potato-haulm is killed off by contact herbicides prior to harvest (i) to facilitate this mechanical operation, (ii) to kill late-germinating weeds before they set seed and (iii) to destroy blight fungus before exposure of the tubers. It is universally agreed that sulphuric acid produces the most satisfactory effect, having an advantage over most other compounds in hastening the decay of the tough fibres of the stem. Its application, however, requires special clothing and machinery. DNOC in oil and, more recently, diquat, are now more often used. Arsenites were at one time favoured but their use has been abandoned in the UK, by agreement between chemical industry and the Government, owing to toxic risk to workers in the harvest.

The falling of leaves of deciduous plants, although assisted by rough weather, is mainly a seasonally-induced positive act of plant growth. A layer of corky tissue develops at the base of the leafstalk which induces weakness and seals off the wound. A specific endogenous substance called abscissic acid is responsible for this development. Where only the seed or, in the case of cotton, the hairs attached thereto, are required, it would be an advantage to cause the leaves to fall prior to harvest. If the cut crop needs drying, the prior abscission of leaves would speed the process. Whether harvest is by hand or by machine, the process is simplified or accelerated if the required plant organ is left outstanding on otherwise naked stems.

Two compounds which have been effectively used to produce the desired clean defoliant action on cotton are calcium cyanamide, CaNCN, and tributylphosphorotrithiolate, $(C_4H_9S)_3P$. The former compound is dusted on and the latter is sprayed as an emulsion. Use is also being made of the naturally-occurring growth regulator abscissic acid.

Prevention of Flowering

It is desirable to prevent crops which are grown for their vegetative parts from flowering as this diverts photosynthetic energy. Disbudding is commonly carried out by hand but can also be achieved by application of chemicals.

An interesting possibility for deflowering agents is as male gametocides to permit ready production of cross-pollinated varieties and thus a hybrid seed which might give higher-yielding plants.

Improvement of Crop Yields

Assuming that a crop can carry out only a specific amount of photosynthesis during a growing season, the problem of yield improvement is essentially one of trying to divert as much of this photosynthetic energy as possible to production of those parts of the plant which are of most value to the grower. For vegetables this may mean prevention of flowering and of formation of side-shoots, and some examples of the use of plant growth regulators to produce these effects have already been given. For root crops it means diversion of the photosynthetic energy away from vegetative growth towards the laying down of energy resources as chemicals such as starch stored in roots and tubers. For crops grown for seeds or fruit and this includes all the vitally important cereals - the aim is to make the stage of vegetative growth as short as possible consistent with the health of the plant and to proceed as rapidly as possible to the reproductive stage. For plants grown for ornaments the aim may be to induce more profuse and prolonged flowering.

Not many chemicals have yet been discovered which produce these effects, and this is certainly the area in which plant growth regulators have the greatest potential in the future.

An interesting compound is gibberellic acid which was first isolated from a fungus, (gibberellus), which attacks rice and produces abnormal elongation of the stems between nodes. The compound occurs to some extent normally in most plants but has now been made synthetically. The effect it produces is the opposite of that produced by the growth retarding plant growth regulators.

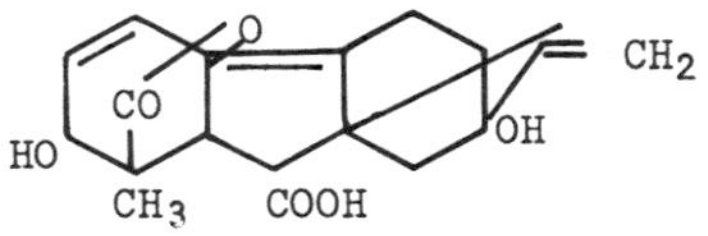

gibberellic acid

Its most important commercial use is in production of table grapes. As a result of elongation of the fruit stalks, more open bunches with larger and more shapely fruits and less susceptibility to fungal attack, are produced. The compound is also used as a seed treatment to hasten emergence of beans, peas, cotton and cereals; to produce bigger, firmer, better-coloured cherries, to hasten their maturity and to prolong their harvesting period, and to counteract the effects of yellowing viruses; to delay ripening of lemons and thus to produce a greater percentage of larger fruits with improved storage lives; to increase the yields and aid the harvesting of hops; to hasten maturing of artichokes and to extend their picking season; to produce longer-stemmed celery and rhubarb; to apply to barley to increase the enzymatic content of malt so that distillers can use less malt in production of grain spirits; to break dormancy of seed potatoes and to stimulate sprouting thus allowing immediate planting of red varieties which normally require 2 to 6 months storage to break their dormancy. It is also used to produce various effects on ornamental plants such as earlier blooming, more profuse flowering and longer flower stalks.

Gibberellic acid is a very safe and extremely non-toxic chemical. It affects only those parts of the plant which are above the soil surface but its effects are transient so that repeated applications are necessary.

Some compounds which have already been mentioned have been used to increase yields. Thus, daminozide can increase yields of tomatoes, and claims have been made that ethyl hydrogen 1-propylphosphate and 1-propylphosphonic acid both may increase yields of potatoes and sugar beets and similar root crops. A compound recently introduced for the same purpose is $HOOCCH_2N|CH_2PO(OH)_2|_2$ (glyphosine). Chloromequat chloride is used to increase tillering and reduce lodging in cereals.

A particular area in which use of chemicals has been successful is in increasing the amounts of specific useful products which can be extracted from plants, in particular, sugar and rubber. A number of compounds have been suggested to increase the amount of sugar in sugar-cane, of which the most successful is methyl 3,6-dichloro-2-methoxybenzoate (disugran) which is applied with a non-ionic surfactant two to four weeks before harvest.

It has long been known that ethylene will stimulate the production of latex by rubber trees. Ethylene is a natural growth regulator given off by a number of fruits during ripening. It is used commercially in fruit storage rooms to hasten ripening of bananas, citrus, melons, pears and pineapples, and also to induce flowering of pineapple plants. Application of ethylene gas to rubber trees is not very convenient so a number of "ethylene generators" have been developed. These are chemicals which can be applied to cuts in the bark and which break down within the tree to give ethylene which stimulates the production of latex. The most successful of these compounds is (2-chloroethyl) phosphonic acid, $ClCH_2CH_2PO(OH)_2$, (ethephon).

An interesting recent observation is that application of the bipyridylium herbicide paraquat dichloride to pine trees greatly increases the formation of oleoresins. It appears that the tree normally maintains a specific pressure of these resins within the cells and that application of the paraquat dichloride increases the permeability of the cells and allows the resin to leak out so the tree produces more to try to maintain the pressure.

Improvement in Photosynthesis Efficiency

Photosynthesis is a very inefficient process from an energy conversion viewpoint. On average, less than 1% of the solar energy falling on a leaf is converted to chemical energy by photosynthesis. Yields per hectare of most crops can be increased by supplying adequate water and sufficient synthetic fertilizers but there is a limit to the yield improvements which can be achieved in this way. Given enough water and fertilizer, the limiting factor to photosynthesis in the field is concentration of carbon dioxide in the air and the ability of the plant to fix it. Glasshouse growers often increase the concentration of carbon dioxide in their houses but this is not a practicable thing to do with a field crop. Nevertheless, it has been shown unequivocally in laboratory experiments that plants do respond to increased carbon dioxide concentrations, although to differing extents according to species, and that the normal level of carbon dioxide in air (0.030-0.035%) is well below the photosynthetic optimum for leaves illuminated with even moderate sunlight. It is thought that this may be a reflection of a much higher atmospheric carbon dioxide concentration on earth at the time plants evolved. However, too great a concentration of carbon dioxide is deleterious, and the optimum concentration is probably 0.5-1.0% for most plant species.

It is often not appreciated how much air a plant has to process to obtain the carbon dioxide it needs. For example, one hectare of maize must assimilate about 22,000 kg of carbon dioxide in a growing season. To do this, it must process about 100,000 tonnes of air and this means that each square metre of foliage must process about 1.5 m^3 of air every day. This presents the plant with an immense engineering problem in mass transfer. For this reason, there are very marked variations in photosynthesis in a field crop consequent in diurnal variations in carbon dioxide concentration. More seriously there are often local gradients of carbon dioxide content in dense stands of vegetation particularly under still atmospheric conditions and this can lead to local depletion of carbon dioxide and a complete local cessation of photosynthesis. Such localized depletion can occur much more commonly than is generally supposed. Considerable increases in photosynthesis have been obtained simply by providing for adequate circulation of air in standing crops.

To apply "fertilizers" which would liberate carbon dioxide is not a practicable proposition since the amounts involved are too great. However, some more subtle approaches may be rewarding. The rate of assimilation of carbon dioxide into the photosynthetic cycle is limited by the rate of entry of carbon dioxide through the leaf stomata, the rate of its passage through various membranes, its solubility in various cell fluids and its concentration near the chlorophyll. The possibility of using chemicals to affect any of these parameters is an exciting one. The degree of opening of the stomata (which are small holes in the leaf cuticle) have a profound effect on photosynthesis and there are a number of chemicals which are known to affect this. Possibly chemicals may be found offset carbon dioxide concentrations in cell fluids.

A further approach to the problem of improving photosynthetic efficiency is to discover chemicals which will inhibit the process of photorespiration. This is the process by which a certain proportion of the carbohydrates which are formed by photosynthesis are converted back to carbon dioxide. It is not, at the moment clear what purpose photorespiration serves in the life of the plant and whether its inhibition would have any damaging effects.

Resistance to Environmental Effects

Several of the plant growth regulators already mentioned are claimed to increase the resistance of various plants to heat, drought and cold. Dinonylsuccinic acid is used on apples, peaches and pears to protect blossoms against frost.

Compounds which prevent the normal opening of the leaf stomata may increase resistance to drought by reducing water loss and may also prevent access of pathogenic fungi or of atmospheric pollutants such as ozone.

General Summary

There are a very large number of biological effects which might be the basis for commercially useful plant growth regulators. However, it must be remembered that, long before the idea of chemical plant growth regulators was conceived, man was attempting to produce plants which grow in the way most beneficial to him by plant breeding. This has been a highly successful approach and high-yielding crop varieties have been developed from poor-yielding natural species. Modern methods of plant breeding and the application of genetic engineering should increase the output of plant breeders and cut down the time scale for establishment of new varieties. Nevertheless, there will still be a place for plant growth regulators. It is very unlikely that all desirable qualities can be bred into one variety of a crop plant and, in practice, improvement in one aspect often induces a disadvantage in another. Thus many high-yielding crop varieties are much more susceptible to disease than the natural stock. The choice may therefore be between breeding for yield and using pesticides to control diseases or breeding for disease resistance and using plant growth regulators to increase yields.

Further Reading

Audus, L.J., Plant Growth Substances, (Hill, London, 1972)

Hanson, L.P., Plant Growth Regulators, (Noyes Data, London & New Jersy, 1973)

Iowa State University, Plant Growth Regulators, (Ames, Iowa, 1961)

Society of Chemical Industry, Plant Growth Regulators, SCI Monograph, No. 31, (London, 1968)

Tukey, H.B., (Ed)., Plant Regulators in Agriculture, (Chapman and Hall, London & Wiley, New York, 1954)

Weaver, R.J., Plant Growth Substances in Agriculture, (W.H. Freeman, San Franscisco, 1972)

Wellensiek, S.J., (Ed)., Symposium on Growth Regulators in Fruit Production (International Society for Agricultural Science, The Hague, 1973)

Chapter 17
APPLICATION OF PESTICIDES

Most modern pesticides are powerful and expensive. Distribution of excessive dosage might lead to undesirable biological side effects and is always economically wasteful. Maldistribution or application at the wrong time may cause crop damage or result in failure to control the pest. It is the business of the field biologist, in the light of his knowledge of the pest and according to the results of many replicated trials, to define the best time and manner of application. In this he must take into consideration the performance of available machinery and the physical properties of the pesticide.

With regard to the latter he must be advised by the formulator whose business is to put the pesticide into a form suitable for application. The formulator must consider not only the requirements as defined by the biologist but also problems of packaging, mixing, stability in transport and storage of the concentrate which is supplied by the manufacturer to be diluted by the user in the field. Both biologist and formulator must keep in mind the probability of the product being mixed with others in the spray tank so that compatibility - chemical, physical and biological - must be examined. It is evident that some compromises must be made between very different, but often conflicting, requirements.

This book is about pesticide chemicals and cannot go into detail of machinery design and operation, but, in this chapter, the types of distribution and deposit which can be obtained will be discussed. They are influenced by the application method and the properties of the pesticide and formulation, which in most cases is diluted before it is applied. In the next chapter, the formulations themselves - i.e. the concentrates as packaged and supplied; will be discussed, together with some devices which are supplied with the means of application "built in".

Some pesticides are used very locally. Seed-borne fungi are often well controlled by dressing the seed with a liquid or dust formulation of a suitable fungicide. A rat-poison is usually put down as a bait in places frequented by rats and not accessible to domestic animals. Houseflies are often successfully controlled by setting out pads attractive to this species and contaminated with insecticide. Materials of construction such as wool cloth, roof timbers or telegraph poles are often treated by a factory dipping or impregnation process to protect them from insect or fungus attack.

With increasing governmental control of the use of pesticides, aimed at reducing unwanted side-effects, localized application is likely to increase in importance. Increasing knowledge and skill can be expected to extend its usefulness. At present, however, by far the major proportion of pesticides produced is distributed more or less uniformly over much larger areas. Growing crops, harvested products and constructed buildings are treated by dusting or spraying. The necessary selectivity on growing crops is partly dependent on physiological and biochemical differences between species and partly on time of application in relation to pest attack and stage of growth. The use of a very general poison in which safety depends entirely on timing is illustrated in the chapter on fumigants.

Most pesticides are sprayed or dusted as dispersions in an inert carrier. In the case of spraying this is necessarily a liquid and, on grounds of inertness, safety and an enormous advantage in cost, this liquid is almost invariably water. In the case of dusts the inert carrier is usually a ground mineral material, with obvious preference, on grounds of transport cost, for a locally available one. Dust bases are therefore very variable.

Dust, a fine-droplet air-blast or aircraft spray can give very good cover of an extensive crop, but cannot be closely localized. It is inadvisable to use these methods for distribution of a selective herbicide in small-field agriculture as there is far too much risk of drift damage to neighbouring sensitive crops. They are suitable only for large areas or where drift would not be environmentally damaging, although it might be wasteful.

In applying soil-acting herbicides to soil before the emergence of crops grown in widely spaced rows, such as sugarbeet and cotton, band-spraying of the seed rows can be practised, leaving the inter-row weeds to the hoe. In a standing crop, such as sugar cane, the spray can be directed to avoid the crop leaves and cover the emerging weed seedlings. It is obvious that only a coarse spray can be so directed.

Dosage and Cover

Most modern pesticides are used in agriculture at a dosage of 1 kg or less per hectare. One kilogram contains 10^9 μg. One hectare (flat) contains 10^8 cm^2. If 1 kg were to be uniformly spread over 1 flat hectare, each square centimetre would therefore carry 10 μg. If the layer could be uniform and had density of 1, its thickness would be 0.10 μ (μ = 10^{-6} metre, or one thousandth of a millimetre, which is about the thickness of a spider's thread).

A well-grown crop has a leaf area of 10-20 times the ground area it stands on. The leaf surface is often very wrinkled - when looked at on the scale which is important. There are hills and valleys many microns high or deep and hairs of several microns thickness. Even 1 kg/ha, a larger dosage than is often necessary with some modern insecticides, would need to be spread to less than 0.01 μ to give complete cover.

A spherical particle of 100 μ diameter has a volume of $6.5 \times 10^{-8} cm^3$. If such particles had a density of 1, 16 of them would be found on each square centimetre if a uniform deposit of 1 kg per actual hectare of leaf surface were applied; say 1 per cm^2 in a fullgrown crop on an hectare of ground, and one on every 10 cm^2 if the dose were as low as 0.1 kg/ha. It is obvious, therefore, that to produce particles so numerous that a small flying insect pest is certain to be hit or a crawling one to pick up a deposit, the particles of pesticide must be much smaller than 100 μ in average diameter.

A 100 μ diameter particle is so much slowed down by air viscosity that it falls at a steady rate of only about 30 cm/sec. It takes only about 1/10th sec time of fall, through about 3.5 cm, to reach this speed. If the particle were projected by some microgun, it would be slowed down in a very short time and distance till it had no trace of its projection velocity left. A 50 μ diameter particle falls at only one quarter of this speed. The droplets would thus be at the mercy of the slightest wind.

The Carrier Function

With a well-translocated herbicide or systemic insecticide or fungicide it may only be necessary to obtain a grossly uniform cover - a few drops on most leaves. Many pesticides, however, require a fine pattern of particles over the leaf surface to be effective. This could be obtained with undiluted pesticide only by using particles so small that they could not be aimed at a small target, but find their way, by wind and gravity, to a large area. It will be seen from the above that one function of the water carrier, in ordinary agricultural spraying, is to provide mass and momentum so that the spray can be directed.

Momentum for aiming the spray can be supplied by air rather than water. This directing force is obtained by high-powered fans which blow the air towards the target, the spray being introduced near the outlet. This method is frequently applied to tree-spraying from the ground when a spray produced by water pressure only would not carry far enough, unless an uneconomic amount of water (which must be carried) is used. The aiming of such "air-blast" spray differs from that of a water-pressure spray, generally used for ground crops, in that the particles in the former are usually much smaller and are carried by the blast towards the target only in a gross way. Some collide with leaves but many drift into the interior of trees or forest and settle under gravity. When a dust is used, it is also distributed in an air blast, or a series of blasts from separate openings on a boom. Wind and gravity, rather than the momentum of individual particles (as in the coarse spray drops from a water-pressure ground sprayer) determine the final pattern within the crop.

Visibility

The main function of the inert dust used as a diluent is to provide visibility. A 50% active dust at 2 kg/ha could do as good a job as a 2% active dust at 50 kg/ha, but it would not be easily visible in the air and hardly visible at all on the crop. The inert diluent must be there to enable the operator to know that he and his machine are doing what is required.

Visibility is also another function of the diluent water used in spraying. Machines and operators are not infallible. Poor spraying can mean wasted chemical and lost harvest. It is therefore necessary not only to arrange that the pesticide can be dispersed and distributed but to be able to check that the intended operation is in fact being carried out. The blockage or wrong setting of a nozzle must make itself at once apparent.

It will be evident that the more reliable the machine, the more skilled the operator and the greater his faith in his machine, the less important is visibility of the carrier. Already good results have been obtained in the U.S.A. by spraying undiluted malathion. Very concentrated DDT formulations have been used successfully on cotton in the Sudan at less than 10 l/ha. Both operations are carried out by aircraft. The wide swathe obtained with very fine droplets enables the job to be done with only few dispersing units, usually rotating discs or cages. These can be more expensive than the simple nozzles of a multiple boom but their performances are more reliable.

The tendency to use this method of "ultra-low volume" spraying has increased in recent years not only for application from aircraft but also for ground spraying. It is less demanding on increasingly expensive and irreplacable tractor fuels because less inert material has to be carried and returns to base for reloading are greatly reduced. It has been made possible by great improvements in the design and technology of spraying equipment.

The Form of the Deposit

Differences in the small-scale pattern of deposit left on leaf or wall surfaces by air-borne dust or very fine spray on the one hand and by a relatively coarse water-pressure spray on the other may be important. The fine particles, having very little momentum independently of the air in which they are carried, tend to follow the air movement around obstacles. The smaller the obstacle, the less the flow of air is deviated and impact of the particle is more probable. In an air stream, therefore, thin stems, hairs and leaf edges collect a disproportionate deposit of fine particles by "dynamic impaction". Relatively little deposit is collected in this way on the central areas of large, smooth leaves. When the air movement slows down as the blast from the machine is dispersed or when the crop or forest canopy has been penetrated, deposition is mainly from slow fall under gravity. Upper leaf surfaces are thus more heavily contaminated than lower.

Coarse droplets are deposited very quickly after their release from the nozzle. They tend not to be collected on fine hairs. A spider's thread passes cleanly through a moderate sized rain drop without collecting any liquid and with very little disturbance. Since the spray transfers momentum to the air, a directed spray-jet disturbs the crop leaves and they may be hit by droplets while in an abnormal posture. Particularly if jets from a boom are set at different angles, a coarse pressure-spray can give better underleaf cover than a fine spray or dust, but the initial cover is still far from complete.

A third function of liquid carrier is to enable the spray drops to spread out when they hit the target. Two hundred litres of water sprayed in drops of mean diameter 300 μ supplies only as many particles as would 1 kg of pesticide alone in 50 μ particles. To obtain uniform cover (which is not always desirable) it is therefore necessary that the drops themselves should spread. Water will not spread at all on the leaf surface of many species, but it can always be made to do so by addition of a good wetting agent. In order that the pesticide should be carried along with the liquid, it must be dissolved or consist of particles very much smaller than 50 μ and this is necessary also to avoid excessive settling of the suspension in the spray tank.

There are three stages of discontinuity in the final deposit from a suspension of powder: (i) a coarse pattern determined by the discrete deposit of drops in low-volume spraying or, after initial complete wetting by high-volume spray, by the structure and posture of the surface - on leaves, for example, the deposit is much denser in the zone in which the liquid collects before dropping off and, on the rest of the surface, there is usually more deposit in the channels above the main veins in the upper surface; (ii) a finer pattern of aggregated particles within this and (iii) a yet finer structure within the aggregates which have been built up by flocculation of the "ultimate" particles, as they are always called, produced in the grinding or precipitation process. Even if the water spreads indefinitely on the surface sprayed, the particle aggregates are usually stranded on the surface within a limited area.

The fine structure within the particle aggregates is probably rarely of biological significance. Any water within the pores of the aggregate is fully saturated with the active compound and an insect walking over a dry deposit will detach aggregates in toto if at all rather than pick up single ultimate particles. If a powder is fine enough to be serviceable for spraying, it will be finer than necessary for maximum biological effect. It is more likely that the size of ultimate particle giving maximum kill of walking insects will be too coarse for satisfactory suspension in the spray tank.

If the active compound is dissolved in the spray water, or in an oil emulsified in the water, it arrives of course in a form which has the potentiality of much more complete spreading. It does not, however, necessarily follow that the final deposit, after evaporation of the solvents, is more evenly spread or more active than that left by a suspension of powder. The oil globules in the emulsion may not spread on a surface already wetted by water, or the oil may evaporate before the water, according to the volatility. The active compound may remain in a super-cooled state for a long time, depending on many factors. Almost always the final pattern of crystals will be much coarser than that left by a suspension of pre-existing crystals.

In the application of pesticides to crops, particularly in temperate climates, resistance to removal by rain-washing is often an important property which may be improved by special additives to the formulation, or may influence the basic choice. An emulsified oil will usually leave a more resistant deposit than a simple wettable powder, because the oil globules, if they coalesce before evaporating, dissolve some resinous constituents of the leaf cuticle which act as adhesives. If an emulsion breaks on exposure, rain-resistance is improved. In the case of wettable powders an adhesive substance, such as starch, is sometimes added. Dusts give good initial cover but poor weather resistance.

Contact action of an insecticide deposit may be reduced by the addition of adhesives to the formulation, because the adhesive is more effective in sticking the insecticide to the leaf or wall surface during drying than in helping the individual particles to stick to the insect's legs. Adhesives make little or no difference to the toxicity of the deposit to a leaf-eating insect.

Evaporation of deposited pesticides is much more important than is generally realized. The compound is distributed in very thin layers or small dots over a very extensive area exposed to wind and sunlight. Attempts have sometimes been made to cut down evaporation by the incorporation of resinous substances. Penetration into leaf or insect is, however, also retarded.

Choice of Type of Residual Deposit

Such conflict of requirements is very frequent in the study of the biological effects of formulation changes. A fungicide deposit provides another example. It is desirable that the deposit should not be washed off by the first shower of rain, but, in order that the active chemical should reach all alighting spores, it is desirable that there should be some redistribution by light rain or dew.

It has already been noted that the deposit of a crystalline substance from solution in an emulsion has a coarser pattern than that left by a suspension of an insoluble powder. The emulsion deposit is, however, generally more rain-fast. More penetration into leaf tissue may have occurred before crystallization. This may make a herbicide more active but also less selective: it may make an insecticide or fungicide more phytotoxic.

Occasionally there is a clear choice on grounds of biological activity. If an active residual deposit of insecticide is to be left on a porous wall surface, a crystalline compound is preferable to a liquid one. A suspension of wettable powder, which is filtered out on the wall surface, is preferable to an emulsion or straight oil solution, which penetrates into the capillary spaces where much of the compound remains uselessly hidden.

Some compromise must nearly always be made and the best formulation will vary with the crop or fabric to be protected and the prevailing climatic factors. The interplay of factors is too complex to resolve in any other way than by numerous replicated trials under realistic conditions. The physical chemist can best contribute by helping the biologist in the design of these experiments.

Biological preference for one formulation or another is therefore much less easy to establish than preferences based on stability, storage, packaging or economic grounds dealt with in the next chapter. These production and commercial factors at present tend to dominate the choice. With increasing interest, particularly in the high living standard countries, in more selective use of pesticides, it is to be expected that increasing attention will be paid to the part that refinements of formulation can play. Until there is a greater appreciation of the possibilities by the user, the formulator cannot play his full part.

Granule Application

Pesticides are sometimes applied in granular form to a crop, sometimes to water and, much more frequently, to soil. There is a technical advantage when a herbicide is intended for action through the soil but is applied in the presence of a standing crop. The dry granules have much less tendency to lodge in the crop. A higher proportion of the pesticide gets to the right place and there is much less risk of damage to the crop leaves. For the same reason, insecticides for control of mosquito larvae in overgrown pools or streams are frequently applied as granules, which are more effective than spray in penetrating the canopy.

There is, however, at least one crop-pest combination where the argument is reversed. The maize leaf clasps the stem, forming a funnel in which granules collect. It is just in this region that the destructive stem-borer makes its entry and this treatment is therefore very efficient. The funnel is not water-tight and a spray is not so effectively collected.

Another situation where granules have a technical advantage is for localized treatment of widely spaced plants subject to root attack by insect larvae. The female often lays her eggs close to the base of the stem and a deposit of granules on the soil around the stem gives very efficient protection. The root-fly which attacks members of the cabbage tribe provides a good example.

Very often, however, granules are used, not for a technical advantage in performance, but purely for convenience. This argument appeals particularly strongly to the small domestic user. A sprinkler tin of dry granules can be ready to hand for treatment of small areas and there is no messiness or need for tedious washing of equipment. Granular herbicides are available in this form for total weed prevention in gravel paths, along wall bottoms, etc., and also (often the same herbicide but used at much lower dosage) for selective weed control in beds of roses, soft fruit or asparagus. In this selective use, the technical advantage of avoiding leaf contamination is also gained.

Insecticide and herbicide granules are made of much smaller size than is customary for fertilizer granules. Fertilizer granules are usually in a diameter range of 1-2 mm whereas pesticide granules are usually in a diameter range of 0.4-0.7 mm.

One reason for this difference is that pesticides are applied at a much smaller rate per hectare. Most of the weight is inert granule base and is, for economy, kept to a minimum consistent with adequately uniform coverage and no excessive scatter by the wind. Fertilizers are applied at higher rates and, as far as possible, their whole weight is made directly useful. They are formed largely of soluble salts so that they disintegrate rapidly in moist soil but need special formulation to avoid stickiness. It is therefore less necessary to produce satisfactory small granules and at the same time more difficult to do so.

An important difference between fertilizers and herbicides, having significance for choice of granule size, is that plant foods are actively sought by the root system. Some herbicides are destructive to root tissue and none are conducive to root growth. Insecticides suitable for soil action are taken up passively by the root. Generally speaking, therefore, uniform distribution is more necessary for herbicides than for fertilizers. Average herbicidal granules number about 5 million to the kilogram. If 20 kg could be distributed with perfect uniformity on a hectare, they would lie about 1 cm apart. If they were of the size commonly used for fertilizers they would lie about 5 cm apart.

Insecticides are often introduced into the seed furrow to protect the roots of the seedlings from insect attack. If the insecticide is systemic some protection of the aerial portion of the seedling, particularly against aphids, can also be achieved. Granules are preferred for this use since the same type of distributor is used for granules and seed and since there is risk with liquid application of upsetting the function of the seed-drill.

Fertilizer is often required in the same situation, relative to the seed, as is insecticide and it is clearly an economic advantage to combine the two - i.e. to have the insecticide impregnated on to the fertilizer granules. Not many insecticides, however, are stable enough in contact with the strongly saline and frequently acid fertilizer. Since aldrin and dieldrin have come into disfavour because of their extreme persistence, combined formulations are now very restricted.

Smokes

A much finer dispersion of stable, low volatile substances can be achieved by condensation of hot vapour than by any process of mechanical scatter of liquid. The vapour must be diluted with air before condensing in order to give a fine-particle smoke and the process must be carried out very quickly to keep down chemical decomposition to an acceptable level. Initially the particles in a condensation "smoke" are less than 0.1 μ diameter but they grow rapidly in a concentrated smoke by collision and distillation. For a given concentration, the light-scattering effect or "obscuring power" of the smoke is at a maximum when the particles are about 1 μ diameter. As little as 25 mg of dispersed material per cubic metre gives a very dense fog.

Dispersion as a smoke can confer on some nearly involatile substances something of the penetrating power of true vapours. Penetration of a smoke into crevices, however, is much more dependent on air circulation than is that of a true vapour, since the thermal diffusion process is very slow. Pick-up by impact is much less rapid from a smoke than from a coarser fog or even a falling spray, because the small particles have too little inertia or momentum. The smoke, however, is more persistent.

Effective smokes of stable substances of not too low volatility, can be made by improvized methods such as leading an oil solution into the exhaust pipe of a tractor or aircraft. The substance is evaporated by the hot gases in the pipe and then condensed as soon as these mix with the air. Special oil-burners are made for more efficient smoke generation in apparatus which can be carried by hand and used in closed spaces, especially against glass-house pests. Entirely self-contained smoke generators are available. They are described in the next chapter. Insecticidal smokes are most used for treatment of pests in glasshouses. Some organic fungicides can also be used in this way, but the copper products are too involatile and elementary sulphur too flammable.

Further Reading

Agricultural Publishing Centre, Proc. 4th Intnl. Ag. Aviation Congress, (Wageningen, 1971)

Bradford, M., Crop Spraying Simplified, (Blackwell, Oxford, 1961)

Deutsch, A.E. & Poole, A.P., Manual of Pesticide Application Equipment, (Oregon State University, 1972)

Hough, W.S. & Mason, A.F., Spraying, Dusting and Fumigating Plants, (McMillan, New York, 1951)

Maas, W., ULV Application and Formulation Techniques, (N.V. Philips-Duphar, Amsterdam, 1971)

Potts, S.F., Concentrated Spray Equipment, (Dorland Books, Caldwell, New Jersey, 1958)

Rose, G.J., Crop Protection, (Hill, London, 1963)

Society of Chemical Industry, Spraying Techniques in Agriculture, SCI Monograph, No. 2, (London, 1958)

Chapter 18
FORMULATION

Concentrates for Dilution

The formulator aims to bring the pesticide into a form convenient to apply, either directly, as in the case of granules or special self-dispensing packs, or after dilution, usually with water. Not only must the product be convenient to use when freshly prepared but also after packaging, transport and storage. In addition to the biological problems discussed in the last chapter, chemical stability, corrosion of containers, sedimentation of suspensions and packing together of powders must therefore be studied. A few frequently occurring problems of this kind will be mentioned but the experienced formulator has long since learned that the next problem, particularly in the case of a solid product, will often be a new and unexpected one.

Concentrates for dilution in water will first be considered. In the spray as applied in the field the active ingredient may be (a) dissolved in the water, (b) dissolved in an oil emulsified in the water or (c) suspended as fine solid particles. The possibilities of (a) and (b) may be restricted by the solubility; (c) is not, of course, directly possible if the active ingredient is liquid. In these respects the properties of the active ingredient may dictate the method of formulation.

If the active ingredient is freely soluble in water, it will be sprayed in form (a) no matter whether supplied as a concentrated solution in water, or in a better solvent, or as a solid. If it is soluble in oil but not in water it can be supplied as a thick concentrated "stock" emulsion ("mayonnaise" type, as it used to be called) or as a "self-emulsifying" (sometimes called "miscible") oil. If the final form is (c), the concentrate supplied must be either a concentrated paste or a wettable powder.

Wettable Powders

Preparation of Free-flowing Powders

The grinding of coarse solids is carried out in many types of mill, ranging from those used in the grinding of grain, which depend on shearing between solid surfaces, to high-speed hammer mills and "fluid energy" mills. In the latter, the premixed, roughly ground material is passed through a quite small chamber in which air is circulated at very high velocity in turbulent flow. The particles are shattered by sudden accelerations and centrifugal forces. This process, generally called "micronizing", is capable of reducing most materials to a very fine powder and is now the most widely used. The micronizing chamber is very compact but high power is necessary to drive the air-stream and, after collection of the powder in a "cyclone", extensive air filters or further cyclones are necessary to avoid contamination. The air can be circulated, if necessary, without exchange with the outside air and, in this closed circuit, it can be replaced by nitrogen or carbon dioxide in processing compounds that are easily oxidized or perhaps form explosive mixtures in air.

Few, if any, organic compounds can be, by themselves, ground to a satisfactory fine powder which does not "cake" on storage. The stickiness of the powder is often aggravated by impurities of manufacture which it may not be possible to remove at a cost acceptable to the consumer. Heat generated in grinding, which can be locally intense, can also fuse together particles of quite high-melting compounds. The wetting and dispersing agents which it is necessary to add tend to increase the problem of stickiness. It is therefore the universal practice to mix in some inert infusible mineral diluent. The function of the mineral diluent is to buffer the organic particles apart. Sometimes a fine clay will suffice but a more absorbent material with an internal "spongy" structure may be necessary. The specially absorbent Attapulgus clays and diatomaceous earths are much used. Some fibrous minerals, e.g. meerschaum, are specially adsorbent, consisting of microscopic tubes.

In some powder formulations, the active ingredient may be one which dissolves completely at spray dilution. The object of fine grinding, which necessitates the admixture of an insoluble filler, is to get the active ingredient into a rapidly soluble form. Dimethoate, 2.5% soluble in water, is an example. At the dosage and volume rate usual in ground-spraying, the compound is wholly soluble. A coarsely crystalline preparation, even if it did not cake in storage, would, however, dissolve too slowly for convenience. Uneven dosing of the crop would result.

If the active compound is itself a mineral substance of low solubility or a rather insoluble organic salt, a filler may not be required, since there is no tendency to stickiness. Single-pack copper fungicides are an example. Copper oxychloride or basic sulphate or cuprous oxide is obtained in very fine-particle form by precipitation reactions and is milled only to break down the aggregates formed on drying. A small percentage of wetting and suspending agents makes up the rest of the "formulation". Wettable sulphur and the zinc and manganese dithiocarbamates can also be formulated in very high concentration, but these compounds introduce a different problem. Both can form explosive mixtures with air. They must be milled in an inert gas or agents must be added to inhibit the build-up of the static electricity that initiates the explosion.

Even a liquid active compound can be successfully put into the wettable powder form if a very absorbent mineral filler is used and care taken not to overload its capacity. Usually, on dispersion in water, the water displaces the organic compound from the hydrophilic mineral and the final spray liquor is really an emulsion in which the mineral powder may contribute to the stabilization.

Generally more troublesome to the formulator than an active ingredient which is always liquid is one with a melting point within the normal temperature range, because this means that change from solid to liquid frequently occurs in storage and the growing crystals can bind the powder together. It is not unknown for storage troubles to arise in wettable powders due to change of crystalline form of the organic compound even when this has a high melting point. The compound may come out of the manufacturing process in a metastable modification and slowly change over to the stable form on storage. New nuclei may be formed which are fed by diffusion in the vapour state so that large particles, often needle-shaped, may grow slowly in an initially satisfactory powder and make it quite unserviceable.

Chemical Stability

Although most compounds are more stable in the solid than the liquid state, because their molecules are held fast in the crystal lattice, the powder contains continuous air channels in which diffusion is far more rapid than in liquid. Consequently, unless the outer package is impervious, a compound sensitive to oxidation may be less stable in the wettable powder formulation. Rates of the heterogeneous reactions have a more complex dependence on time and temperature than those of reactions in solution.

Chemical stability must always be proved by storage of the wettable powder, even when the evidence for stability of the pure substance is satisfactory. Some clays contain strongly acid centres which may catalyse a decomposition reaction. Some compounds may be intolerant of an acidic and others of an alkaline diluent. One compound, e.g. diazinon, may be satisfactorily formulated with chalk as a filler: for another, e.g. dimethoate, chalk may be a highly undesirable constituent. Obviously knowledge of the reactions of the active compound, particularly of acid or base catalysed hydrolysis, can be a useful preliminary guide to the choice of mineral filler.

Physical Stability

Physical stability must always be proved by storage trials. These should include subjection of the sample to vibration and compression. Temperature fluctuation, since it gives rise to local diffusive distillation within the sample and also to partial solution and recrystallization, is particularly prone to produce caking. The tests should therefore include taking the sample from one extreme of the probable range of storage temperature to the other, as frequently as the thermal capacity and conductivity permit.

A powder settles into a container with a large proportion of included air. The actual solid itself occupies generally between about 40 and 60% of the total volume, though extremely "fluffy" solids such as light magnesium oxide may occupy as little as 20%. The "bulk density" of the wettable powder is therefore subject to variation with slight changes in composition and in handling and packing operations. The powder will frequently settle down on storage. The formulator must keep these variations as small as possible. Technically, they may appear to matter little, if the powder is measured out by weight or packed in suitable units, but variation of bulk density can be an important commercial disadvantage. The customer does not like to receive a half-full container and the packaging department does not like to find that a new formulation cannot be got into a stock of labelled containers.

Performance

It is easier to make a wettable powder which can give a good suspension only by "creaming" before dilution than one which disperses rapidly and easily when scattered on a water surface or rinsed through a filter-basket. By "creaming" is here meant the process of stirring the powder to a stiff, smooth paste with a little water before adding the bulk of dilution water. Pockets of dry powder shear more easily than the stiff paste between and are therefore reduced in size and eventually eliminated. With the very wide range of wetting agents now available, however, it is nearly always possible to make the powder dispersible quickly and completely in the full bulk of water and this should always be the aim of the formulator. A good wettable powder should disperse without agitation in a few seconds when scattered on the surface of water in a large beaker and very little should fall to the bottom without dispersing.

An insoluble powder must remain suspended for an adequate time. Obviously it cannot remain in suspension indefinitely since the filler at least, and usually the active compound, is heavier than water. Suspensibility is usually tested by a W.H.O. (World Health Organization) standard method. The suspension is held at rest in a 250 cm^3 measuring cylinder for half an hour and the upper 90% of the liquid drawn off and analyzed. The ratio of active compound found to that which would be present in 90% of the volume initially is defined as the suspensibility and should exceed 80%. This is not a very severe test. If the top 90% has an average concentration of 80% of the initial, the bottom 10% has a concentration 2.8 times the initial. This variation may be acceptable for an insecticide or fungicide, but, with a herbicide, overdosing with the first offtake from a spray load could lead to crop damage. For the operator, time as well as chemical costs money, and he has therefore a good reason to spray out his load quickly, but he should advisedly have some means of agitating the contents of his tank on long runs.

The powder may settle too quickly for one or both of two reasons. The "ultimate" particles may not be small enough. More often, they may loosely aggregate together - "flocculate" - after dispersion. The first fault, unless the particles have grown during storage, can be cured by finer grinding. The second requires alteration of the amount or composition of the suspending agents included in the formulation. The agents included to make the powder rapidly wettable are not usually adequate to protect the suspension. Suspending agents are usually of much higher molecular weight than the wetting agents. They include lignin sulphonates from waste-liquors of wood-pulp refining, sulphonated naphthalene formaldehyde condensation products, proteins and gums. Selection of the best compound can be made only by trial.

Although prolonged storage of the diluted suspension is not usually necessary, it may sometimes be unavoidable if spraying is interrupted by adverse weather or mechanical faults. It is therefore desirable that the settled powder should be easily resuspended by agitation. To meet this requirement some compromise may be necessary because the cake, which eventually forms at the bottom of a tank on long standing of a very good - not flocculated - suspension, is usually much more difficult to disperse than a more quickly settled flocculated precipitate.

Packaging

For many uses, wettable powder formulations are preferred at the merchanting and user level because they are considered less subject to loss during transport and handling, and can be packed in cheaper containers. The industrial chemist, experienced in the manipulation of liquids, regards the liquid state as by far the most convenient for measuring, piping and mixing. The preference for powder formulations may seem illogical to him but it has to be accepted as the purchaser's choice.

An advantage often claimed for the powder is that it can be packed in cheaper containers, from which it is not all lost if these are broken in transit, but it must be remembered that a poorly packed powder may be vulnerable to oxygen and water vapour permeating this container inwards and to evaporative loss occurring through it outwards. Perhaps the commonest of all mistakes made in this subject is to believe that a solid is involatile. Compare a little naphthalene wrapped in paper, lubricating oil in an open dish and some ether in a well-stoppered bottle. What the solid does do is to stop the contained air from moving. Evaporation from the interior of the powder must take place by diffusion. Consequently a large package of powder loses its active ingredient much more slowly than a small one.

Emulsifiable Oils

Composition

Many water-insoluble pesticides are supplied to the user in the form of an emulsifiable concentrate. This is a solution of the pesticide, together with oil-soluble emulsifiers, in a solvent which is mainly a hydrocarbon oil. The formulation should disperse as a fine emulsion when simply poured into water, but gentle agitation is, of course, necessary to ensure that the emulsion is uniformly distributed throughout the contents of the spray tank.

The formulation of emulsifiable oils has been greatly facilitated by the commercial development over the last 20 years of non-ionic emulsifying agents in which the hydrophilic portion of the molecule consists of a polyethylene oxide chain. A typical example is

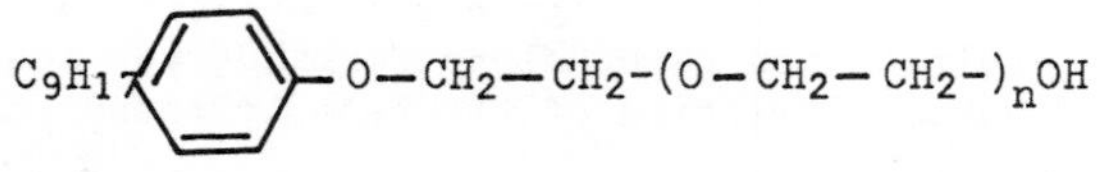

where _n_ is distributed rather widely about a mean value which, for emulsification of oil in water, is around 12. An advantage of these compounds over the ionic types is their much greater oil solubility. To get an effective concentration of soap or sodium high alcohol sulphonate into the oil, it is usually necessary to include a more polar solvent and perhaps even a little water. Chemical instability may result.

Another advantage of the ethylene oxide type emulsifiers is that their behaviour is little influenced by saline impurities which vary in amount and composition in water from different sources. The hydrophilic property is the combined effect of a large number of moderately hydrophilic ether groups which form hydrogen bonds with the water. The hydrophilic property of the ionic agents is usually concentrated in a single ionized group which is much influenced by other ions, particularly multivalent ions of opposite charge.

There is now available a very wide range of non-ionic emulsifiers having various bridging groups between the paraffinic and hydrophilic groups. Some have both an ethylene oxide chain and an ionic group in the same molecule. In the example illustrated, a branched nonyl alcohol of petroleum origin is condensed with phenol, substituting the ring mainly in the 4 position, and ethylene oxide is then condensed with the phenolic OH group. Ethylene oxide can also be condensed, with rather more difficulty in control of the reaction, with the OH groups of a primary fatty alcohol or of a polyhydric alcohol already partly esterified with a fatty acid. It may also be condensed with a carboxylic acid, an amide or an amine. Compounds in which a carboxylic ester group forms the bridge are rather sensitive to alkaline hydrolysis.

Bearing in mind that the number of ethylene oxide molecules condensed in the chain can be increased almost indefinitely, it is evident that there are very numerous possible chemical structures. Also very important are the "block" co-polymers of ethylene and propylene oxides, in which considerable lengths of the chain consist of one kind of unit only. The extra methyl groups in the portions of the chain made up of propylene oxide make these regions predominantly hydrophobic.

No example seems to be known where the use of only a single, pure emulsifying agent can give a satisfactory self-emulsifying oil. A mixture of rather different emulsifiers seems to be necessary. Even a mixture of similar non-ionic compounds, but with two widely different mean numbers of ethylene oxide molecules in the chain, can be successful when neither alone would be. More commonly a mixture of non-ionic and anionic emulsifiers is used, around, say 4% of the former and 1% of the latter, choosing for the anionic constituent one with a branched hydrophobic chain, which generally forms more soluble salts.

Generally speaking, the more hydrophobic the emulsifying agent, the greater its tendency to produce water-in-oil rather than oil-in-water emulsions. The best system for producing a self-emulsifying oil is usually found to be near the limit for good oil-in-water emulsifiability. Some authorities consider that a water-in-oil emulsion is first produced when the two liquids meet and is then inverted on further dilution. Frequently, the larger globules in a dilute emulsion prepared by pouring a self-emulsifying oil into water are found to be multiple - i.e. finer globules of water inside a larger globule of oil.

Performance

The detailed mechnism by which an emulsifiable oil breaks up into globules in the 1-10 μ range on simply pouring into water is by no means clear but it is certain that the process is not strictly spontaneous. If water and an emulsifiable oil are carefully layered, the lighter over the heavier, and left at rest for a long time, there will be found only a narrow zone of emulsion and the rest will not now emulsify on gentle shaking. It seems therefore that the agitation caused by pouring the oil into the water is necessary for the process. Deliberate extra agitation, however, short of the high velocity agitation of an emulsifying machine, makes little difference to the degree of dispersion.

The fat globules in milk rise to form a cream layer but, in this layer, the fat is still dispersed and can be stirred back into the skim milk. This "creaming" is very different from the "breaking" of an emulsion, when the globules coalesce to give a coherent separate phase, but breaking is accelerated by creaming since the globules are forced into contact by gravity. The larger the globules and the more their density differs from that of water, the more rapidly will the emulsion "cream". In many pesticide emulsions the oil phase is heavier than the water and the "cream" concentrates downwards. Since the active compound is usually heavier than water and the solvent lighter, it is sometimes possible, but may not be economic, to adjust the density of the solution to that of water. Otherwise creaming can be prevented only by agitation, but, in the dilute emulsion formed by a good emulsifiable oil, the globules are usually fine enough to remain well suspended for an adequate time.

There is sometimes a good case, on biological performance grounds, for the emulsion to be an unstable one that breaks easily, since the oil layer is then held preferentially by the plant surface. There is no point in using an over-coarse density-adjusted oil phase for this purpose, since the spray nozzle itself breaks up the oil phase further. Instability in the spray tank is undesirable since the separated oil phase of a broken emulsion cannot, like the cream, be easily resuspended. If instability is required, it should develop after the spray has been exposed to air. Instability is achieved in old-type tar oil winter washes for fruit trees by using alkaline soap as the emulsifier. The soap is reduced in effectiveness by uptake of atmospheric carbon dioxide after exposure.

Dilute emulsions, formed easily from a self-emulsifying oil, are not of very high stability. They do not remain dispersed as long as do emulsions produced in an emulsifying machine with careful selection of the stabilizers, particularly when these include macromolecular agents such as casein Generally, in fact, good emulsion stabilizers are not facile emulsifiers. A stable emulsion can often be made unstable by adding a good emulsifier. An emulsifiable oil which gives satisfactory emulsion at, say, 1 in 100 dilution will often be much less stable at much lower dilution, say 1 in 10. The self-emulsifying oil can, however, give emulsions of adequate stability for spraying purposes where it is not necessary to keep the diluted liquor for long periods.

Solubility Limitations

The emulsifiable oil is much more restricted than the wettable powder in the number of compounds for which it is suitable. Many pesticides are of too low a solubility to be supplied economically in this form. Not only is the cost for solvent, packaging and transport too high, but the excessive oil often gives rise to crop damage when used in agriculture. The choice of oil is necessarily restricted because it must have a very low water solubility; otherwise the oil itself will be dissolved on dilution and the pesticide, usually less soluble in water and now with no separate phase to dissolve it, becomes supersaturated and crystallizes out in an uncontrolled manner. It is unwise to risk the use of a supersaturated spray liquor. It may give rise to no trouble in two or three successive spray loads but further loads can crystallize quickly once nuclei have been formed. Under-dosing, poor control and nozzle blockage can result.

The limitation of choice to oils of very low water solubility is more restricting than is often realized. Many modern pesticides have moderately polar molecules for which rather polar solvents are optimal. Ketones are usually much better solvents than hydrocarbons, but acetone is completely miscible with water. Methylethyl ketone and cyclohexanone are around 10% soluble in water. Chloroform is often an exceptionally good solvent and is only about 0.5% soluble in water but it is expensive and frowned upon by medical authorities. The choice is in fact limited to crude aromatic hydrocarbon mixtures from coal-tar or petroleum sources. Methyl naphthalenes in the 230-250°C b.p. range are much used in formulations of the organochlorine insecticides.

It may be permissible to boost the solubility to some extent by the addition of more polar, and therefore more water-soluble, solvents without incurring risk of crystallization in the spray tank. Some relaxation is general and arises from the fact that the marketed concentrate must remain a solution at the lowest temperature likely. Spraying, on the other hand, is usually carried out well above this temperature. It may therefore be permissible to add a proportion of ketone to an aromatic solvent to keep the active compound in solution at -5°C, provided it would remain in solution at 10°C if the ketone were removed. Further relaxation is permissible where the active compound is soluble in water at the spray dilution. An example is provided by a dimethoate concentrate in cyclohexanone. For the recommended dilution, emulsification need only be transient, to hasten the process of solution, all constituents being dissolved in the water when this dilution is reached.

Stock Emulsions

Emulsions are sometimes marketed as such - in concentrated cream form - to be diluted in the field. A cheaper product can be made this way because less emulsifier is needed. With the rapid advance in recent years in the availability of good emulsifiers for the oil concentrate, the stock emulsion has receded in importance. It is less satisfactory in storage, more demanding in container choice, because generally more corrosive, and is in any case restricted to hydrolytically stable compounds. Probably, however, the strongest reason for manufacturers' preference for the oil concentrate is the greater certainty with which shelf life can be tested and assured. The product is a homogeneous solution in which only chemical changes are possible and these follow a simple pattern. "Accelerated" (high-temperature) storage tests can give a reliable indication. In the stock emulsion, on the other hand, much more complex and less predictable physical changes may also occur.

Solution Concentrates

The simplest of all formulations is the solution concentrate, a homogeneous water-miscible liquid merely requiring dilution in the spray tank. An aqueous solution will be used where possible, but it is sometimes necessary to add a better, but water-miscible, solvent in order to reach an economically acceptable concentration of the active compound. For this purpose, ether-alcohols, such as butyl cellosolve (butoxyethanol) or diethylene glycol (2,2'-dihydroxyethylether), are preferred to methanol or acetone as they have lower flash-points. It is important, where possible, that the concentrate should not incur the higher freight-charges imposed on flammable products.

The number of pesticides that can be formulated in this simple way is limited by two factors - solubility and hydrolytic stability. With the exception of the herbicide amitrol (and the insecticides octamethylpyrophosphoramide and dimefox which now have only very limited use), the solubility factor limits these formulations to salts of active acids or, in the case of diquat and paraquat, of active bases. The active acids, however, include 2,4-D, MCPA and other "hormone" herbicides which are one of the major classes of herbicides in terms of world tonnage used.

The inactive ion of the salt may have to be chosen for solubility as well as cheapness. In the case of MCPA an aqueous concentrate can be made containing up to 250 g/l as the sodium or potassium salts (or preferably a mixture), but, in the case of 2,4-D, only 2% can be achieved and an amine salt must be used. The ammonium salt itself is not soluble enough and dimethylamine or ethanolamines are mainly used as the neutralizing bases. Mixed bases give a higher solubility than single bases. Solubility advantage can therefore be secured along with economic advantage by the use of unfractionated bases. Since high solubility in the residual deposit assists leaf penetration, there is no conflict in this case between biological and packaging requirements.

In terms of inanimate chemistry all these compounds are very stable and there is no problem on this account in the storage of aqueous concentrates. The phenoxyaliphatic acids have all a rather limited life in fertile soil. There is generally no significant residue left in soil after 3 or 4 weeks in the summer. TBA, dicamba and picloram are much more persistent but there is no dangerous residue, of the first two at least, 12 months after application of the dosage used in selective weed control. The decomposition in soil is entirely due to microbiological action.

A problem peculiar to the aqueous concentrate is the precipitation of tarry matter when the concentrate is diluted. Pesticides must be cheaply produced. An unnecessarily high degree of purification adds to the cost. Despite the saline nature of most concentrates, their organic content usually increases the solvent power of the product for other organic substances. It sometimes happens that some resinous by-product of manufacture is held in solution by the concentrate and not removed by filtration. On dilution, this by-product can come out of solution and lead to filter- or nozzle-blockage. Tests for filterability after dilution should always be carried out in the formulation laboratory. The addition of a suspending agent or "protective colloid" is often adequate to avoid the tar problem, but further purification during manufacture of the pesticide itself may be necessary.

An aqueous concentrate need not contain surfactants. The diluted emulsifiable oil or wettable powder has always at least moderate wetting properties due to the content of emulsifiers and wetting agents necessary for suspension. The diluted aqueous concentrate may have good or poor wetting powder according to whether surfactants are added specially for this purpose. It is possible, therefore, to meet a wider range of biological specifications. In the formulation of dinoseb for spraying the pea crop, for example, minimal wetting is desirable since undamaged pea leaves then reflect the spray and reduce risk of chemical injury to the crop.

At the manufacturer's end, water-based concentrates have limitations which are clear and easily established. At the user's end, they would seem to be the simplest of all formulations, but their very apparent simplicity has often led to trouble. The user expects to have to cream a powder or otherwise coax it through a filter basket. He expects to have to stir an emulsifiable oil. Because the aqueous concentrate is water-miscible, he is inclined to hope that it will mix itself. It is only miscible in the sense that there is no phase boundary between the concentrate and the water. If a heavy concentrate is just poured into a tank full of water it will sink, only partly diluted, to the bottom. Trundling the machine over a rough farm track is surprisingly ineffective as a means of gross mixing of the contents of a full tank. A heavy over-dosing of the first part of a field and inadequate dosing of the rest has often resulted. The concentrate should always be poured into a half-full tank and the rest of the water then added quickly. The tank agitator, if there is one, should be used. If there is no agitator, a minute's work with a stick is a small price to pay for a good harvest and a cheap insurance against serious damage and ineffective weed control.

Corrosion

Corrosion of containers must always be considered in the choice of formulations and is generally a greater risk with aqueous concentrates than with anhydrous formulations. Even in this respect the aqueous concentrates of salts of "hormone" acids are satisfactory. They are generally made slightly alkaline and do not attack mild steel when the container is kept closed. A combination that must be avoided is excess ammonia and zinc-containing metals. Galvanized containers should always be avoided for pesticide storage, and hydrogen evolution is exceptionally rapid when aqueous ammonia is kept in them. Even the diluted product will attack brass filter gauze very rapidly.

It is worth mentioning that metal pesticide containers are very often corroded from the <u>outside</u>. They should obviously be stored away from foodstuffs and this often means that they are kept on a damp floor, which may well have had fertilizer spilt on it. Even the acids from organic manures can be very aggressive. The containers should always stand on a clean, dry floor.

Dusts

The mineral bulk of a dust should be composed of rather coarser particles than are necessary in a wettable powder. Ideally they should also be more uniform in size. Too fine grinding or too wide a range of particle size results in a "claggy" product which does not flow well through the regulating gate of the dusting machine. When an over-fine powder is dispersed in an air-blast, the wind-borne particles are mainly aggregates of the ultimate ones and often larger than are produced from a more free-flowing, coarser dust.

Ideally the smaller particles of the active pesticides should adhere to the coarser particles of the dust base so that the operator will know that, wherever the dust is visible, there will also be pesticide. This is not, however, as important in practice as might be expected; there is not much tendency, if all particles are below about 20 μ diameter, for significant "winnowing" according to particle size to occur. In perfectly still air, of course, differential settling would occur, but the free rate of fall is less than 2 cm/sec and, except under extreme inversion conditions, the turbulence of the air frequently lifts all particles up or brings them all down much more rapidly. Any small air volume does not deposit its contents until it has become very stagnant close to the ground or within the crop and then all particles are deposited.

The ideal dust is certainly not often produced. A dust must be much more dilute than the wettable powder because dusting machinery cannot generally discharge evenly less than about 10 or 20 kg to the hectare. Two to five per cent of active ingredient is usual. High transport costs dictate that some locally available mineral base be used and, as dust is sold in a cheap market, the cost of this base is often more important than its quality. Both the final formulation of the dust and the means of applying it are subject to much improvisation. Many a peasant farmer has dusted his crop with a mixture of wettable powder and road dust mixed on the floor and shaken out through his wife's discarded - or not discarded - stocking.

There are, however, many efficient dusting machines available and a large area can be quickly and cheaply treated. A very common practice is for local formulating companies to blend a wettable powder with some local inert dust. The wetting agents present in the wettable powder serve no useful purpose in this case but they are not detrimental and there is convenience and economy in using the one product of the pesticide manufacturer for both final products of the local formulator.

Aerosol and other Automatic Dispensers

A very fine spray can be produced by discharging through a small orifice a liquid which is contained in a pressure vessel above its boiling point. The issuing liquid is shattered more completely than in a normal spray nozzle operated under pressure, by a combination of three factors: (i) a small discharge orifice and thin issuing stream, (ii) a liquid of very low viscosity and (iii) the actual boiling of the liquid immediately the jet enters the air.

The method is expensive. Only small packages are feasible as they must withstand high pressure in storage. Moreover the need for an issuing stream of small cross-section limits their use to small spaces or areas. The jet is under control of a finger-button. The packages are non-returnable. They are suitable only for use in domestic premises, aircraft cabins, amateur glasshouses and small gardens where their great convenience and "instant" availability outweigh their high cost. Usually the cost of the active compound is only a small fraction of that of the total package.

The droplets are in the 10-50 μ diameter range, which can be obtained in large-scale operation only by an air-blast machine or a very high-speed rotating distributor. This range is particularly effective against flying insects. It can also yield a good deposit after penetrating fairly dense foliage, but this is controllable only in still air conditions.

The aerosol insecticide dispenser has followed the development of similar packages for other uses - dispensers of disinfectants, deodorants, hair-lacquers, cleaning and polishing products. The manufacture and filling of containers is, in fact, normally carried out in factories specializing in these packages. The self-contained automatic dispenser can be adapted, as with the other above-mentioned products, to discharge other forms than sprays. A common alternative is a foam. A lower pressure and larger orifice is used and the liquid product is made much more viscous and has foam stabilizing agents incorporated. The dissolved - or emulsified - propellent expands when the pressure is released, giving a very visible and voluminous foam.

Foam products are made for dispensing cleaning materials and shaving soap. The method has been applied to herbicides for small garden use. A "spot" treatment of herbicide can be quickly given to rosette weeds in lawns or, with more care, to troublesome perennials such as bindweed in herbaceous borders. The advantage of the foam is that it "bulks out" the very low dosage necessary of a high concentration product so that the deposit is clearly visible and the operator knows what weeds he has already dealt with. It is desirable in this case that the foam should collapse after a few minutes in order to improve contact and minimize blowing in wind.

Because automatic pressurized dispensers must have a very small orifice in a nozzle system which cannot be dismantled, the fillings must be scrupulously free from suspended impurities. This necessitates a special standard of stability in storage and of freedom from corrosion. Often a specially purified active ingredient may be necessary.

Fluorochloro carbons such as difluorodichloromethane (b.p. $-28^{\circ}C$) and 1,1- and 1,2-dichlorotetrafluoroethanes (b.p. $-2^{\circ}C$ and $4^{\circ}C$), made in the U.K. by I.C.I., are the most common propellants for domestic dispensers, having the advantages of only slight odour, low toxicity and non-flammability. They are also better solvents than the flammable butane, but the latter is used in foam formulations and, with addition of a better solvent, in aerosol formulations for outdoor application. With advance of container technology and development of protective coatings, propellents such as methyl chloride, potentially more corrosive, but cheaper, are becoming more widely used.

For packages discharging a foam, carbon dioxide dissolved in a water-based liquid is sometimes used as an alternative to the more general butane emulsions.

Smoke Generators

Smokes may be conveniently produced from specially fabricated smoke generators. These are available in various sizes. They are set into operation by ignition and are provided with a delay composition so that the house can be evacuated before the smoke composition itself comes into action. Generators of this type were first developed during the Second World War for military remote signalling purposes for which it was desired to have smoke plumes of different colours. It was found possible to make such smokes from quite complex organic dyestuffs. Indeed, rather unstable compounds can, surprisingly, survive treatment in a well-designed generator. Even the highly inflammable dinitro-ortho-cresol can be produced in smoke form, but is not (on grounds of toxic hazard) in commercial use.

In the self-contained smoke generator the compound to be vaporized and condensed is mixed with an oxidant and combustible material capable of generating a large amount of hot gas of non-oxidizing balance. Sodium chlorate and a solid carbohydrate is the most common mixture, the hot gas consisting of water vapour and carbon dioxide with a small proportion of carbon monoxide. Thus a mixture approximating to $2C_6H_{10}O_5 + 7NaClO_3$ giving $10H_2O + 9CO_2 + 3CO + 7NaCl$ is preferable to the mixture $2C_6H_{10}O_5 + 10NaClO_3$ giving $10H_2O + 12CO_2 + 3O_2 + 10NaCl$, but excessive CO can lead to ignition of the escaping gas.

The insecticide is protected from destructive oxidation (i) by the fact that sugars are exceptionally reactive with chlorate, (ii) by the use of a slight deficiency, as above, in the amount of chlorate necessary for complete oxidation of the carbohydrate and (iii), if necessary, by the intimate mixture of chlorate and carbohydrate being formed into a coarser powder which is then mixed with the insecticide.

The design and filling of such smoke generators has to be carefully worked out by experience and accurately controlled. To avoid explosion the discharge orifice must not be too small. Rate of burning and danger of ignition of the issuing gas on contact with the air can be controlled by the degree of compression of the filling and incorporation in it of retarding agents such as ammonium chloride.

Dilution of the vapour with air before condensation, so that a fine smoke is produced, is achieved automatically by the turbulence set up by the high velocity of the jet of hot gas.

Azobenzene for mite control, lindane and parathion for general insect control, DDT, several of the bridged diphenyl acaricides, captan and karathane fungicides and various proprietary mixtures are obtainable in smoke generators. Even pyrethrins can be put into smoke form without undue loss.

Granules

Granules can be made by several methods. Most dusts will aggregate to coarser particles in a fairly uniform way if tumbled in an oblique rotating drum - a concrete mixer, or the more refined pill-coating machine - while a liquid, preferably containing an adhesive substance, is sprayed in at intervals. The process is more reliable if the initial mixture contains a proportion of coarser particles to act as nuclei. Pastes, if rubbed through a sieve before drying, especially if they contain a self-setting component such as plaster of Paris, are easily broken down into fairly uniform granules after drying.

These methods, which require experience and "know how" for success, are well adapted to the manufacture of granules in which the active pesticide is incorporated at the start and present throughout the bulk of the finished granule. Most pesticide granules, however, are made by impregnation of dissolved pesticide into preformed, porous, inert granules or by sticking the powdered pesticide on to the outside of preformed inert granules, which, in this case, need not be porous. The last method is the most common and requires least research to bring into operation with a new pesticide. Suitable base material is sieved ground mineral matter - usually limestone or brick waste. Sieved crushed nutshell or corn-cob chips are also much used. Walnut shell base is already industrially available in the right size range as an abrasive for air-blast cleaning of internal combustion engines. The coating is accomplished by tumbling the base granules in a rotor, together with a suitable admixture of powdered pesticide and mineral dust, during addition of a water solution of a gum, a soluble cellulose derivative, high molecular weight polyethylene oxide or other adhesive.

There has recently been some production of "microgranules" with particle sizes in the 100-300 μm range. These can almost be considered as a coarse dust and they are usually distributed with air-flow assistance in a special machine, although the product is much more free-flowing than most dusts. Absence of very much smaller particles is the main reason for "clagginess" and bridging of orifices with ordinary dust formulations and this fine component must be eliminated from microgranules since the advantage claimed is reduction of drift hazard.

In this respect microgranules are a parallel and alternative development to controlled drop application (CDA) made possible by the development of spinning disc applicators. In both developments the object is to put the maximum proportion of the applied active ingredient in the most effective particle size. A much narrower particle-size spectrum can be obtained than with hydraulic spray nozzles and there is therefore greatly reduced wastage in unnecessarily large drops in order to avoid drift of undesirably small ones. Both developments are now being tested under different conditions against different targets and some results appear to be conflicting but there is little doubt that both will find a useful place in the near future.

The dynamics of impaction and spread and the strength of adhesion will be very different between liquid and solid particles. The spherical residues of evaporation of spray drops with a heavy load of wettable powder have very little adhesion and form a very inefficient means of contaminating foliage. Microgranules must be flat or formulated so as to become sticky.

According to the method of manufacture and the solubility of the pesticide there is scope for variation of the rate of release of the pesticide from the granule. Obviously, if the pesticide has very low solubility in water and is contained within a very close-packed mineral matrix, it may take a long time to diffuse out, even if the granules are shaken in free water. Equally obviously, if the pesticide has a high solubility and forms only a thin envelope around the granule base, it will disperse in free water almost immediately. In moist soil, however, on to which most granules are distributed, the difference in action may be very much less. In one sense the pesticide in the second case is "available" immediately to any organism touching the granule. If, however, it is necessary that the pesticide should diffuse away from the location of the granule, this process is so much slower than that occurring within normal granules that the latter has little influence. If the formulator is to exploit logically the potential variation he can introduce, he must be well guided by the biologist about where, and how well distributed, the pesticide should be.

Delayed release has been successfully exploited to secure prolonged treatment of flowing water. Here the external resistance is low and mixing efficient so that very large "granules" can be used, which simplifies the technical problem. A more extreme form is the thick plastic strip for slow release of a volatile insecticide (usually dichlorvos) into the air of a room or store.

Generally any homogeneous body containing pesticide, filler and thermoplastic material will release its volatile or soluble content at a rate decreasing with exposure time. If the resistance is provided mainly by a surface layer or skin this tendency is reduced. Various other devices are being explored to make release rate more uniform including the chemical binding of some pesticides to functional groups in polymer chains. Release then depends on breakage of the bonds, usually by hydrolysis, but, of course, this method is only possibly with some pesticides. Controlled release has achieved much more success and technical refinement in the pharmaceutical field than in agriculture for two very powerful reasons - a much higher cost per unit is acceptable and the product carries out its function in an aqueous environment at constant temperature and pH.

Baits

Baits may consist of palatable granules widely scattered in high dosage as, for example, in the control of immature "hopper" locusts, or used in a much more localized manner. Local baits may incorporate attractant substances or be shaped or coloured in a manner found to be attractive to the pest. In such applications the behaviour of each pest must be exploited. It is not possible to deal with this subject in a general way from the point of view of formulation. It is hoped that enough has been said to illustrate the forms in which the pesticide itself can be used and the problems likely to be encountered in meeting the particular requirements of the biologist - or rather of the pest as interpreted by the biologist. As always, storage tests for continued efficacy are an important feature of the formulator's work. The more complex the formulation and the more it must meet specific requirements not fully understood, the more must reliance be placed on actual performance against the pest in question, after realistic periods of storage under practical conditions.

Use Problems

The manufacturer cannot anticipate all the problems which the user of his product may meet. He can help the user avoid some mistakes by adequate and clear instructions. There is need for more general instruction in some basic rules of use of pesticides. It is well known that instructions are often not read. The formulator should try to make his product serviceable even if not used quite in the way intended, but in doing so he may have to make it fall short of the best performance if used correctly. Formulations are often blamed for faults which lie in inadequate maintenance of distributing machinery.

The commonest faults of this kind are:

1 Excessive frothing in the spray tank. It is difficult to use wetting agents, which may be essential for good biological effect, without to

some extent stabilizing foam, but, to create the foam, air must have been introduced into the liquid. Only the wettable powder can itself introduce air. The commonest way in which unwanted air gets into the tank contents is through a leak on the suction side of a pump used for return-to-tank agitation. It is more profitable to tighten a gland than to blame the formulation.

2 Filter or nozzle blockage. A bad formulation can cause this trouble. More often it is caused by a dirty water supply or a machine brought out of haphazard winter storage and not cleaned out. Water may be fit to drink but not fit to spray. The clear waters of the trout-stream - which does not often flow through agricultural land - may contain insect parts, ribbon diatoms and seed hairs, which species of poplar scatter in abundance in the spraying season. The proportion is not high, but it all builds up on the filter.

3 Inadequate mixing of concentrate and dilution water. Mixing of emulsifiable concentrates or wettable powders is often done with needless waste of energy, but the stirring of a solution concentrate is often neglected. To invert the tank of a knapsack sprayer two or three times is quite adequate, but it is quite essential.

4 Inadequate washing out. Corrosion products and dried formulations can produce filter-blocking scale in pipework. The time to wash out a sprayer is immediately after use, particularly if a formulation of insoluble pesticide has been used. One should not blame the next formulation used if this elementary precaution is not taken.

Further Reading

Society of Chemical Industry, Formulation of Pesticides, SCI Monograph, No. 21, (London, 1966)

Society of Chemical Industry, Physicochemical and Biophysical Factors affecting the Activity of Pesticides, SCI Monograph No. 29, (London, 1968)

Tahori, A.S., (Ed)., Herbicides and Fungicides: Formulation Chemistry, (Gordon and Breach, New York, 1972)

Van Valkenburg, W., Pesticide Formulations, (Marcel Dekker, New York, 1973)

Chapter 19
FUMIGANTS

Fumigation

In practical pest control the word "fumigation" is restricted to operations where a special method of application is required or permitted because of the high volatility of the toxicant employed. There are numerous special situations in which fumigation is used, e.g. hydrogen cyanide against rabbits, moles and wasps, but the major operations are four: (1) the treatment of empty transport containers, grain stores, warehouses, glass-houses, etc., to clean up residual pest populations between storage or cropping uses; (2) the treatment of grain or other harvested products prior to or during storage; (3) the treatment of soil to destroy pests, particularly nematodes difficult to control by other methods, and (4) disinfection of animal skins, etc. by quarantine authorities.

These different situations call for different properties in the chemical. The first calls for no selectivity. Ideally the building should be sterilized completely, but it is important that there should be, after an acceptable period of ventilation, no corrosive, toxic or tainting residue. The third might be highly demanding in selectivity if, for instance, it is desired to kill nematodes in the soil under a growing crop. Only to a very limited extent, with dibromochloropropane in citrus, has this yet been found possible. For the most part soil fumigation is done in the absence of a crop. Complete sterilization (including the killing of weed seeds) is therefore sought, but the toxic effect must, of course, be of limited duration.

A very important area nowadays for fumigation is aircraft. Cargo and passenger planes move continually from one part of the world to another in very short times and offer opportunities on a scale never approached in the past for transport of viable insect pests from one region to another. "**Disinsection**" of aircraft is now standard practice.

In practice, therefore, the first and third situations call for similar biological effects. In respect of physicochemical properties, however, they are very different. All parts of the empty warehouse are freely accessible to the main air space and distribution is assisted by stirring fans. The air space of the soil, however, is in close contact with an enormous area of adsorbing and chemically active surfaces: agitation is generally impracticable and molecular diffusion must be relied on for distribution of the toxicant.

The second situation is the one where choice of chemical is most restricted by the demand for no residual toxicity, no detectable residual taint and no effect on viability of seed. Physically, this situation is intermediate between the other two. A large surface is involved but it is smaller than in the soil case and there is no free water. Diffusion is often relied on but it is practicable in suitably designed storage bins to use forced air circulation. In fact numerous such bins are in use for the drying of grain freshly threshed from the standing crop.

All fumigation operations depend for their success on rapid diffusion or mixing of the toxic vapour into all parts of the air-space to be treated. It is a necessary corollary that the space to be treated must be sealed off from the outside air. Advantage is taken of sites which are naturally nearly closed, such as rabbit burrows and wasps' nests and of the fact that grain stores, being designed to prevent ingress of birds, mice, etc., are also closed habitats. Many buildings which are desirably disinfested by fumigation are, however, much less easily sealed and it should be mentioned that the use of fumigants has in recent years been greatly helped by another branch of the chemical industry - the plastics industry. Field clamps of stored vegetables, piled sacks of grain in arid regions, houses and even food factories have been completely covered by polyethylene or polyvinylchloride (PVC) sheeting to permit successful fumigation. Sheets laid on the soil surface, held down by a ploughed slice of soil around the edge, make practicable the use of methyl bromide sterilization. This compound diffuses very rapidly in soil and uneconomic dosage would be required if free escape were permitted.

Since most fumigants are gases at room temperature or liquids which can rapidly build up a high pressure of heavy vapour there is considerable possibility of layering of concentration. With care and experience this tendency can be relied on to assist mixing, since deep but localized penetration can be allowed to occur by convection before forced mixing is commenced. Even in soil, when using methyl bromide, it is found that deeper penetration occurs when the application is made in spots than when made in lines. Least penetration occurs from a surface application where there is rapid initial spread under the sheet. The extent of convective penetration increases with the vapour density. It is much greater with methyl bromide than with the much less volatile ethylene dibromide. It does not occur with hydrogen cyanide (mol wt 27) nor significantly with formaldehyde (mol wt 30) as even the pure gases differ very little in density from air. The former is the only lighter-than-air fumigant in use.

The fumigation of stored products, although having to meet more exacting standards, can be carried out by more sophisticated methods than fumigation of soil. The high value of the products allows more expenditure and their generally dry nature simplifies handling and permits less stable products to be used than in soil.

Methods of filling large plastic containers have been evolved in which the product is completely sealed, containing only its interstitial air which is rather less than half the total volume. If no external air can gain access, there is only enough oxygen to oxidize about 0.4% of the carbohydrate of the grain to carbon dioxide. It is evident that pest insects will be completely asphyxiated before much of the grain is attacked. There may be problems of taint, varying according to the product stored, and seed grain slowly loses its viability under anoxic storage, but the method is cheap once the plastic containers are obtained.

Much grain is treated with insecticide before it is put into storage. Forced air fumigation in the drying towers is one method. Spraying on a conveyor belt with an insecticide of acceptably low toxicity and not necessarily volatile - pyrethrum or malathion - is another. Two devices can improve the efficiency of transient fumigation treatment and, since they reduce the amount of toxic substance used, they help to achieve an acceptably low residue. The first is the introduction of carbon dioxide, up to about 5% of the air content, along with the toxic vapour. The increased efficiency is probably due to increased respiration. The second is the reduction of pressure of the whole air content. If a completely filled container is used, there is no need for strong walls, the compressive strength of the bulk grain being sufficient to resist distortion. There is some doubt about the mechanism of the improvement by "vacuum" treatment and it is not uniformly effective with all pests and toxicants. When a pile of grain sacks or cotton bales has the air partially removed before fumigant is introduced and pressure is then restored, the fumigant is driven into the interior of the units more effectively than if diffusion alone is relied on.

Grain, as might well be expected, is more vulnerable if stored at a high moisture content, particularly to initial fungus damage which can then create "hot spots" attractive to insects. Moisture content is particularly important for direct effect of the chemical on viability of the seed. The dry seed is relatively resistant but the chemical must be removed by ventilation before the seed is sown. A moisture content of around 12-16%, depending on the kind of grain, is critical.

Fumigation of viable plants for quarantine purposes makes the most difficult demands on selectivity. The plants should be in a dormant phase, so maintained by low temperature (but not frozen) and in as dry a condition as is consistent with their survival. Excess soil should be removed to make the pests as accessible as possible to the fumigant. Hydrogen cyanide or methyl bromide are the chief agents used. The plants should be washed after exposure to cyanide but maintained dry and freely ventilated after exposure to methyl bromide.

Fumigants

Probably because all the fumigants have small molecules gaining access via the tissues adapted to gas exchange, they show little biochemical selectivity. They are used in high concentration and their use has therefore special danger. The operators must have special training and equipment. The problems of use will be illustrated with special reference to the two most widely used compounds - hydrogen cyanide and methyl bromide - and a summary of the important properties of other commercially used compounds follows.

Hydrogen cyanide is the most effective fumigant for empty buildings or containers, being the most rapidly and generally lethal of all readily available, very volatile substances. It is very soluble in water and easily adsorbed on protein and cellulosic fibres and reacts rapidly in the adsorbed state. It has therefore poor penetrating properties and finds little use in the treatment of stored grain and none in the treatment of soil.

The high chemical reactivity of hydrogen cyanide ensures satisfactorily low residues, the hydrolysis product, ammonium formate, easily oxidized to the carbonate, being quite harmless at practical levels. Safety with regard to residue is helped also by the behaviour of hydrogen cyanide itself as a toxicant. It is the classic example of a very potent but essentially acute poison, long-term, low-level intake having no significance.

It will be seen that the danger of cyanide lies in its very rapid effect at high concentration. Handling the concentrate therefore requires proper equipment and training. It should never be undertaken by untrained workers and it is of special importance that no operator with the concentrate should be alone. A man overcome by cyanide can recover completely if he is at once dragged away from the source.

Anhydrous hydrogen cyanide boils at 26°C and is therefore supplied in pressure cylinders. It is an endothermic compound and can explode by a chain reaction initiated from an alkaline source which may be a metal oxide contamination in the container. For this reason it is always supplied containing about 0.2% of anhydrous oxalic acid in solution, but, even so, should not be stored in a hot place. It is sometimes supplied adsorbed on to discs of solid adsorbent and can be generated on site by dropping the relatively safe sodium salt into excess of fairly strong sulphuric acid (if the sodium salt solution is only partly acidified, a black polymer is precipitated, from a reaction between the free acid and its anion).

Methyl bromide (b.p. 4.5°) is easily vaporised but can be packaged in light-wall pressure vessels. It is recommended for empty buildings at about 40 g/m^3 which is about one-fifth of the dose of hydrogen cyanide, but it is effective in both stored grain and soil at a comparable overall concentration. It owes this advantage to its much lower affinity for polar substances. The volumetric partition ratio between water and air is only 4:1 at 20°C compared with ca 300:1 for hydrogen cyanide. In consequence a lower proportion is removed from the diffusible continuous vapour phase and held back in discrete phases. It is, moreover, less reactive, undergoing rapid hydrolysis only in media more alkaline than are met in stored vegetable products or soil. In dilute solution in water, buffered at pH 7, its half life is about 3 weeks at 20°C. A lower rate of decomposition therefore occurs during a more rapid transport process and a higher proportion reaches remote sites. It must disappear from buildings and soil by diffusion and ventilation. It tends, of course, to be retained in fatty substances and sufficient time must therefore be allowed for oily seeds to lose the toxicant before they are allowed to take up moisture for germination.

Methyl bromide has the disadvantage that its toxic action is more insidious than that of hydrogen cyanide. Its maximum permissible concentration is only twice as great and it is undetectable by smell at a concentration which can produce serious effects after a few hours' exposure. For these reasons it is normally supplied containing a few per cent of chloropicrin as a warning agent.

While ethylene dibromide (b.p. 131°) is used in open agriculture by simple injection into soil, methyl bromide is introduced under temporary impermeable sheets since it would otherwise escape too quickly. This is not because its vapour concentration is 100 times as great, since normal rates of both evaporate completely in the soil air space (5 cm depth of air over an acre would accommodate 20 kg of EDB as vapour). In dry sand the loss would be determined only by rate of gas diffusion which does not greatly differ between compounds. The greater retention is due to greater partition in favour of water (45:1 compared with 4:1) and greater adsorption on soil solids.

Chief Other Fumigants

Compound and b.p.	Partition Water/Air 20°	Reactions	Chief Use	Special remarks
Sulphur dioxide gas	80	Oxidised on surfaces	Empty buildings	Residue corrosive
Ethylene oxide 14°	200	Addition to $-OH_2$ $-NH_2$ etc	Stored grain	Not seed grain Explosive hazard
Methyl formate 31°	300	Hydrolysis	Furs and skins	-
Methyl isothiocyanate 35°	300	Hydrolysis and oxidation	Soils	Generated in situ from sodium N-methyl dithiocarbamate
Formaldehyde gas	500	Oxidation	Surface soil against "damping off"	-
Carbon tetrachloride 80°	1	Stable	Assists penetration of grain	Suspected carcinogen Inhibits explosion of carbon disulphide
Chloropicrin 112°	18	Hydrolysis	Soils	Added to cyanide or methyl bromide as sensory warning
Carbon disulphide 46°	1	Oxidation	Soils, esp. against fungi	Danger of explosion: wide concentration limits

Chief Other Fumigants (Continued)

Compound and b.p.	Partition Water/Air 20°	Reactions	Chief Use	Special remarks
Phosphine gas	-	Oxidation	Stored grain	Generated in situ by moisture on aluminium phosphide
Ethylene dibromide 131°	45	Stable	Soil, harvested fruit and vegetables	Suspected carcinogen
Ethylene dichloride 84°	30	Stable	Grain, in mixtures	-
Dichloropropene mixtures 78°	ca 20	Stable	Soils esp v nematodes	-
Dibromochloropropane	ca 50	Stable	Soils esp v nematodes	With care can be used in soil near live citrus

Further Reading

Hough, W.S. & Mason, A.F., Spraying, Dusting and Fumigating Plants, (McMillan, New York, 1951)

Munro, H.A.U., Manual of Fumigation for Insect Control, Agricultural Studies, No. 56, (FAO, Rome, 1961)

Chapter 20
PEST RESISTANCE

Resistance of Insects and Mites

In any species of a living organism individual susceptibility to poisonous substances will vary considerably. So, in large populations of insects and mites, there will always be a proportion of individuals who are naturally resistant to a particular pesticide and who will survive doses of it which are lethal to the rest of the population. If generation time is short and treatment with the same pesticide is repeated frequently the resistant strain survives and increases in numbers and eventually comprises the bulk of the population. Resistance can develop very rapidly if a steady and drastic selection pressure is exerted. Resistance is, however, less likely to develop when the kill achieved by each spraying is poor, treatment is intermittent and generation time is long.

Some species will be effectively resistant when they first encounter a particular pesticide. Resistance may be naturally present in one strain of a species even though all other strains are susceptible. More generally, resistant strains arise when the whole population has been selected by killing off the susceptible individuals. In all these situations the organism has an intrinsic potential to develop a genotype which can cope successfully with the toxicant. It matters little whether the genes necessary for resistance already exist in high frequency or low frequency or whether they are developed by mutation during exposure to the toxicant. The ultimate result in every case is a population which can withstand the toxicant more effectively.

At one time biologists believed that resistance genes were extremely rare in wild populations which had never been exposed to pesticides. Now, however, the general view is that most natural populations of any pest have sufficient genetic diversity to develop resistance to any pesticide, and that whether this happens, and the speed at which it happens are determined by the ecology of the pest, by the nature of the selection pressure, and by the genetic nature of the system confering resistance. Very little is in fact known about the factors which govern the development of resistance in the field and this is an area of knowledge in which much more information is needed if pesticides are to be used in the safest and most effective manner.

The first example of a pest species developing resistance appeared in California in the 1920's. It had become the practice to control a scale insect in citrus orchards by enveloping the individual trees in mobile tents into which gaseous hydrogen cyanide was introduced. After some years it was no longer possible to control the pest without so greatly increasing concentration or exposure time that the trees were directly damaged. About the same time in Queensland strains of cattle ticks evolved which were no longer controlled by the arsenical dips which had been in use for many years.

Despite these known examples, DDT was at first hailed, during the Second World War, as the miracle insecticide which would solve all insect pest problems and it came as a new surprise to many people when resistant bodylice appeared in Korea, resistant mosquitoes in India and resistant houseflies in many parts of the world. Resistance is now very widespread. Nearly all established insecticides have induced resistance somewhere in some of the 200 and more species about which there are clear records. At least 104 species of insects of public health importance are now resistant to one or more pesticides. In many localities wild strains now fail to respond to applications of quantities even a hundred or more times greater than were previously adequate.

Mites, because of their short generation period, are particularly prone to develop resistance rapidly. During the period 1952-1973 20 different acaricides were recommended for use in apple orchards by the U.S. Agricultural Extension Service. Of these, 15 have now been discontinued because of development of resistance. In Australia, the cattle tick has become resistant to practically all the chemical control agents which have been introduced during the past twenty years.

Resistant insects are sometimes compared to the fabled human opium or arsenic eaters, capable of handling heroic doses of these poisons. There is the very important difference that the insect <u>population</u> becomes resistant by selective breeding, whereas the habituated human has adapted himself as an <u>individual</u> by slowly increased intake. There is no known adaptation of the individual insect to pesticides. The pest insect is normally short-lived in an environment made hostile by many factors other than the pesticide. It will either survive its single contact with an insecticide, be killed by it, or be so incapacitated that it will succumb to some other enemy.

Resistance is now closely studied in many laboratories. Not only is it desirable to know the performance of a new insecticide against strains already resistant to established insecticides, but also how quickly a new strain, resistant to the new insecticide, can be developed. Resistant strains are produced by breeding under what is called high insecticide pressure. The environment of the breeding stock is contaminated in some standard way which kills the majority of the insects before they reach the breeding stage. The progeny of the survivors have usually rather greater chance of survival and the degree of contamination is increased for each successive generation. In this way special stocks have been bred which are, in some instances, several thousand times more resistant than the original.

The mechanism of resistance can be morphological, behavioural or biochemical. By morphological is meant the development of some structure in the insect which reduces the dose transferred to the vital site. For example the cuticle may become considerably thickened or the legs become less hairy and so have less tendency to retain contaminating particles. By behavioural is meant a change in the behaviour of the insect which brings it less into contact with the insecticide. For example insecticides are often applied to internal wall surfaces and kill flies and mosquitoes by contact. Contact is made most efficiently by walking, less efficiently by resting on the surface and not at all in flight. Therefore a change from restless walking activity to an alternation between flight and rest makes the insecticide less effective.

Such modifications have undoubtedly arisen and are rather general in their effects. The physical properties of the insecticide, particularly melting point, crystal shape and vapour pressure, are more important for morphological or behavioural resistance than are differences in chemical reactions. Morphological or behavioural changes rarely give rise to a very high degree of resistance.

Resistance could also arise if the insecticide fails to reach the site of biochemical action, or does so at a slower rate. This could happen as a result of (a) failure to penetrate the cuticle in sufficient amount (b) an increase in the rate of excretion of the pesticide (c) storage of the pesticide in inert fat deposits. It has been shown that movement of some insecticides into and through the cuticle of susceptible strains of houseflies is two to five times more rapid than for resistant strains. There is some evidence that DDT-resistant larvae of tobacco budworm may have higher sclerotin levels in the cuticle than susceptible larvae and that this may bind a higher proportion of the applied DDT. A number of instances of diazinon resistance have been correlated with binding to tissues. BHC-resistance in one strain of housefly has been associated with a threefold increase in cuticular absorption.

The above effects are essentially biophysical. However very high degrees of specific resistance are almost always the result of additional biochemical effects, in particular, an increase in the ability of the insect or mite to detoxify the pesticide by means of some enzymatic reaction. It appears that enzyme systems capable of detoxifying most insecticides exist naturally in insects but generally at too low a level to be an effective protection to the organism. Those individuals which appear to possess a high degree of natural resistance merely possess the detoxifying enzyme systems at a higher level of activity. This is determined by particular genes and so it can be seen why a severe selection pressure may rapidly increase the genetic purity of the resistant strains.

Study of the genetics of insect resistance reveals a complex situation. When insecticidal pressure is removed, an isolated laboratory colony becomes generally more sensitive again through succeeding generations, but resistance is much more persistent in some strains than others. If allowed, in the absence of insecticide, to interbreed with the original stock, the colony loses resistance more quickly but some resistant individuals may be found after many generations. The resistance is usually a dominant characteristic but may be associated with other characteristics less favourable to survival in an environment free from insecticide. Under field conditions, where associated undesirable characteristics would be more important, one would expect reversion to type to be more rapid.

Generally a strain selected under pressure from one insecticide will show resistance, as would be expected, to a closely related compound (e.g. DDT and methoxychlor), but not to unrelated compounds. In this respect the cyclodiene compounds behave as closely related and distinct from DDT, with BHC usually coming into the cyclodiene group. Resistance to organochlorine insecticides does not usually confer resistance to organophosphorus insecticides, but resistance to organophosphorus insecticides often gives a very pronounced cross-resistance to organochlorine insecticides. The reason for this is not known. Resistance to organophosphorus insecticides also often confers resistance to carbamate insecticides but this is not surprising as they both have the same toxic mechanism, namely, inhibition of cholinesterase. Multiple resistance - i.e. simultaneous resistance to a number of unrelated insecticides - can be induced in strains subjected to pressure from all these insecticides, but it can also appear when pressure has apparently been from one insecticide only. However, in such cases, it often happens that the resistance is not of a very high order and that a small increase in dosage of the insecticide can produce an acceptable level of kill. Such resistance is, therefore, most probably morphological or behavioural in origin.

However, cross resistance is a very complicated problem because insects with a high degree of resistance induced by pressure from one specific insecticide and almost certainly biochemical in origin are often found to be resistant to unrelated insecticides.

An example is the pattern of resistance found by exposure of houseflies to DDT, γ-BHC or dilan (= DDT with the three aliphatic Cl atoms replaced by H, NO_2 and CH_3 or C_2H_5). In strain 1, DDT resistance was associated with cross-resistance not only to methoxychlor but also to the quite unrelated pyrethrins. This strain remained susceptible to γ-BHC, dieldrin and thanite. Strain 2, almost completely resistant to γ-BHC, displayed slight cross-resistance to DDT only. Strain 3, resistant to dilan, was resistant also to DDT and, slightly, both to dieldrin and γ-BHC.

By interbreeding strains of houseflies resistant to organochlorine and organophosphorus insecticides it has been possible to study the genes that confer resistance to them. These genes are situated on chromosomes 2,3,4 and 5. Some are dominant, some are recessive and others intermediate. All appear to exist as alleles; the different members of the allelic groups confer resistance of different intensities. Three major genes appear to control resistance to organochlorine insecticides: Deh a dominant on chromosome 2, kdr and kdr-o recessive genes on chromosome 3 and md an intermediate dominant on chromosome 5. Deh controls dehydrochlorination and confers resistance to those organochlorines that are readily dehydrochlorinated. kdr and kdr-o confer resistance to knockdown by DDT and are thought to diminish nervous sensitivity. Several genes are known to confer resistance to organophosphorus insecticides. Houseflies can resist insecticides from most chemical groups by means of more than one breakdown mechanism. Often several such mechanisms can coexist within the same insect. Individual mechanisms can prolong: (a) protection against most insecticides of a given group, e.g. most of the chlorinated insecticides, (b) protection against a limited number of compounds within a group, e.g. only against DDT and TDE, (c) activation against insecticides or different groups, e.g. diazinon and DDT, or (d) act as modifiers that manifest themselves only when other mechanisms are present. The mechanism-delaying penetration of insecticides controlled by the dominant gene Pen is an example of (d). Synergists may be too specific and so increase rather than decrease resistance, e.g. sesamex acts as a synergist to organophosphorus compounds degraded by oxidases but with genes which control a modified carboxyesterase and gene gst which controls glutathione-S-transferase, sesamex acts as an antagonist and confers increased resistance.

Much chemical and testing effort is devoted not only to the search for insecticides which will defeat existing resistance, but in the hope that some will be found much less prone to induce resistance. That this may be possible has been shown in a manner extremely interesting to the chemist, but probably without commercial future, by the finding that as small a change as the substitution by deuterium (D) of the lone H atom in the centre of the DDT molecule produces a compound fully effective against DDT-resistant strains of *Musca* and *Aedes*. In the case of *Aedes*, many generations under laboratory pressure from deutero-DDT have increased the resistance by a much smaller factor than in parallel experiments with normal DDT. Resistance is due to the existence of an enzyme called DDT-dehydrochlorinase which is specifically active in dehydrochlorinating the molecule. It would seem that the activation energy for removal of DCl is just sufficiently greater than that for removal of HCl to put the reaction out of reach. More practical, but not very potent, compounds which have shown no tendency, in laboratory rearing programmes, to induce resistance are the bromoacetic esters referred to below, particularly the 2,4,6-trichlorophenyl ester.

Cross-resistance is the exception rather than the rule, but it has usually been found in laboratory tests that a strain already resistant to one insecticide can develop resistance, under pressure from a second insecticide, more quickly than the original strain. It almost seems that insects can "get into the habit" of breeding resistance. The opposite phenomenon to cross-resistance is, however, known. A strain of *Drosophila* highly resistant to DDT was found to be more sensitive than the normal strain to bromoacetic esters. Similarly, it has been shown that BHC resistant flies are more susceptible to knockdown by halohydrocarbon narcotics such as chloroform than are non-resistant flies. This is a predictable effect based on biophysical considerations. It is possible that alternate use of insecticides negatively correlated with regard to resistance could help to ameliorate the resistance problem.

Resistance in Relation to Manner of Use

Until compounds have been discovered to which insects cannot develop resistance and which are also economic, increasing attention must be given to the strategy and tactics of use of existing insecticides. It is evident that certain general conditions are necessary to induce resistance. *If* it could be arranged that the individuals in a wild population were *either* subjected to so high a dose that they had no chance of survival *or* were not subjected to the insecticide at all, no selection could occur. Even were the dose in the first group not adequate, selection would be improbable were there a sufficiently well-mixed population of individuals not exposed, because interbreeding would eliminate the resistant characteristic, if associated, as is generally the case, with some characteristic disadvantageous in a normal environment. Resistance development is most favoured by subjection of all the individuals in an island population to a significant, but not always lethal, exposure. A population is effectively an island one if it occupies an area so large, in relation to the mobility of the species, that interbreeding with an outside population is unimportant. Widespread use of an insecticide leaving a persistent deposit, active by contact, provides the most suitable environment for the development of resistance.

This is why this phenomenon became of such great importance after the introduction of the stable organochlorine compounds. Persistence must not, however, be over-emphasised, since the earliest resistant strains were induced by treatment with the most transient of poisons, hydrogen cyanide.

It is widespread use throughout an island community that is particularly likely to create resistance. The extremely persistent organochlorine compounds have been the most effective inducers of resistance, not so much directly because of their persistence, but because this property assists their distribution throughout the environment. Their low cost and apparent safety to mammals also, of course, promoted carelessly extensive application in the early days, which greatly aggravated the problem.

There are several modifications of strategy which have been advocated to avoid the development of resistance. The first is moderation, to the point of apparent inefficiency, in the use of the insecticide. If, instead of aiming to control the pest on all the land carrying the particular crop in a selected area, there were left a well-distributed pattern of untreated pockets, interbreeding of the survivors from the treated areas with the untreated population might prevent resistance developing. The pest would be less well controlled at first and the competing companies in the chemical industry would not find it possible to advise the apparently inefficient use of their products. Government legislation would be essential. If it were necessary to sacrifice not only pockets of crop on one holding but complete holdings in a peasant community, a political problem would at once arise. It should be emphasized that, for success of this plan, some areas should receive at least the full dosage and others none. A reduced dosage throughout the area is more likely to aggravate resistance build-up.

Another plan is to arrange for alternation in the pesticide used - say an organochlorine for 2 years, then an organophosphorus for 2, and so on. The plan would be most likely to succeed were the compounds known to show no tendency to cross-resistance but rather the reverse - the strains resistant to one being more sensitive to the other - as in an example, unfortunately using one rather feeble insecticide, quoted above. Such an arrangement would be difficult to enforce in a community of separate farmers. Obviously, trade in insecticides can be more easily controlled by Government than can their use in the field. If both insecticides had a rather short storage life, control of supplies might be adequate.

A third plan, or perhaps more a combination of details, in which education and propaganda would be the chief activity of Government, is the more cautious and selective use of pesticides. This would include the discouragement of widespread prophylactic use of insecticides leaving a persistent residue, an improved system of warning of potential pest build-up and the encouragement of localized application rather than overall spraying.

Together with all these measures a greater use of biological control and conservation of predators and parasites is desirable. The aim should be a total system of pest management involving the wise use of selective pesticides with full consideration, as far as knowledge permits, of the interactions of species.

Effects on Predators, Parasites and Population Balance

Apart from differences in individual susceptibility to a particular pesticide within a species, there may be gross differences in susceptibility between one species and another. In the extreme, some species may be totally unaffected by the dose applied. If this is so, then application of the pesticide will remove all the predators and parasites which normally hold the resistant species in check. Even if the species has no predators or parasites the effect of removing competition for food, light, water, etc. may be sufficient to allow the resistant species to become dominant. The total effect of using the pesticide may, therefore, be to create conditions in which a previously minor pest can flourish. A notorious example of this is the red spider mite of fruit trees. Prior to the introduction of DDT this had been only a minor pest because it was held in check by predators and by competition. As it is a mite, not an insect, it was not affected by DDT which killed practically all the fruit tree insects. As a result of widespread and injudicious use of DDT in orchards red spider mite became, and still is, a major pest to fruit growers.

It might appear from this that the desirable aim is always to discover insecticides which attack only the specific target pest and nothing else. This is a good aim if you can be quite sure that the target pest is the only species which is, or could be, a pest to that particular crop. Experience in commercial glasshouses where biological control methods using introduced predators of the pest insect have been widely investigated. This method of control is highly specific to the particular pest but this specificity can, in practice, be a disadvantage if some minor pests are present which can become major pests themselves if competition from the original major pests is removed by highly specific biological agents. In such circumstances, better overall results can be achieved by using a general insecticide which kills major and minor pests alike.

Nevertheless where a pest is mainly kept in check by predators or parasites any reduction in populations or activities of these predators or parasites could lead to an enormous increase in a pest problem. Long-term effects on population numbers may completely outweigh any short-term advantage gained by application of insecticides.

The above discussion illustrates the need to consider pest management as a total problem and not just to use insecticides in isolation. The correct choice of pesticide and of its method and timing of application demands an extensive knowledge of the ecology of the pest and of its relationships with other species.

Agricultural pest insects are necessarily herbivorous, that is, they feed on plant tissue although, in the case of some sucking insects they do far more damage by transmission of virus than by direct consumption of the crop. The predators and parasites of agricultural pest insects are carnivorous. The ideal agricultural insecticide might be one that could make this very broad distinction. The biochemical processes of herbivorous and carnivorous insects may have some characteristic differences dependent on the biochemical differences between the two sorts of feed. So far no selectivity based on this difference is known, nor, to the writers' knowledge, deliberately sought. The difference is obscured by several factors. Insects of many families, while structurally closely related, have adapted themselves to very different feeding situations. For example, the family Syrphidae, of the order Diptera (flies) includes some of the small hoverflies characteristically predators of aphids, but also feeders on the tissues of growing plants and fungi and scavengers of decaying organic matter. Habits of feeding and movement do not therefore follow genetic relationships. Biochemical processes tend to be more closely coupled with genetic relationships than with feeding habits.

The differences between pest and beneficial insects most likely to influence their response to insecticides arise therefore from the habits themselves and the different opportunities they create for exposure to the insecticide. Most herbivorous insects are (in the feeding stage) slow-moving and hidden, either because they burrow into the crevices between leaves, form tunnels within leaves or stems, or produce leaf-curling or gall-formation. While parasites are also hidden, they need to be more mobile than their hosts in the egg-laying adult stage because they have to seek out a smaller food source. Predators are necessarily more mobile because they must seek out small prey to devour in numbers.

A spray directed generally into the air, regardless of season, would therefore be likely to hit a higher proportion of parasites and predators than of pests. A persistent surface deposit active by contact with walking insects is, by and large, more effective against beneficial than against pest insects. That it often serves, in fact, a very useful crop-protective function depends a great deal on timing and placement. No matter how chemically stable the compound, it is bound to be most active on crop foliage during and shortly after spraying. During spraying, accessible insects can be directly hit. After some days there will be some loss by physical factors of weathering and dilution of the overall surface concentration of deposit by growth of the crop.

The success of a persistent general insecticide applied at the right time is usually achieved despite reduction of the parasites and predators. A low dosage of some very selective insecticide can reduce the pest population to a level which can be held in check by the remaining parasites and predators. A non-selective insecticide can achieve a more complete clean-up but has had to brush aside the help it might have had from beneficial insects. As these decline and as resistance in the pest perhaps develops, higher dosages are found necessary. In any plan of integrated control, the moderate dosage of selective insecticide will always be preferred.

The systemic insecticides show a high degree of selectivity by automatic correct placement, quickly retreating within the plant tissue and becoming directly available only to leaf-eating or sucking insects. If parasites or predators are killed, it is because they consume already poisoned pests. Three factors safeguard them to some extent. A sucking pest already affected is less attractive than a healthy one. Some decomposition of the insecticide occurs in the body of the pest. Finally, and in fact at present most important, water-soluble systemic compounds are intrinsically more toxic to aphids and spider mites than to most other insects. This is, of course, a disadvantage when the pest is a leaf-eating or stem-boring one, and is not planned. If an effective systemic of a wider spectrum of action appears, there may be more problems with beneficial insects.

The generally greater mobility of beneficial predators and parasites than of pests has another important effect on the balance as affected by insecticides. If the beneficials are eliminated, and the pests reduced, in a local treated area, the beneficials from the surrounding untreated area can invade the treated area and clean up the residual pest. It therefore pays to leave pockets or strips untreated or to treat these strips at different times. Similar strip-harvesting of a crop suitable for this method - e.g. alfalfa (lucerne) in California - is another practice recommended by the protagonists of integrated control, in order to conserve the mobile beneficials.

An advantage claimed for insecticidal control measures moderated in some way to leave predators and parasites in action is that the development of resistant strains is made less likely. Although the residual population of pest may contain a higher proportion of resistant individuals, the parasite or predator will kill these off. There is some dispute whether this statistical argument is valid without some factor of discrimination. It is, for instance, likely that a resistant individual, being more healthy than its neighbours after ingestion of insecticide, would be sought out by a parasite, but clear experimental evidence is hard to obtain.

If it seems unreasonable that a species can develop resistance to an insecticide but not to a predator or parasite, it should be remembered that the insecticide is new but the natural enemy very old. Evolution has been going on many million times longer than exposure to insecticides. Doubtless many parasites have long ago become extinct because their hosts successfully developed resistance. Many species were preyed upon to extinction by a too successful predator. There are now left the species which have achieved a precarious balance in a world of ancient hostilities. Insecticides are a new factor in selection. After they have been used for many million years and most have been discarded, the situation may become much more stable.

Resistance and Predator Suppression

It may now be apparent why resistance and effect on natural enemies have been included in the same chapter. In the field it is often difficult to distinguish which has produced an observed decline in the effectiveness in an insecticide. The biologist must make laboratory observations on a pest population brought in from the field and tested in isolation before he can say with certainty that resistance has appeared. The effects tend also to be induced by the same defects in chemical properties or mode of application - excessive persistence and too widespread application.

The history of the cabbage-root-fly (*Erioischia brassicae*) in the U.K. provides an interesting example of resistance and predator suppression. It seemed easy to attack with a stable soil-applied insecticide. It could do considerable damage to the crop and could not be effectively controlled once the damage (plants wilting due to root destruction) was evident. General soil treatment in advance of planting was adopted. The practice brought about increase in the population of the fly through slaughter of its beetle predators. Damage in fields where no insecticide was used increased but good control could still be obtained by soil incorporation of dieldrin. There then appeared, in some areas of intensive *Brassica* cropping, strains of fly which were no longer controlled by this insecticide. Other compounds, such as diazinon, intrinsically less effective against the original strain, are holding the pest at present, but the grower is on balance in a worse position than before the first use of insecticides. Some authorities contend that had treatment been confined from the beginning to seasons when a bad outbreak was expected and had the treatment been localized by dipping, drenching or granule application to individual plants this undesirable development could have been avoided.

Resistance in insects and other arthropods is not necessarily detrimental to man's interests. It is most strikingly evident in rapidly breeding flies, aphids and mites, but parasitic and predatory flies, wasps and mites have also been shown to be capable of developing resistance. One of the problems of biological control is that, while it is very cheap and successful against some pests, it may have made no progress against other pests of the same crop. If insecticides are called in to deal with the latter, they upset the biological control of the former. The chemist may be able to produce an insecticide harmless to the parasite or predator. Alternatively the biologist may be able to breed a strain of parasite or predator sufficiently resistant to an established insecticide.

The biological situation is immensely complex. There is no space in this book to deal with it more adequately, nor would the writers be competent to do so. They may already, in the interests of brevity, have made generalizations unacceptable to the expert. Their object has been to show that the use of insecticides is necessarily a subject of increasing complexity. The search for new insecticides is therefore one of increasing difficulty but also a stimulating challenge. The future research chemist in this field will meet endless frustrations if he is not prepared to let his biological colleagues teach him some of their problems. He must be prepared to give advice as well as take it and will then find that the pesticide field is so full of interest that the frustration is less important.

Whatever the future, the world of agricultural pests no longer lies at the feet of the chemist with the powerful, stable, general insecticide and the engineer with the long spray boom. The fact that agriculture in many of the developing countries employs much more hand labour than in the highly mechanized countries may make them better able, with adequate study of the problems, to use pesticides more intelligently. They have more need of them and can secure higher yields if the pest problems are surmounted. They are in a position to profit well from the mistakes made in the U.S.A. and northern Europe.

Resistance of Fungi

In general, a high degree of resistance to "conventional" protective agricultural fungicides, which do not penetrate to any significant extent into plant tissue, has been seldom observed in the field. Even when attempts were made to induce resistance in the laboratory by successive transfers of fungal mycelium to agar medium containing increasing concentrations of the fungicide, there has been little success. Resistance has, from time to time, been reported to chlorinated nitrobenzenes, dicloran, hexachlorobenzene, dodine and organomercurials. These examples of resistance are very limited and when one considers the long period during which protective fungicides have been used commercially the number of cases in which resistance has become a problem for practical disease control is very low and is restricted mainly to fungicides used on a limited scale for special purposes. In all cases the acquired resistance disappeared after transfer of the fungus to a medium free from the fungicide.

The situation is quite different with regard to systemic fungicides which penetrate and move about in the plant tissue. Very soon after compounds of this type were put into commercial use there were dramatic examples of acquired resistance. In 1968, dimethirimol was introduced to control cucumber mildew but, by 1969, resistance had developed. Even though use of dimethirimol was discontinued, resistance to it was still widespread in 1971.

A similar example is of use of benomyl at 0.5 ppm to control heart-rot in cyclamen. Very rapidly a resistant strain emerged which was not killed by 1000 ppm of the fungicide. Emergence of resistant strains took place very quickly in these two examples because of the artificial, confined environment of the glasshouses. Resistance to systemic fungicides in the field has taken longer to develop but reports are now increasing rapidly and concern practically every systemic fungicide which has been introduced commercially.

It is not yet clear how large a problem this is going to be in practice because little is known yet about the epidemiology of resistant strains and their competetive ability. It is probable that populations will become fully resistant only in those cases where the resistant strains have a large and continuous selective advantage over the susceptible strains.

However, to reduce the emergence of resistance, it is desirable that more than one systemic fungicide, with different biochemical mechanisms of action, should be applied either together or alternately. The use of such fungicides should be restricted as much as possible and methods of application should be chosen to reduce selection pressures to a minimum. The possibility should be borne in mind that a systemic fungicide may upset the total balance of microorganisms in the soil or on the plants and that competing fungi or even fungi which produce antibiotics effective against the pathogen may be inhibited. This, as with insecticides, could cause new disease problems to arise.

The interesting question why resistance has developed rapidly to systemic fungicides but not to protective fungicides is probably answered by the fact that protective fungicides all act by interfering with the general energy processes and protein synthesis of the fungus, which are not determined by single genes, whereas most systemic fungicides act by inhibiting specific enzyme reactions which are often carried on a single gene. Plant breeders produce strains of crop plants which are resistant to various fungi but, invariably, new strains of fungi emerge which overcome this resistance. One of the most notorious examples is the wheat rust (Puccinia graminis) which has produced successive new strains which have successively rendered dozens of new wheat varieties, which were initially resistant, susceptible. This suggests that "natural" resistance also depends on the production within the plant of natural antifungal agents which act like systemic fungicides by inhibiting specific enzyme systems in the fungi but which are, because of this fact, prone to induce resistance. The occurrence of such natural antifungal agents - the phytoalexins - has been demonstrated and some have been isolated. More knowledge about "natural" resistance would assist understanding the problems of acquired resistance.

Resistance of Weeds

Compared to insects and fungi, plants have long generation times and their seeds are mobile or stay in a dormant state in soil for long periods. Evolution of plants is thus a comparatively slow process and even if genes carrying resistance were produced by herbicide application they would be continuously diluted by non-resistant individuals. For these reasons, acquired resistance of plants to herbicides is rare. One of the few examples is an annual weed common in sugar plantations (Erechtipes hieracifolia) which can no longer be controlled by 2,4-D on plantations where this herbicide has been used extensively for several years. Common groundsel has been reported as showing acquired resistance to simazine and atrazine. There have been many reports of "difficulties" in weed control because of decreased effectiveness of herbicides but it is not clear whether any of these reported cases involve true genetic resistance.

There obviously is a potential for induction of resistance to herbicides into plants and weed scientists are alive to this possibility. If resistance could be induced artificially by pressure from successively increasing doses of herbicide this could be made very positive use of to produce strains of major crop plants which were more resistant to herbicides and thus improve the margin of safety of selective herbicides.

Agronomists do talk about "resistant" weeds but they invariably mean weeds which are naturally resistant as a species not weed populations which have acquired resistance. Most herbicides are deliberately designed to be "selective", that is, to kill all weeds but not to damage the crop. Complete selectivity is, however, unlikely to be achieved. So long as there is one plant present in a particular environment which is naturally resistant to the herbicide then, no matter how small a proportion of the normal flora this plant forms, it will, if competition for water, light and nutrients is removed by destruction of all other plants, take over all available ground-space and become a major weed pest. This is because nature abhors a vacuum and always fills any available ecological niches. As an example of this tendency, 2,4-D and MCPA have been used for 30 years to control weeds in cereals. Broad-leaved weeds have been almost completely eliminated but the space has been taken over by the wild oat, which is now the major weed pest in cereal cultivation. Many herbicides to control wild oat have been introduced during the past decade but, when this has been eliminated, some other plant resistant to currently used herbicides will emerge to "take over". This situation can be avoided, or at least mitigated, by "ringing the changes" on herbicides or by using mixtures of herbicides.

Resistance of Vertebrates

The most important example in terms of its practical consequences has been the development of resistance to warfarin in rats and mice. Resistance to warfarin is known to occur in four species, the Norway rat, the roof rat, the mouse and man. In man, transmission of resistance is by a single autosomal dominant gene, but, in rats and mice, more than one gene is involved. It is interesting to note that resistance can be observed in species such as man which have relatively small populations and relatively slow reproduction rates and long generation times.

Other examples of resistance in vertebrates include pine mice which have acquired resistance to the organochlorine insecticide, endrin, as a result of increased rate of excretion of hydrophilic metabolites, and two strains of mosquito fish from the Mississippi which are resistant to the organochlorine insecticides aldrin and dieldrin. Resistance to organochlorine insecticides has also been found in three species of frog. DDT-resistant mice have been artificially produced in the laboratory, and it appears that this resistance is due to preferential deposit of DDT in the fatty tissues and consequent reduction of levels in sensitive tissues.

Further Reading

World Health Organisation, Insecticide Resistance and Vector Control, WHO Technical Report Series, No. 443, (Geneva, 1970)

Chapter 21
SAFETY OF PESTICIDES

Introduction

Crop protection and pest control chemicals have been deliberately developed to be toxic to some living organisms and this is the reason for their commercial utility. There is a unity between all forms of life, so accidental ingestion of pesticides by humans or animals might produce adverse effects if they were very poisonous. In this case, there would be a possibility of health risks to the operatives who are actually engaged in handling and spraying them. There is the possibility of hazard to children, when the carelessness of adults permits them access to such chemicals, and the possibility of their being deliberately used for suicide or murder. Since crop protection chemicals are applied to plants which will produce edible crops and, in some cases, to the crops themselves, there is the possibility that small residues of the chemicals might remain in the crop until it is eaten by the consumer and that, if this happens, it might be deleterious to the consumer's health either acutely or in the long term. Furthermore, since crop protection chemicals are sprayed widely over large areas, there is the possibility that they may drift on the wind and that small concentrations may build up in the atmosphere at large. They might be hazardous to beneficial insects such as bees, to wild animals and birds which feed in the crop, and to creatures which live within the crop or in the soil beneath it and, thus, indirectly to wildlife which feeds on those creatures. Chemicals which fall on to the soil can be washed down into it by rain and eventually find their way into lakes and rivers and thence into estuaries and harbours, where they might adversely affect fish and other aquatic life. All these are possibilities, the risks of which have to be weighed against the benefits which the pesticides produce. What is clear is that it is as unthinkable for the community to do without modern pesticides as to do without modern medicines. Present-day agriculture in developed countries is dependent on crop protection chemicals as much as it is on the internal combustion engine. No one would want to return to the hunger, malnutrition, disease and discomfort of a primitive existence. However, our goal must be to employ crop protection and pest control chemicals in agriculture, horticulture, public health and amenity and recreational areas as part of the total management of the environment for the long-term benefit and survival of mankind, and to minimise any possible risks arising from their use.

Right from its start in the 1940s the modern pesticides industry has been aware that there were risks and has taken steps to eliminate or minimise them. In the early days, farmers were not made sufficiently aware of the possible dangers of crop protection chemicals and sometimes used them indiscriminately and not too wisely, but the standard of knowledge has risen greatly in recent years. However, there was always the possibility that unscrupulous manufacturers might not give sufficient consideration to the safety of their products or that careless growers might handle and use them in an unsafe way, and it was to prevent situations like these that Governments had to act by means of legislation and regulatory controls to protect the consumer and the community as a whole.

A major responsibility of the Government of any country is to try to ensure that adequate supplies of nutritious and wholesome food are available at reasonable prices to its inhabitants, and it must encourage and promote any technology which can help to achieve this aim. At the same time, it has the responsibility for trying to ensure that application of the technology does not present unacceptable or unreasonable risks to the community or to the quality of our lives. It is inevitable that introduction of any new technology must bring with it the possibility of some new risks, and the aim of Government must be to enact legislation which will minimise these risks without seriously detracting from the benefits.

In the U.K. a system of voluntary rather than legislative control has been developed based on the Agricultural Chemicals Approval Scheme and the Pesticides Safety Precautions Scheme. The Approval Scheme is concerned mainly with establishing that the product is efficacious for the purpose for which it is intended. Government representatives have access to trial plots and records of a manufacturer who seeks approval for a new product, and may arrange for independent trials to be carried out in government experimental stations. Approval is for a particular use or set of uses and entitles the manufacturer to use the official approval mark on his labels, the wording of which must be agreed.

The Safety Precautions Scheme agrees conditions of use to ensure that hazards to users, consumers and to wildlife are reduced to an acceptably low level. It is administered by the Ministry of Agriculture, Fisheries and Food, acting in association with other government departments and calling on various expert committees for advice.

In the U.S.A. control over pesticides began with the Federal Insecticide Act of 1910 which laid down certain labelling requirements designed to ensure that the products were both effective and safe. These requirements were greatly extended for all pesticides by the Federal Insecticide, Fungicide and Rodenticide Act of 1947 which required that all pesticides should be registered and could not be used in any application without obtaining a "label" specifically for that use. Such a label has to carry adequate instructions for use and warning statements about possible injury to men, animals and plants. The scope of the act was increased by the Federal Environmental Pesticides Control Act of 1972. This now requires that a pesticide shall be registered if (a) its composition is such as to warrant the proposed claims for it (b) its labelling and other material required to be submitted comply with the requirements of the act (c) it will perform its intended function without unreasonable adverse effects on the environment (d) when used in accordance with widespread and commonly recognized practices it will not generally cause unreasonable adverse effects on the environment.

The administration of the act is in the hands of the Environmental Protection Agency set up in 1970, which has an Office of Pesticide Programs specifically to deal with crop protection and pest control. They are required to make an analysis of the risks and benefits before making any decision on registration of a new pesticide or amendment or withdrawal of the registration of an established product. That is, they are required (a) to make inferences about future impact of the pesticide which are justified by existing scientific knowledge (b) to determine possible harmful risks and to estimate their magnitude as to compare these with the benefits to society which may result from use of the pesticide.

The Environmental Protection Agency have control not only over registration and use of pesticides but also of biological methods of crop protection and pest control such as release of predators or use of pathogenic microorganisms. Also they may consider whether new varieties of plants produced by the plant breeders might contain naturally occurring chemical substances which might have harmful short-term or long-term effects on men or animals. In particular, where varieties have been bred to be resistant to a particularly disease and it is believed that this resistance is due to increased production of natural fungitoxic substances, then the effects of these on man and animals need to be studied.

It should also be realised that the registration requirements are strict and apply not only to new pesticides but also to established pesticides which can be called to account and have their registrations amended or cancelled if new evidence comes to light of possible risk in their use. The procedure then is for EPA to issue a "rebuttable presumption" that the registration should be cancelled, and the manufacturer then has six months in which to bring evidence to convince EPA otherwise. At the moment "rebuttable presumptives" have been made against 47 established pesticides. One consequence of this is that pesticide companies are having to spend a greater proportion of their research and development budgets - currently estimated at about 25% - just to protect their existing products, which reduces the effort available to develop better and safer new products.

The strict registration requirements pose problems for innovation and development of new pesticides but these have been resolved by the issue of experimental use permits.

With regard to control of use, rather than manufacture, of pesticides the Federal Environmental Pesticide Control Act of 1972 provided that pesticides shall be classified for "general use" or "restricted use", and chemicals in the latter category can be applied by or under supervision of a certified applicator.

The National Environmental Policy Act requires a full statement of environmental impact for every major federal activity which may "significantly affect the quality of the human environment". Obviously any crop protection or pest control programme which was planned over a whole region or area rather than just on an individual farm would come into this category.

Most other developed countries favour a mandatory system such as that of the U.S. It is not pertinent in this book to argue the relative merits of voluntary or mandatory systems. The voluntary system in the U.K. has led to co-operation and open communication between all interested parties but it is possible that it could operate effectively only in a comparatively small country such as the U.K. which does not use pesticides intensively.

The evidence which must now be provided makes a very impressive list entirely contradicting a rather common popular picture of the pesticide industry irresponsibly scattering untested substances over the population.

The cost of providing the necessary evidence of safety may be anything from

£50,000 to £1 million and falls entirely on the company which wishes to market the new product. It would be difficult to give a precise figure in any one case even if the company concerned were willing to disclose its full accounts, because some of this cost is incidental to other research costs. The total cost of discovery and development of a new pesticide may exceed £10 million when the cost of extensive field trials and of the "wasted" basic work on unsuccessful compounds is included. One unfortunate result of this high cost is that is has become increasingly unprofitable for the manufacturer to develop a compound for a restricted and specialized market. It is obvious, however, that compounds with a specialized action or use are, on the whole, those least likely to give rise to side-effect problems. Also, because of their multiplicity and diversity, they are less likely to create real cumulative risk to the consumer than a few very widely used compounds. One effect of present safety legislation is to force the manufacturer to develop only those compounds which have a wide spectrum of activity and a large market, and this may, ironically, be the least safe state of affairs.

Safety to Wildlife

In 1963 a U.S. journalist, Rachel Carson, published a book, "Silent Spring" in which she dramatized certain errors which were made in the early years of modern crop protection because farmers were inadequately warned about the inherent hazards of distributing highly biologically active compounds over wide areas. All the evidence of damage from pesticides to wildlife relate to one group of pesticides, namely, the organochlorine insecticides. This is not to say that other pesticides could not, if misused, have damaging effects on wildlife, and it is essential that a constant watch be kept to detect early warnings of any such damage and thus to avert it. This is done in the U.S.A., for example, by the National Pesticide Monitoring Program. There is no evidence that any such damage has occurred, and the extent of ecological and environmental information which nowadays has to be supplied to regulatory and registration authorities makes it very unlikely that it would occur.

It is important to emphasize here a difference about which there is a great deal of confusion in the minds of the public, namely between damage to individuals in a species and damage to a species as a whole. Individuals in a species are constantly prone to death from many natural causes, they are hunted by humans but the species as a whole survives. Accidental isolated emissions of pesticides into the environment have resulted in death of many individuals in some species but have given no reason to believe that any permanent harm has been done to those species. Even the large-scale misuse of certain pesticides described in "Silent Spring" have had no permanent effects on any species. Silent Spring has turned out, in fact, to be Noisy Fall - the birds sing in greater numbers and there is no evidence of harm to the health of the population who are more healthy and better-fed as a result of crop protection and pest control.

The safety precautions to ensure no hazard to wildlife have already been outlined and comprise extensive ecological and environmental tests and toxicological studies on non-target species, coupled with continual monitoring. A particular region in which great care needs to be exercised is water. In all developed countries there are legislative controls over the discharge of chemicals in effluents, such as the Federal Water Pollution Control Act

Amendment of 1972 in the U.S.A. Factory effluents have, without doubt, been a source of contamination of the environment with pesticides over a protracted period in the past. There is, however, at present no legislative control over possible run-off of pesticides from land into watercourses, rivers or lakes except that given by common law which provides for redress if actual damage is caused to person or property.

The extensive environmental and ecological work which now has to be carried out on any new pesticide makes it very unlikely that incident involving widespread destruction of wildlife could occur again unless a user grossly disregarded all of the manufacturer's instructions. The safeguard against such misuse - which is mainly caused by ignorance - is the increasing education of farmers in safe and effective ways to apply crop protection treatments. The only safeguard against misuse from lack of concern is the legislative sanctions which are applied to any body who commits an anti-social act, and laws against misuse of pesticides should be stringent. This should apply not only to people who cause excessive damage to wildlife by flagrant misuse but also to people who discard unwashed pesticide containers or leave pesticides in unlabelled and improper containers within reach of children.

The danger of large scale episodes of wildlife destruction has, therefore, been largely removed. It is doubtful whether such incidents had any permanent long-term effects on wildlife populations and, in most cases, recovery has been much more rapid than ecologists anticipated. It remains to examine the question whether pesticides could have any more subtle effects on wildlife which could have lasting effects on populations of mammals, birds and beneficial insects.

The situation with regard to pesticides and effects on wildlife can, therefore, be summarised as follows. Because of ill-considered application, a number of incidents of destruction of wildlife occurred in the early days of use of organochlorine compounds - although not in the U.K. - but there is no evidence that these incidents have had permanent long-term effects on population numbers. There is a suspicion that, because the organochlorine insecticides are very persistent and tend to accumulate in biological food chains, the reproductive capacity of some predatory birds may have been affected, but this is by no means proved. There is no evidence that any other pesticides have produced long-term effects on wildlife populations.

A rational attitude based on these conclusions would be to continue to require adequate environmental and ecological studies as a necessary part of registration requirements, to keep a constant watch for any adverse effects on wildlife, to supervise carefully any application of pesticides over large areas, particularly from the air, and to avoid the use of highly persistent pesticides when some acceptable alternative is available. This last point is prompted by the impossibility of ever conclusively proving a negative. Although there is no evidence that persistent pesticides are doing harm through their low-level presence in the environment we cannot be sure that no harm, however small, is occurring nor that they will never be concentrated to harmful levels by some living organisms, so that, when use of persistent compounds can reasonably be avoided, it should be. The decision must, in each case, be made on a cost-benefit assessment. If use of a particular persistent pesticide is essential for protection of a particular crop and there is no effective alternative it may be that a certain amount of risk of environmental damage may have to be accepted by society as the price it pays for the benefits it receives. It has been said that 'the perfect is the enemy of the merely good'. To refuse to use pesticides which are of overall benefit to mankind because they have some shortcomings would not be sensible if it led to food shortages while the search for the perfect pesticide was in progress.

A great difficulty is the complexity and uncertainty of environmental and ecological work. A vast amount of data have been collected, and is being collected, on the concentrations of pesticide residues present in various parts of the environment using highly sensitive analytical techniques which are often capable of detecting as little as one part of pesticide per million million, equivalent to 1 gram distributed through 1 million tonnes. The concentrations of pesticides found are generally very small, but there is no method at present available by which this vast amount of data can be interpreted in terms of a quantitative risk to any species of wildlife. A practical answer may be to maintain constant ecological watch for any detectable effects on wildlife populations and, in the absence of these, to assume that any risk is hypothetical.

Safety to the Consumer

The consumer wants food which is pure and wholesome, of high quality and attractive appearance, in as great a variety as possible, at reasonable prices and in good supply. The general increase in affluence of people in developed countries over the past thirty years has made it possible for them to get what they want since they can afford to pay for it. Apart from any of the other requirements mentioned above, the matter of purity and wholesomeness is one on which the consumer is very sensitive, as the fear of mysterious and secret poisons in the food he eats is a very deeply-rooted and atavistic one which prompts an exaggerated response if any suggestion of contamination is made. In previous centuries, widespread adulteration of foodstuffs was common commercial practice, but legislation and public inspection in developed countries have steadily brought this under control, so that any gross adulteration of food is now very rare. Legislation and general education in food handling and hygiene have reduced the likelihood of bacterial contamination of food, although there is still room for improvement in this area as local instances of food poisoning are still much too common. Nevertheless, public fears of gross adulteration or bacterial contamination of foodstuffs have been largely allayed. In recent years, however, the consumer has been made aware of what is to him a more insidious threat of the possibility of residues of pesticides in his foodstuff which he thinks might make him acutely ill or, more alarmingly, have long term effects on his health.

Government authorities responsible for registration of pesticides have, very properly and rightly, made provision of sufficient acceptable data on residues an essential requirement for permission to sell. In the U.K. only arsenic, lead and fluorine have statutory tolerance limits. However the Food and Drugs Act of 1955 prohibits the sale of food which is not of the nature, substance or quality demanded by the purchaser and also prohibits the sale of food unfit for human consumption. The Pesticides Safety Precaution Scheme always makes recommendations on the maximum allowable residues in foodstuffs when it approves a new pesticide. In the U.S.A. tolerance limits for pesticides in foodstuffs are established by the Environmental Protection Agency and authorized by the Federal Food, Drug and Cosmetic Act.

Manufacturers have always been concerned about possible residual toxic effects of their products and have carried out toxicological studies of their effects on test animals and field studies of the amounts of pesticides actually present in harvested crops. With increasing public concern, these tests have been made more and more comprehensive, so that nowadays a new pesticide is subjected to an extensive and searching toxicological testing programme. This programme includes study of acute toxic effects from doses of various sizes: 90 day and two year continuous feeding to dogs and rats - after which the animals are humanely killed and every organ of their bodies closely examined histologically for any departures from normal; studies of reproduction over three generations of rats to ensure no effects on embryos and no transmitted effects to offspring; and studies of general toxicities to birds, fish and shellfish.

At the same time it is a requirement for registration that analytical methods for a new pesticide shall be developed, capable of detecting and measuring 0.1 parts per million in meat, 0.01 parts per million in food crops and 0.005 parts per million in milk. Determinations then have to be made of residues present after harvest in all crops to which the pesticides may be applied and, in the case of animal feeding stuffs, of any residues in the meat or milk.

There has been a certain amount of confusion in the minds of consumers about residues. This is partly because the layman does not always appreciate that there are no poisonous or non-poisonous substances, only poisonous or non-poisonous amounts. The enormous increase in the sensitivity of analytical methods has increased this confusion. If the only analytical method available for a particular pesticide could not detect more than 1000 parts per million, then a sample containing 900 parts per million would be designated by the analyst as 'non-detectable', and the consumer would feel reassured. If then a method was developed for detecting 0.01 parts per million, a sample containing 0.03 parts per million would show up positive even though this is 30,000 times less than the amount in the previous 'non-detectable' sample. The consumer then feels uneasy on a general principle that 'if it is there at all, it must be doing some harm'.

This illustrates the crux of the pesticide residue problem. Everybody agrees that harmful residues should not be present, but what amounts should be specified as the maximum allowable for safety? We are up against the impossibility of proving a negative and of being able to say for certain that a particular amount of residue might not cause some harm to somebody, somewhere, sometime. Likewise, it is not realistic to specify that no residues should be present, because this can likewise never be proved, as the most that can be said is that the substance is 'not detectable' by the most sensitive analytical methods available. Registration authorities are forced therefore, into specifying amounts, that is, laying down finite limits for residues. The 64 thousand dollar question is, 'What is a safe amount?' To specify 'non-detectable by the most sensitive analytical methods is unrealistic because modern analytical methods are so sensitive that the amounts involved are fantastically small - 0.01 parts per million is equivalent to one teaspoonful in a rail tanker wagon. From an overall cost-benefit view, the banning of a pesticide, because it could not achieve such low limits, might have large economic repercussions on the production of one type of food, without any compensating benefit, since this action may have been taken against a non-existent danger. What then is a reasonable amount of residue to specify which will not expose the consumer to an unreasonable risk? What is an unreasonable risk?

A generally accepted basis is that the maximum permitted amount of residual pesticide in a foodstuff is determined from three factors: (1) the smallest dose, expressed in parts per million, which produces detectable harmful effects in experimental animals; (2) a safety factor which is usually 100, but may be less if a great deal is known about the physiological and pharmacological effects of the pesticide; (3) a food factor based on the proportion of the particular foodstuff in an average diet. Thus if the minimum harmful dose is 10 parts per million, the safety factor 100 and the food factor 0.2, the permitted residue level will be 10 divided by (100 x 0.2), that is 0.5 parts per million. The safety factor is intended to be large enough to compensate for any differences between man and test animals, for any variations in susceptibility due to age, health or personal idiosyncrasies, and for any differences in eating habits such as the person who easts an inordinate amount of a particular food.

The problem of determining safety is a problem of interpreting data obtained on experimental animals under laboratory conditions in terms of an actual hazard to people in practical use. Nobody is sure how to do this. Is there, in fact, any evidence of harm caused by pesticide residues in practice? In developed countries regular 'market-basket' surveys are made in which Government analysts purchase various foodstuffs in the normal way in shops and supermarkets and analyse them for pesticide residues. Such surveys invariably show that most samples contain well below the acceptable levels of residues - for instance, the Government analyst in the U.K. has reported that residues were, in general, of the order of a few parts per thousand million in the various foodstuffs. These findings are confirmed by the extensive monitoring programmes of the U.S. Bureau of Foods. In the few cases where abnormal levels of pesticide residue could be detected, these were traced either to human error or human greed. For instance, a recent case in the southern United States in which millions of broiler chickens had to be destroyed because they contained more than the acceptable amount of an organochlorine insecticide, resulted from the fact that seed treated with this substance had been improperly obtained and sold as animal feed by unscrupulous operators.

The very low levels of pesticide residues found in practice are not surprising when it is realised that the acceptable residue levels are not the amounts which normally occur, but the maximum amounts which could result from use of the pesticides in a proper way. They serve, therefore, as a means of detecting improper use of pesticides. Instructions for use and applications are drawn up by the manufacturers so that, if these are followed, there is no danger of undesirably high residue levels.

The general conclusion is, in the words of the late Dr J. M. Barnes, Director of the U.K. Medical Research Councils' Toxicological Research Unit, 'that there is not a tittle of scientific evidence that pesticide residues in human food present a real threat to health'. The organochlorine compounds are the only pesticides which are detected regularly and the general levels observed are well below the accepted safety levels. Other pesticides are detected only very rarely and then in very tiny amounts. Even the very widely used organophosphorus insecticides are detected in only about 1 in 70 samples of food tested, and even then at exceedingly low concentrations.

There remains the contention that is sometimes put forward that, although the trace amounts of pesticides actually present in foodstuffs will not cause any acute damage to health, they might have subtle long term chronic effects. In this connection, cancer is a very emotive word. It is very difficult to refute this type of argument because it involves conclusively proving a negative.

On the question of cancer it is certainly true that some chemical substances have been unequivocally shown to promote the formation of malignant tumours, but none of these are used as pesticides. There is, however, a very much larger number of very common chemical substances which can be induced to promote tumours in susceptible experimental animals by implanting massive amounts of them under the skin. These include such common substances as gold which has been worn in contact with human skin for many years without any recorded ill-effects. It is difficult to assess what meaning experiments of this type have in relation to an actual risk in practice. Some pesticides have been suspected of producing tumours in animal experiments carried out under these highly artificial conditions but this is not necessarily a good reason for restricting their use. A careful study of nationwide mortality data throughout the U.S.A. over the past 30 years has shown no alterations which could be attributed to use of pesticides.

It is obviously desireable for growers to use pesticides which, on the basis of toxicological investigations on animals, are of as low toxicity as possible so that the safety margin is wide and, preferably, they should not be persistent but be broken down rapidly in practical conditions to harmless substances. These requirements are more than ever being taken into account in development of new crop protection chemicals.

It should be remembered that actual pests, parts of pests, insect excrement are aesthetically objectionable to the consumer and may be a health hazard. So may seeds of other plants in grain. A much more serious problem is that the by-products of some fungi are very toxic. In particular, the substance aflatoxin produced by some fungi is one of the most potent cancer-producing agents known. It has been reported that a toxin produced by potato blight may cause birth abnormalities in children. The risks to health which might arise from using pesticides therefore has to be weighed against the risks to health of not using them.

Safety of Farm Workers

In the early days of modern pesticides some deaths of spraying operators occurred because some substances used at that time, particularly DNOC, were much more highly toxic than most pesticides used nowadays, and their dangers had not been communicated to the personnel concerned. As a result, the Agriculture (Poisonous Substances) Act of 1952 was introduced in the U.K. and provides for provision, use and maintenance of protective clothing and equipment and for the keeping and inspection of records. Special regulations are made with respect to certain specified pesticides. The Act is enforced by inspectors of the Ministry of Agriculture, Fisheries and Food, and both employer and employee must comply with its provisions. As a result the safety record of pesticide use on farms in the U.K. has been excellent and there have been no fatal accidents to spray operators since the Act came into force. The Act applies only to "employees", so a farmer or other self-employed person could conceivably use pesticides without taking suitable precautions, but it is unlikely that they would do so.

Similar protection of workers is provided by legislation in most developed countries, for example, by Occupational Safety and Health Act in the U.S.A. It is estimated that about 12 deaths per year occur in the U.S.A. from occupational exposure to pesticides. Aerial application of pesticides poses particular safety problems not only for pilots flying through spray but also for "markers" on the ground. In the U.S.A. in 1970 there were about 275 accidents during aerial spraying of pesticides involving about 30 deaths. In that country aerial application is regulated by the Federal Aviation Administration.

Safety of Industrial Workers

The worker in the chemical industry is protected from hazards inherent in the processes which they are operating by a complex mass of legislation. In the U.K. these comprise all the various Factory Acts which lay down requirements for safe design of plants, safe methods of working, availability and supply of adequate protective clothing and safety training. Factory inspectors, with unlimited powers of access, ensure that the provisions of these Acts are carried out and the trade unions keep a constant watchful eye on safety. Most far-reaching of all, the recent Health and Safety at Work Legislation in the U.K., requires that the toxicological hazards, both from acute exposure and from exposure to very small amounts over a period of years, of any chemical substances with which industrial workers might come into contact shall be established and communicated to the workers and that adequate precautions shall be taken to guard against them. Similar legislation exists, or is being introduced, in most developed countries, such as the Occupational Safety and Health Act in the U.S.A. The possibility of workers being harmed as a result of their own negligence or carelessness can be minimised only by constant safety training and competent supervision. All large manufacturers of chemicals now regard the safety of their workers as a top priority and safety training and control is an essential part of their operations. Before any new plant is put into operation extensive hazard studies are made to visualise any accident which might happen and how it could be prevented.

The possibility of harm not only to workers in a factory but also to people resident in the neighbourhood from accidental failure of chemical plant and resultant dissemination of toxic substances into the air can be reduced to a minimum only by ensuring that chemical plant is adequately designed for the job, with built-in "belt and braces" safety precuations, and operated properly. One or two incidents in recent years suggest that there is still room for improvement in this respect by some manufacturers.

Safety in Transport and Distribution

Safety in transport and distribution is covered by a multiplicity of acts which regulate the carriage, transport and storage of dangerous substances. The important matter of labelling and of adequate packaging are dealt with in the U.K. by the Farm and Garden Chemicals Act of 1967 and by the Pharmacy and Poisons Act of 1933 and all the various rules made under it subsequently. The latter act also restricts the classes of people to whom certain pesticides may be sold.

In the U.S.A. transport and storage of pesticides are regulated by the Federal Environmental Pesticide Control Act, the Poison Prevention Packaging Act, the Hazardous Materials Transportation Control Act and various other acts administered by the Department of Transportation. All incidents in which a pesticide is unintentionally released during transport or storage have to be reported to the Office of Hazardous Materials. In 1973 it was estimated that about 200 such incidents occurred, 35 of which caused injury to humans but no deaths. Since about 1.3 M tonnes of formulated pesticides were transported 517 million tonne-miles this would seem to be an adequate safety record. But all accidents should be prevented if possible and the Environmental Protection Agency have recently set up a Pesticide Accident Surveillance System to investigate accidents involving pesticides more thoroughly.

Responsibility of Manufacturers and Farmers to the Public

If, despite the various regulations covering their manufacture and use, pesticides are allowed to escape from a factory or drift during a spraying operation and, as a result, cause harm to any person or to his property, then the remedy lies in common law. The principle in the U.K. was laid down by Blackburn and approved by the House of Lords, as follows "The true rule of law is that the person who for his own purposes brings on his lands and collects and keeps there anything likely to do mischief if it escapes, must keep it at his peril, and if he does not do so, is prima facie answerable for all damage which is the natural consequence of its escape". There is no need to prove negligence, the fact that injury has occurred is sufficient. This is just the same principle which applies if you allow a bonfire in your garden to set light to your neighbour's house. The principle that the person who has a pesticide in his possession is to be held responsible for any damage it causes applies not only to manufacturers and farmers but also to distributors, wholesalers, retailers, transport organisations and anyone else who may handle and keep a pesticide, including the householder who keeps pesticides in his garden shed. Of course, prevention of harm is far better than provision of recompense if harm occurs, but this can be achieved only by widespread and thorough instruction of all who handle pesticides in the possible damage they might cause if mis-used and in ways to prevent this. The possibility that accidents may occasionally happen no matter how much care is taken can be insured against by all who handle pesticides and this should be a legal requirement to ensure that any person harmed does receive financial compensation.

A particular hazard to the public is discarded pesticide containers which have not been decontaminated properly. This is, in fact, the largest source of injury from pesticides in the U.K., unfortunately, mostly to children. It is an offence under the Dangerous Litter Act of 1971 but it is often not easy to pin-point the culprit. The related hazard of adults putting pesticides into unintended containers such as lemonade bottles and leaving them within reach of children is totally reprehensible and can be stopped only by education and safety propaganda.

The suggestion is often heard nowadays that the pesticide manufacturer should be held responsible for all harm caused by pesticides no matter how it is caused or who causes it. This is as ridiculous as requiring the car manufacturer to pay for damage caused by a drunken driver or for a cutlery manufacturer to pay compensation to a victim stabbed by a maniac with a table-knife. The common law principle would seem to be the sensible one and extends not only to actual harm to person or property but also through the common law relating to torts such as negligence and nuisance.

Further Reading

Baxter, W.F., People or Penguins, (Columbia University Press, London & New York, 1974)

Biros, F.J., Pesticides Identification at the Residue Level, (American Chemical Society, Washington, D.C., 1970)

British Agrochemicals Association, Pesticides: A Code of Conduct, (London, 1968)

Busvine, J., Insects and Hygiene, (Methuen, London, 1966)

Edwards, C A., Environmental Pollution by Pesticides, (Plenum Press, London & New York, 1973)

Faust, J.D., Fate of Organic Pesticides in the Aqueous Environment, (American Chemical Society, Washington, D.C., 1976)

Goring, C.A.I. & Hamaker, J.W., Organic Chemicals in the Soil Environment, Volumes I & IV, (Marcel Dekker, New York, 1972)

Green, M.B., Pesticides: Boon or Bane? (Elek Books, London, 1976)

National Academy of Sciences, Contemporary Pest Control Practices and Prospects, (Washington, D.C., 1975)

Rosen, A.A. & Kraybill, H.F., Organic Pesticides in the Environment, (American Chemical Society, Washington, D.C., 1976)

Tahori, A.S., Pesticide Terminal Residues, (Butterworths, London, 1971)

Chapter 22

ALTERNATIVES TO PESTICIDES

The problem of crop protection and pest control has been with man ever since he turned from a nomadic hunting life to practice agriculture and horticulture in settled communities. Wherever a particular plant is growing in abundance there will be other forms of life which can use it as food. Monoculture, that is, cultivation of serried rows of the same plant over a wide area, can be maintained only by a never-ending struggle against competing plants, pests and diseases. Most insects, fungi and bacteria have evolved to live specifically on one source of food and to infest one particular type of plant or animal. In a natural environment, any plant which can be a host for one organism will be surrounded by other types of plants which cannot and so, for any particular organism, the supply of food will be limited and this, together with the presence of natural enemies of the organism (predators) limits reproduction and population growth. A large area covered by plants of the same kind standing close together offers abundant food and no barriers to inhibit movement of any pest or disease for which that type of plant can be a host. The attacks can therefore build up exponentially throughout the whole crop. The same is true for large stores of a commodity such as grain, the existence of which can enormously increase rodent populations, or for large herds of animals or massive populations of men jammed into cities. These all provide conditions under which infestations and epidemics can run riot.

The foregoing is an oversimplification, as population dynamics are a very complex problem which cannot be discussed in detail in this book. It is clear that natural communities of animals, plants, insects and fungi have a much greater stability - commonly referred to as "the balance of nature" - than the artificial communities produced by agriculture and horticulture. In them, there are rarely enough of any particular pest to destroy all the host plants and, in general, population numbers are considerably below those that the host plant could carry without serious damage. The artificial communities are inherently unstable and prone to epidemics.

Man has attempted to use both technological and non-technological methods for crop protection and pest control. Examples of the latter are scattered in ancient and mediaeval literature. The Romans held an annual feast - the Robigalia - to try to protect their cereals from mildew and rust. Ecclesiastical courts in the middle ages solemnly tried insects and excommunicated them. An Elizabethan writer recommends making love to a weeping maiden in the middle of an orchard, when the caterpillars will drop off the trees - presumably from surprise and astonishment. There is no evidence that any of these non-technological methods had any degree of efficiency.

Man has used his developing technological skills to assist him in the fight against pests and diseases. The possibility of using chemical substances occurred quite early to him and a papyrus of 1500 BC records formulae for preparation of insecticides against lice, fleas and wasps. One of the main sources of man's early interests in chemistry was his search for means to relieve human ailments and it was natural for him to seek similar remedies for the pests and diseases of his plants and animals. However, it was not until comparatively recently that he had any conspicuous success with the chemical approach. In the last thirty years his success in this field has been remarkable, as the present book shows.

Man had, however, prior to the development of modern chemical pesticides, evolved three methods of crop protection and pest control which were successful to varying degrees. These are:- (1) cultivational practices (2) plant breeding (3) biological control methods. In the remainder of this chapter application of these three techniques to control of weeds, fungi and insects will be summarised. This will necessarily be a very short account of a complex topic and so subject to shortcomings and accusations of superficiality, but the reader who wishes to know more can consult many excellent detailed books on the subject.

Weed Control

Weeds have, in the past, been controlled by careful preparation of the seed-bed by ploughing and other cultivational practices, and by mechanical weeding and hand-hoeing. The latter operations are not possible in cereal crops for which, until the advent of herbicides, there were no effective methods for controlling weeds once they had emerged. Modern trends towards reductions in crop rotations, precision drilling, continuous cropping and high plant densities for many crops have made traditional methods of weed control difficult to apply, and there is scope for development of new techniques of cultivation.

Legislative control of the quality of crop seeds and regulations specifying maximum permitted contents of weed seeds in them have helped, in modern times, to reduce the spread of weeds in this way. Dissemination of air-borne seeds has been reduced by control of weeds in non-cropped areas, particularly by prevention of their seeding.

Biological control by introduction of insects which feed specifically on certain weeds has some applications. The most successful example is control of the prickly pear in Australia by the cochineal insect, *Dactylopius tomentosus*, and by the moth, *Cactoblastis cactorum*. This technique is applicable only to long-term control of a single dominant weed which is present over large areas of uncropped land. It is not suitable for rapid control of mixed weed infestations in particular crops or for the high level of control needed in arable crops. It is essential to be sure that the introduced insect will not attack related plants of economic importance nor produce any adverse effects on the natural ecological balance in the area. The potential environmental hazard of introducing such an agent might be much greater than that of any herbicide. A related technique which has had some success is the use of weed-eating fish, such as the grass carp, for control of aquatic weeds. This fish does not breed in the UK so could not cause ecological upset: it can be regarded as an environmentally safe persistent herbicide.

For the foreseeable future, weed control will depend mainly on herbicides, which play an essential part in modern intensive cultivation in developed countries. They are versatile, labour-saving, fuel-saving and convenient, and they give substantial economic returns for their cost. Their application by light tractors or from the air causes less disturbance to soil structure than compression by heavy tractors or continuous trampling by manual workers, and this can be an important factor in maintaining maximum fertility. In those very large areas of the world which are subject to wind or water erosion, use of herbicides to avoid ploughing and other massive disturbance of the soil can be of great value. For these reasons, herbicides comprise by far the largest proportion of all pesticides sold, and the demand for them is increasing rapidly. Fortunately, as has already been pointed out, they are generally not very poisonous and, as far as is known, they have no adverse effects on the environment and they do not induce resistance in the weed species. Possibly their overenthusiastic use might eliminate some potentially useful wild plants.

The trend for the future will be to regard weed control merely as a part of the total management of farm and regional agricultural systems. Development of this systems approach will involve balanced programmes of rotation of crops, rotation of herbicides, use of combination and sequential treatments, and an integration of chemical control with cultural practices.

Disease Control

Until the introduction of fungicides and bactericides, mankind had been singularly unsuccessful in controlling plant diseases which were often the cause of total loss of crops and consequent famine. This crop loss still occurs in developing countries and, even as recently as 1970, 800 million bushels of corn were lost in the U.S.A. in an epidemic of Southern corn blight. This was an example of the damage which can be caused to susceptible varieties grown over large areas when weather conditions favour rapid spread of a virulent fungus. The total financial loss of about $2 thousand million was paid by the consumers of corn products since the price of corn rose from $1.28 per bushel to $1.60 per bushel as a result of the epidemic.

Build-up of populations of soil-borne diseases have, in the past, been prevented by systems of crop rotation, but there has been little direct protection against air-borne diseases. In modern times, there has been a greater appreciation of the importance of farm hygiene in limiting the spread of diseases. Removal of rubbish heaps, diseased tissue, dead leaves, fallen fruit, etc. which could be sources of infection is most desirable. It is interesting to note that, as long ago as 1873, Le Baron suggested that pigs should be allowed to run free in orchards to eat fallen fruit and thus prevent infection. Contaminated manure has often been a carrier of diseases. Legislative control over seed quality and restrictions on sale of infected seed have helped control infection, and plant quarantine laws have helped prevent spread of diseases from one country or state to another.

For many years it has been realised that external factors such as weather, soil condition, application of fertilisers, time of sowing, etc., could greatly affect the incidence of diseases. Thus, there is evidence that less damage is caused by yellow rust to wheat growing on heavy, deep, moist soils than to wheat growing on light, shallow, dry soils. An effect of soil aeration is suggested by the observation that red rot of sugar cane is common on impervious soils but rare on porous soils. Differences in soil drainage may account for variations in incidence of root rot in peas. It has been suggested that susceptibility of potato blight is increased by heavy nitrogenous manuring. Early-sown winter wheat is more prone to *Helminthosporium* disease than is a late-sown crop.

Studies of the life cycles of fungi have revealed that some can live on alternative host plants to the crops which they usually attack and that some need alternative hosts in certain stages of their development. Eradication of the wild host-plants may effect a significant degree of control. A well-known example is wheat rust for which barberry is the alternate host. This connection was recognised long before the reason for it was understood, and a law was enacted in Massachusetts in 1760 for destruction of all barberry bushes.

There are very few examples - mostly in diseases of trees - of successful commercial control of fungal diseases by insects or other fungi. Apart from fungicides, the most important method for preventing loss of crops from diseases is by breeding resistant varieties of crop plants. This was first suggested in 1815 by Knight who advocated cultivation of fungi resistant cereals. A practical difficulty is that it is not often that the variety which has the greatest resistance to disease turns out to give the greatest yields or best quality. Where this is the case, cost-benefit considerations may make it more advantageous to the farmer to grow the variety which gives the biggest yield and to protect it, if need be, by fungicides. It may take twelve years, or more, to develop a new variety for commercial use and a variety which has been bred during this time specifically for resistance to disease may have to compete with other new varieties bred during the same period for improved yield and quality.

In the past, resistance of a plant variety to a certain disease has generally depended on one gene. Sometimes this is sufficient to give a permanent solution to a particular disease problem. Resistance of potatoes to wart disease depends on a single genetic factor and the disease has been diminished in the UK by legal restrictions on cultivation of susceptible varieties. More usually, however, the disease develops new strains which break down the resistance of the plant, often quite quickly and suddenly. In most natural disease populations there are strains capable of overcoming inbred resistance in plants and it is, generally, only a matter of time before these strains become dominant. For instance, the yellow rust of wheat has continually developed new strains which have successively overcome the resistance of a range of new varieties. Resistance can also be broken down by external factors, e.g. if the plants are transferred to different soil or climatic conditions. This widespread failure of major gene resistance has led plant breeders to turn from using race-specific resistance ('vertical' resistance) to non-race-specific resistance ('horizontal' resistance). This approach coupled with developing knowledge of 'genetic engineering' offers more hope for the future, and plant breeding will certainly have an important future contribution to disease control. If this is to be achieved, it is necessary that as many varieties as possible of the major crops of the world and of their wild progenitors shall be collected and conserved in 'gene banks' to give the breeders sufficient basic material to work with.

The use of resistant plants is one of the most economical methods of disease control. Once a resistant variety which gives as good quantity and quality as susceptible varieties has been commercially established, there is no annual outgoing for the farmer for crop protection. However, resistance may break down after a time so a succession of resistant varieties may be needed.

However, as it is unlikely that a plant bred to be resistant to one particular disease will be immune to other pests and diseases, there are obvious limitations if the crop is likely to be attacked by a variety of these. In the process of natural selection varieties evolve which have a good average resistance, since the plant has to fight not only one but many enemies. It can happen that breeding of resistance to one particular disease increases susceptibility to other diseases. As with weeds, future management of diseases will depend on a judicious use of resistant varieties, cultural practices and fungicides in combination with each other and as part of a total system of farm management.

Insect Control

Traditionally, cultivational practices have been relied on for control of insects and nematodes, particularly the rotation of crops to prevent build-up of populations of soil pests, adjustments of sowing and harvest times, and removal of weeds, rubbish heaps, etc., which could be overwintering refuges. Thus, oats sown before the end of February generally escape damage by fruit fly, whereas those sown in April are often severely attacked.

Breeding of varieties of crop plants resistant to or tolerant of certain pests has not been as well exploited or as successful as breeding for resistance to diseases, so there is considerable scope for future developments. It is believed that many plants, such as the yew and laurel, are naturally resistant to a variety of pests, because they contain natural repellent chemical substances. This offers the possibility of a new type of chemical approach to control.

Since it is mainly insecticides which have been suspected of producing undesirable environmental effects, it is in the field of insect control that most attention has recently been given to biological control, although this type of control was, in fact, used before insecticides became generally available.

One technique involves culture and release of millions of male insects which have been sterilised either by radiation or by treatment with suitable chemicals so that they compete with normal males for mates but fail to fertilise the females, so the pest population diminishes. World progress on application of this technique has been disappointing. One problem is the prodigious number of male insects required. It has been calculated that, if there is a natural population of one million virgin females in an area, then release of two million sterile males in each of four successive generations will theoretically reduce the population substantially to zero. This is obviously not a method which can be applied on a single farm, but must be applied to the total region in which the insect occurs, or to an area which can be geographically protected against an inflow from outside. Thus, it had a spectacular success in eradication of the screw-worm from the island of Curaco. The campaign against the Mexican screw-worm in the south-west USA necessitated release of 4 thousand million sterile males over an area of 85 5000 square miles. It is a long term technique which cannot be used for rapid control of a sudden infestation. Also it is most effective at low pest population levels and cannot deal with major attacks. It is a technique which is really effective only with female insects which mate only once.

If this technique is to be used extensively in the future, economical methods of commercial mass production of sterile insects will have to be developed. A deep understanding of the reproductive processes and habits of the insects which are to be controlled will also be needed. A major problem is that the insects to be sterilised usually have to be reared on the crop plant. It has been estimated that, in the Mediterranean area, rearing of sufficient olive fruit flies to provide effective control of the natural pest population would require a substantial proportion of all the olive trees in the area to be devoted to this purpose. It is therefore unlikely that the sterile male technique will be used on its own in the future. It is more likely that infestations will first be reduced to low levels by conventional insecticide treatments and then kept under control by sterile males.

A much more important method of biological control of insects depends on introducing a predator or parasite of the pest into the affected area. One of the earliest applications of this technique was the control in 1889 of the cottony cushion scale on citrus trees in California by the Australian lady-bird beetle, Vedalia cardinalis. This method was highly successful. It has been suggested that birds might be used for biological control of insects, since they naturally devour large quantities of insects and insect larvae, but this idea has never been practically developed. An attempt was made in the USA to use bats to control codling moth but, by and large, control of insects has been by other insects. Most success has been with the use of imported predators or parasites to control imported pests, particularly in Australia, and difficulties are often encountered when attempts are made to control an indigenous pest.

The method has had some success in Southern Europe but none at all in Northern Europe. In the UK, the only examples of successful control have been in the highly controllable environment of the glasshouse. In the USA, up to 1956, 485 biological agents had been released against 77 species of pests. Of these agents, 95 became established and exerted some control over 22 of the pest species. Very large numbers of the predators or parasites are required. Thus, in Canada, between 1916 and 1956, one thousand million specimens of 220 species of parasites and predators were liberated against 68 pest species. Production of these numbers presents considerable difficulties, particularly if the parasite has to be reared on the host plants.

A related technique is the use of pathogens, that is, fungi, bacteria or viruses or their toxins which infect or poison the insect and cause its death. This is not an easy method to use since it is very susceptible to changes in external factors such as weather conditions. One of the most successfully used pathogens is Bacillus thuringiensis, which has proved effective against a number of species of caterpillar.

Control of insects by predators, parasites or pathogens can, if successful, be a cheap method of crop protection. Biological control agents also have the advantage of being highly specific in that they affect only the target pest and have no direct effects on non target species. This specificity can, however, be a disadvantage when certain minor pests which are controlled incidentally by chemical pesticides used against major pests become major pests themselves when competition is removed by biological agents highly specific to the original major pests. This is an example of Nature's tendency to fill any available ecological niche.

Use of biological control agents requires a great deal of care and extensive background studies. It is also a technique for use by specialist operators and not one which can easily be applied by the average farmer. It is applicable to large regions rather than individual farms and is a long-term measure which cannot be used to give rapid control of an unexpected infestation. It is not usually effective at low population densities of the pest, so it is normally necessary to maintain the pest population at a minimum level by artificial infestations to prevent the biological control agents from dying out. Expert judgement and constant observation are required to ensure that the right balance is maintained and this is very difficult to achieve in regions with unpredictable weather patterns.

Biological control is not without the possibility of environmental risk, since introduction of an alien life-form into an established environment is bound to have some effect on the equilibrium of species in that area. The possibility always exists that an introduced insect might become a pest of some economic crop. Thus, the larvae of blister beetles imported into the Philippines to control louse eggs gave adult insects which attacked lucerne. An introduced pathogen might change and become infectious to man or animals. It is therefore essential to undertake extensive studies before any new system of biological control is introduced, just as it is with a pesticide.

What then is the future for non-chemical methods of crop protection and pest control? Traditional methods based on cultivational practices, such as crop rotations and manipulation of sowing times, depths of planting, plant spacing, size of fields, sowing patterns, etc. will continue to be utilised and new methods such as direct drilling, minimal cultivation and modification of soil structure will be developed to meet the requirements of modern farming systems.

In temperate countries with rapidly changing weather patterns, such as the UK, pest and disease attacks on outdoor crops tend to be sporadic and rarely reach epidemic proportions. The instability of the environment tends to militate against use of sterile male and predator/parasite methods. The non-chemical method which has most future utility in these countries is development of resistant crop varieties. Plant breeders, aided by modern developments in genetic engineering will make an important contribution to future crop protection. In particular, the possibilities of plant breeding for resistance to pests, as distinct from diseases, have hardly begun to be exploited. However, this is a very long-term task.

In countries in which large areas of a single crop are grown, especially those in tropical or semi-tropical regions with regular weather patterns, sterile male, predator/parasite and pathogen methods may have utility provided that commercial mass production of the required organisms is feasible and that problems of registration requirements for use of predators, parasites and pathogens can be solved. It is implicit in such methods that they should be applied over whole regions, not individual farms, and so they will need to be applied by Government agricultural departments and financed on a national or regional basis from public money, as for public health or education. Use of such methods requires that pesticide usage over the region shall be fixed in advance and strict legislative control over use of pesticides by individual farmers in the region will have to be introduced. The modern farmer is working against a background of increasing land values and farm rents, high cost of marketing and fuel, shortage of labour and demands from the processed food industry for substantially undamaged crops. He will oppose any control measures which are labour intensive or which impose unacceptable constraints on his farming operations and he will wish to retain freedom to use the most economical methods to safeguard the quality of his crops. He considers that, if protection of the environment is a national interest, it should be paid for by the nation and that he should not be left, as at present, to foot the bill.

It seems clear that, within the constraints of the need for increased production of food, increasing costs of production and demand for high quality, the farmer will require crop protection and pest control methods which are cheap, highly effective, easily applied and quick acting and that chemical pesticides will be the mainstay for many years to come. However, all the non-chemical methods of crop protection and pest control will have their part to play, alongside pesticides, in the future. Pointless arguments over the respective merits or demerits of the various methods serve no purpose. What is needed is widespread adoption of the concept of pest and disease management by a carefully balanced use of all available methods in a way that is most effective against the pest or disease and least harmful to its natural enemies and to the environment. The decision on the precise way in which the various techniques should be investigated in each pest and disease situation obviously depends on a comprehensive cost benefit study of that situation. The aim should be to keep pest populations continuously below a level at which unacceptable damage to the crop is produced - the so called 'economic threshold'. This level is not easy to define because there is a continuous relationship between pest numbers and crop yields which is influenced by environmental and climatic stress on both the pest and the plant. The fact that arable crops are short lived and nowadays often do not follow one another in set rotations may make a sustained policy of pest management difficult. A psychological barrier is the desire of the farmer to see an immediate and effective kill and a practical difficulty is the current demand of consumers and food processors for substantially complete freedom from blemish or damage.

It has already been pointed out that successful application of many non-chemical methods of crop protection and pest control demand a very detailed knowledge of the ecology, habits and life-cycles of the various pests and diseases. The amount of research work required is prodigious and it is pertinent to ask who will do it and how it is to be paid for? Although most large companies engaged on discovery and development of pesticides now devote considerable research resources to studying the problem of integration of chemical and non-chemical methods of crop protection, much of the basic knowledge required is more appropriate for academic than industrial research. The inescapable conclusion is that, if it is in the nation's interest to achieve efficient integration of chemical and non-chemical methods into total systems of pest and disease management, then the research work will have to be paid for out of public money, that is, by the taxpayer and consumer.

A related problem for Government is that of disseminating to farmers all the detailed knowledge required for such systems of pest and disease management and of assisting them with their application, or of taking the responsibility for that application when this is needed over whole regions. Complex systems of pest and disease management would make far too great demands on the average farmer's time and technical knowledge. The main burden will fall on Government agricultural advisory and extension services, but it is likely that these will be augmented by emergence of a class of 'plant doctors' who will take full responsibility for the health of a farmer's crops. These may be independent consultants, comparable to veterinary practitioners, or more likely, members of the staff of the larger pesticide manufacturers. It is logical that the pesticide industry should shift from being a purely manufacturing industry to being a service industry and that it should offer to its consumers a total pest and disease management service on a contract basis. From the grower's viewpoint, this would ensure safe operation and maximum financial return, from the food processors' viewpoint it would ensure maintenance of quality standards and defined crop protection programmes, from the consumer's point of view the danger of accidental misuse and introduction of toxic residues would be minimised and from the environmentalists' point of view all operations would be supervised by trained biologists and ecologists. For the manufacturer, it would ease the present economic difficulties of discovery and development of pesticides by giving him the added value of a service and this, in turn, would facilitate the introduction of a wider range of effective, environmentally safe pesticides for both major and minor crops by improving the financial expectations of such projects. It is possible that a comprehensive cost-benefit study of crop protection and pest control from the point of view of the community as a whole may point to a move in this direction as an approach to optimisation. Many of the larger pesticide manufacturers have moved a long way in this direction by providing extensive expert technical service to farmers, although their representatives may need to acquire considerably more biological expertise in order to cope with future pest management programmes.

Further Reading

Apple, J.L. & Smith, R.F., (Ed)., Integrated Pest Management, (Plenum, New York, 1977)

Baker, K.F. & Cook, R.J., Biological Control of Plant Pathogens, (W.H. Freeman, San Fransisco, 1974)

De Bach, P. (Ed)., Biological Control of Insect Pests and Weeds, (Chapman and Hall, London, 1970)

Gilbert, L.I., (Ed)., Juvenile Hormones (Plenum, New York, 1977)

Hall, S.A., New Approaches to Pest Control and Eradication, (American Chemical Society, Washington, D.C., 1976)

Huffaker, C.B. (Ed)., Biological Control (Plenum, New York, 1977)

International Atomic Energy Agency, Sterility Principle for Insect Control and Eradication, (Vienna, 1971)

Price Jones, D. & Solomon, M.E., (Ed)., Biology in Pest and Disease Control, (Blackwell, Oxford, 1974)

Chapter 23
FUTURE TRENDS

For the foreseeable future, pesticides will remain the main weapons in the farmer's armoury of defences against pests and diseases. Over the past thirty years, the developing technology of crop protection and pest control has become highly effective and now provides methods for preventing or reducing the damage caused by most economically important pests and diseases. Application of this technology has proved very financially rewarding to farmers, and this is the main reason for widespread use of pesticides and for the remarkably rapid growth of the pesticides industry. It has also made a major contribution to the substantial increases in yields per hectare of many crops, particularly of cereals, which have been achieved during that period.

Human populations have increased steadily and are likely to go on increasing, but the amount of arable land available in each country remains the same, or actually decreases as cities and towns expand to accommodate increased numbers, as agricultural land is taken for motorways, airports, shopping plazas, parking lots and the like, and as affluent societies demand that more land be reserved for their amenity and recreation. If these populations are to continue to be provided with their present variety and quantity of foodstuffs at reasonable prices and if they are not to be faced with the need to spend much higher proportions of their incomes on food than they do at present, then yields per hectare will have to be increased still more and, to achieve this, use of crop protection technology will have to be intensified, not restricted. With growing populations the human race simply cannot afford to allow substantial proportions of the food they grow to feed insects or nourish microorganisms nor can they permit unwanted plants to compete with essential crops for growing space, nutrients and water.

The costs to the community of discontinuing the use of some major pesticides have been calculated by the United States Department of Agriculture and have been referred to in Chapter 2. For example, if the cereal herbicide 2,4-D were no longer used, the direct costs of U.S.A. farm production would increase by about $ 300 million per year, which the consumer would have to meet by increased food prices, and would necessitate 20 million more hours work from farmers and their families, without additional income, just to maintain current levels of production. A practical demonstration of the economic significance of crop protection was provided in the U.S.A. in 1970 when failure to control the Southern corn leaf blight cost the consumer $ 2 billion in increased food prices.

As well as crop protection, control of insect pests which debilitate animals by the irritation they cause or which actually transmit lethal diseases to them, is an important factor in maintaining adequate supplies of food, as well as being an humane objective. It is also essential to protect public health and to alleviate human suffering by controlling insects which carry disease to man. It is desirable to prevent the economic losses and dangers to life caused by "biodeterioration", which comprises the effects of attack of materials by living organisms of a wide range of types. Economical losses can range from rotting of timber and textiles to increased fuel consumption by ships consequent on fouling by algal growth. Safety hazards may arise from causes varying from collapse of wooden buildings undermined by termites to fires started by rupture of gas mains by the sulphuric acid produced by anaerobic bacteria or by rodents gnawing through electric cables.

The need for crop protection and pest control is now generally recognized and accepted but there is considerable public controversy about the methods which should be used and about the significance of unwanted or hazardous side-effects which they might produce. This matter has already been discussed in Chapter 21 and the various arguments will not be repeated here. It appears inevitable that, no matter what the validity may be of these arguments and of the accusations made against pesticides, public opinion, spurred on by the mass media, will demand tighter controls over the registration of pesticides and greater restrictions on their use, and that politicians will respond to these demands. Affluent societies may be willing and able to pay the increased food prices which will inevitably result in order to feel a greater sense of security against possible or imagined risks. If, however, controls and restrictions reached a point where they resulted in actual food shortages, then public opinion might undergo a sharp reaction, as it probably would also if insect-borne diseases such as malaria once again became prevalent in areas such as the southern states of the U.S.A. where they were originally indigenous. Even if controls and restrictions did not result in actual food shortages, they might reduce the amount of foodstuffs which major exporting countries such as the U.S.A. had available for sale abroad, and this could have economic, political and international consequences which would be unwelcome to the Governments of those countries.

Crop protection and pest control must be continued and intensified in the future. What alternatives are there to pesticides? Biological methods are very successful in some situations and can be economical, but they have limitations and are not without possible risks. Cultivational practices, improved by modern agricultural techniques and machinery, will certainly have an important part to play in the future. Plant breeding has a substantial contribution to make, although development of varieties which give greater yields, which is the main objective, can bring problems if, as is often the case, varieties which are more productive are also more susceptible to pests and diseases.

None of the alternatives to pesticides can, by themselves, provide solutions to all crop protection and pest control problems, particularly solutions which are both practicable and economical. If there is any reason to believe that certain pesticides may present an unacceptable degree of risk of toxic hazards or adverse environmental effects then the alternative which will cause the least complications and fit in best with established agricultural practices is to replace them by other pesticides which have been proved to be free from residual or environmental effects or for which the risks of these is substantially less. Discovery and development of new pesticides to replace any older pesticides which are considered to present an unacceptable degree of risk, will be major future targets for the pesticides industry, as will still further increases in the very great safety and freedom from risk of most modern products. It is now generally accepted by the industry that persistence in the environment is an undesirable property and that short-lived compounds are most unlikely to cause long-term environmental effects, so the aim will be to develop pesticides which do the job required of them and are then decomposed or degraded to harmless products. Society must, however, accept that pesticides which are not only effective in preventing crop losses but also meet the most stringent requirements of freedom from risk will be more expensive and that society will have to meet the increased costs. This may present some political problems because people are very often greatly in favour of desirable objectives provided they personally do not have to pay to achieve them. Furthermore, if the only known method of controlling a particular serious pest or disease of a major crop is to use a pesticide which involves some degree of risk then society should carefully consider the social, political and economic implications of a decrease in the amounts of that crop and of the effects on the well-being of the farming community before demanding withdrawal of the pesticide in question.

There is considerable future scope for us to learn how to use existing pesticides so as to produce maximum benefit with minimum of risk. This will require much greater understanding than we at present possess of the ecology and population dynamics of pests and diseases so that reliable "early warning" systems can be instituted to permit optimum timing of crop protection operations and to avoid unnecessary or excessive applications of pesticides. In this respect, improvements in formulations and in application methods and equipment have an important part to play.

However, all this knowledge will be useless unless it can be imparted as practical advice to farmers and to those who actually carry out the crop protection and pest control operations. For pesticides, as for most technologies, the risks lie not so much in the intrinsic nature of the technology but in the way it is applied, and the greatest problem is how to deal with human ignorance, apathy, carelessness, stupidity or greed. There is a massive task of education to be undertaken, the burden of which will fall largely on Government agricultural advisory and extension services and the costs of which will have to be met from public revenue.

All knowledgeable and informed people who are concerned with crop protection and pest control are now agreed that we should move towards the concept of "pest management". This requires that we develop as extensive a range as possible of chemical and non-chemical weapons for the continual battle against pest and diseases, acquire the biological knowledge required to decide how each may best be used and then provide the farmer with information on their relative utilities and limitations and with expert advice to help him select the best combination for each situation. If this aim can be achieved the farmer will have a range of options of which use of pesticides is only one, albeit the most effective and generally applicable one, but which are not mutually exclusive and which may, according to circumstances, be used singly, in combination or successively.

Future trends in non-chemical methods of crop protection and pest control have been described in the previous chapter. In the remaining sections of this chapter we shall discuss future trends in herbicides, fungicides and insecticides, in formulation and application technology, and towards "pest management", and assess the possible future effects of all these on the chemical industry.

Future Trends in Herbicides

Herbicides present very little risk of toxic residues or environmental damage. Most of them have low toxicities not only to men and animals but also to birds, fish, insects and microorganisms, and are therefore unlikely to cause adverse ecological effects on any species. They are normally applied well before harvest and often before emergence of the crop or even before sowing, so the risk of toxic residues is very slight. The current tendency is to phase herbicides which persist for long periods in the environment out of agricultural use (although highly persistent chemicals may be used for long-lasting weed control in non-agricultural situations) and to replace them with compounds of shorter life.

During the past 30 years a wide range of herbicides have been developed and there is now available to the farmer a reasonably effective and economical solution to most weed control problems. Many existing herbicides which have been in production for a long time and which are now free from patent restrictions have become comparatively cheap because of scale of manufacture and competitive production and many more commercial products will move into this category during the next 10 years.

Since current herbicides are both effective and safe it is pertinent to ask whether there is any incentive to try to discover and develop new products especially since these would almost certainly be more expensive and farmers might be reluctant to pay considerably more for purely marginal advantages. There are, however, a number of future trends which may justify a continuing search for new herbicides.

Widespread use of some standard herbicides has created major weed problems by altering competition and balance of populations; for instance, a major weed problem in cereals is now other graminaceous plants. Further changes in weed flora can be expected, and there will be problem weeds in particular areas and particular crops. As long as there is one weed species resistant to a particular herbicide, there is the possibility that this species may become a local problem, or even a national problem, when competition is removed. The economics of production of various crops may change in response to changes in local, national, and international economics or to changes in the nature of consumer foodstuffs. New farming techniques may call for new techniques of weed control; for example, if the dwarf wheat hybrids are introduced their cost will necessitate precision drilling and this, coupled with their growth habit, will mean less competition to weed growth from the crop and greater need for effective weed control. There will be scope for more residual herbicides which do not present any pollution problem and for pree-mergent herbicides which are more highly selective so that they can overcome environmental variations and can be used over a very wide range of soil and climate conditions. New herbicides may make essentially new techniques of cultivation possible in the way that the bipyridylium herbicides have engendered the minimum tillage techniques, which are so useful. New varieties of existing crops may be introduced and also, possibly, completely new types of crops may assume economic importance either for food or for industrial uses, and these may all present new weed problems which need new solutions.

The rapidly increasing cost of agricultural land will necessitate maximum land utilization; and as the limit to increasing yields by breeding new varieties is being approached, herbicides will have a large part to play in allowing even higher plant densities by providing the weed control needed to conserve water supplies and facilitate harvesting. Herbicides are a substitute for machinery and labour, both of which are becoming more expensive; and labour will get scarcer as the world-wide drift from the country into the cities is unlikely to be reversed. Population pressures provide another reason for maximum effective use of agricultural land as expansion of urban communities into the countryside demands more houses, shops, motorways, airports, etc. Herbicides will assume a greater importance in the control and improvement of grassland, particularly in relation to the need for increased efficiency in meat production, and also in control and management of commercial timber. At the same time, affluent societies will demand more recreational grassland, forest and water but what they want is not nature run riot, so amenity management will assume greater importance, especially as tourism is now a major part of many national and local economies; and this will bring new weed control problems.

The farmer is demanding, and will continue to demand greater convenience in use and application of herbicides. Logistics are very important to him and he may decline to undertake an extra operation at a busy time even if it might increase crop yield. The food processor is calling for more stringent standards of quality and freedom from impurities and taint. The increasing tendency towards home-saved seed to offset rising costs of commercial seed calls for stricter weed control in seed crops if the total weed problem is to be contained. There is a move towards zero weed tolerance not only in horticulture but also in agriculture, one reason being the very high cost of inspection of processed crops. Last, but not least, the public will continue to demand higher standards of quality and appearance and, at the same time, freedom from residues in the crops and from hazards to the environment in which the crops are grown.

All these trends suggest that there will be continuing need for discovery and development of new herbicides. These will be mainly herbicides of the conventional type because of their proved effectiveness and economy. Some novel chemical approaches to weed control are being investigated such as use of compounds which interrupt winter dormancy of weed seeds in the field so that they emerge at times when climatic conditions are unfavourable for their survival, and use of seed-treatments which reduce the time between sowing and germination of crop seeds so that these emerge quickly and can compete more effectively with later-emerging weeds. Novel chemicals of these types impinge on the field of plant growth regulators. The future will without doubt see many advances in the use of chemicals to change the ways in which plants grow so as to make them more convenient to cultivate or harvest or to make them produce more of the types of growth which are commercially useful. Although these effects do not come within the category of crop protection and pest control the modifications in plants produced by growth regulating compounds could, in some cases, result in less susceptibility to pests and diseases and greater ability to compete with weeds.

Future Trends in Fungicides

There is no doubt that the standard protective fungicides such as sulphur, copper preparations and dithiocarbamates are cheap and, in a great many cases, adequately effective. For this reason, they are likely to continue to be widely-used and to constitute the major part of the fungicide market. With rising development costs and the increasing difficulty of obtaining registration for new compounds search for complex organic protective fungicides is not a financially attractive target for the pesticide industry.

Foliar-applied protective fungicides are effective only against air-borne fungi. The main line of defence against soil-borne fungi is likely to continue to be seed-dressing. The main seed-dressing fungicides of the past, the organomercurials, have come under a cloud because of the hazard of persistence of mercury in the environment. Mercury seed-dressings actually present very little hazard and introduce very little mercury into the environment - much more comes from cremation of corpses with mercury fillings in their teeth - but public opinion is unselective so their use will probably be discontinued. There are opportunities for future research to develop new fungicides effective against soil-borne diseases and which can be used either as seed-dressings or by incorporation into the soil at the time of sowing.

One of the most notable events in pesticide science in the past decade has been the development of the systemic fungicides which can enter the plant and combat the disease within it. The future will see discovery and development of many more compounds of this type. The systemic fungicides so far discovered are most effective against those fungi which do not invade the plant very deeply, such as the mildews and the rusts. Hopefully, new compounds will be produced which can deal with those diseases which penetrate the plant tissues deeply, such as the blights. Also, because resistance to a particular systemic fungicide is likely to develop comparatively quickly, there is need for a wide range of systemic fungicides so that any one compound does not have to be used over a number of seasons in a particular locality and so that their effectiveness can be maintained by restrained use and rotation.

A great deal of work has already been done, and more will be done in the future, on the vast number of naturally occurring fungitoxic substances in plants the presence of which accounts to a considerable extent for the fact that most plants are resistant to most fungi. No such compound has yet found commercial use but this may only be a matter of time and we may possibly see the development of a range of natural antifungal agents for treatment of plant diseases similar to the antibiotics used in human and animal medicine.

Future research may also produce synthetic compounds which are not fungicidal in themselves but which will provide indirect control of plant diseases by stimulating endogenous production of the natural antifungal agents in the plant or, possibly, by altering the morphological characteristics of the plant so that it is more difficult for the fungus to penetrate, or by changing the biochemistry of the plant in some way which makes it less able to support the fungal growth.

Future Trends in Insecticides

It is use of insecticides which has generated most public concern about possible adverse effects of pesticides on the environment. When they were first introduced, the organochlorine insecticides were, without doubt, misused as a consequence of ignorance on the part of farmers and public health authorities and over-enthusiasm by industrial salesmen. Incidents of the type dramatized by Rachel Carson in "Silent Spring" did occur. The organochlorine insecticides are the only pesticides for which there is any evidence of long-term environmental effects as distinct from transient damage caused by a local accident or misapplication, but, as a result of public concern about organochlorines, all insecticides have come to be regarded as suspect.

Particular attention has, therefore, been given to possible alternatives to conventional insecticides as methods of insect control. There are good reasons for doing this, quite apart from satisfying public opinion. Production of many crops requires the assistance of pollinating insects, and the health and fertility of the soil is dependent on the activities of its micro-fauna. Interference with either of these might have deleterious effects on crop yields. In most agricultural environments there is a balance of insect populations, of predators and prey, disturbance of which may lead to a resurgence of pest populations after control measures have been applied, which could exceed the original infestation, or to an emergence as major pests of insects which had previously been kept in check by natural enemies or by competition for food sources. In the comparatively new agricultural environments of monoculture insect populations tend to be metastable and any imbalance which is induced in them can rapidly become exaggerated, whereas long established mixed environments tend to be stable and any population imbalance in them is self-correcting. The economic and environmental consequences of non-selectivity in insecticides have, therefore, the potential to be more serious than those of non-selectivity in fungicides or herbicides. Finally, there is the problem of development of resistance to insecticides. It now seems probable that, no matter which insecticide you select, there will be some individuals within every species of insect which have a natural resistance to that compound, and that the selection pressures of repeated applications lethal to most of the individuals in a species will rapidly make these resistant strains dominant and thus render that particular insecticide of little further value for control of that particular pest.

Much attention has, therefore, been given to the methods of biological control of insect pests outlined in the previous chapter. Given the right circumstances these can be both spectacularly successful and cheap but they have considerable limitations and do not, at the moment, provide a complete practical answer to all insect control problems.

Because of the limitations of biological control methods much attention has been given to improving conventional insecticides. Until recently, there have been, apart from one or two isolated compounds, only two main chemical classes of insecticides, namely, the organochlorines and organophosphates.

The organochlorines vary in their toxicity to men and animals. Some, like dieldrin, are highly toxic whereas others, like DDT, are virtually non-toxic and extremely safe to use. Unfortunately, the organochlorines are not broken down by natural microorganisms and have therefore persisted in the environment, virtually for ever, and have become widely distributed in small amounts, producing adverse effects on some non-target species, although there is no evidence that permanent damage has been done by them to any species as distinct from individuals in that species. Because of their lipid solubility they have accumulated in the body fats of men and animals and we now all have a certain amount in our bodies. There is no evidence that this does us any harm whatsoever but, on general principles, it is undesirable.

The other main group of insecticides, the organophosphates and related carbamates, depend for their activity on inhibition of acetylcholinesterase, an enzyme which is essential to the proper functioning of transmission of nerve impulses to muscles. As this system is common to both insects and animals many organophosphorus insecticides have a high mammalian toxicity. Many, however, have been developed which are very safe to use because the particular compounds can be broken down and rapidly detoxified in the animal body but not in the insect body. By the same taken, organophosphorus insecticides have been developed which have a very high degree of selectivity towards certain insects.

Organophosphates and carbamates break down rapidly in the environment and are therefore not persistent and present no long term residual effects. However, resistance of a number of major insect pest species has developed to many of the most widely-used organophosphates and there is a tendency for an initial resistance to one particular compound to develop into a generalised resistance to all compounds of the group.

The potential of the new synthetic pyrethroids has yet to be assessed. Certainly they appear to have very low toxicities to men and animals and are not persistent in the environment. Nevertheless, because they are general insecticides, they may pose problems of adverse effects on beneficial and non-target insect species, and their comparatively high toxicities to fish may restrict that use near rivers, lakes and watercourses.

Organophosphates, as has been said, act on an enzyme system common to both animals and insects. Much thought has been given to the differences in physiology and biochemistry between animals and insects and to ways in which these differences might be exploited for selective control of insect pests.

Arthropods have no internal skeleton. Their shape is maintained by a horny outer casing incapable of continued growth. Characteristically, at regular intervals, they produce a new soft integument inside the old one, which is then split and discarded. The new integument hardens after expansion. These moulting operations separate each stage of development (instar) from the next. Although successive instars may differ mainly in size, sometimes a considerable change of structure is involved. Such change of structure (metamorphosis) is particularly significant in insects where only the final instar is sexually mature. In the lepidoptera (butterflies and moths) there are four instars of the larval (caterpillar) form followed by an immobile non-feeding fifth instar (pupa) during which the final winged insect is formed.

This type of development and metamorphosis has no counterpart in vertebrate animals and is controlled by two hormones which are specific to insects. One, the so-called "moulting hormone" (MH) controls the process of development of a new integument and shedding of the old at each instar which takes place without change of structure. The other, the so-called "juvenile hormone" (JH) controls the process of metamorphosis. The chemical structures of both have been determined. MH is a steroid, ecdysone, and JH is methyl trans, trans, cis-10-epoxy-7-ethyl-3,11-dimethyl-2,6-tridecadienoate, although there may be minor variations in structure between different species.

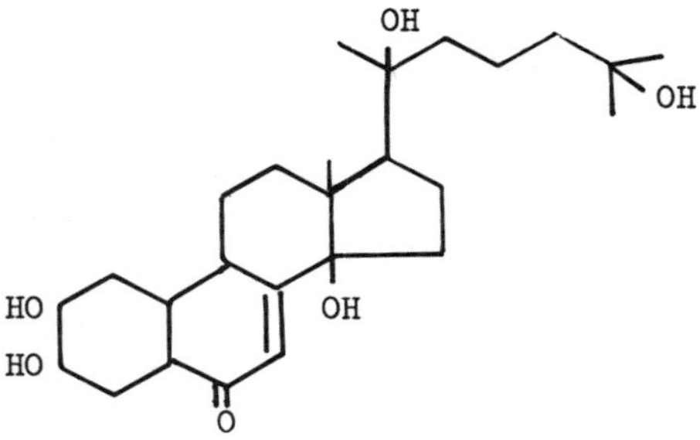

Moulting Hormone (MH)

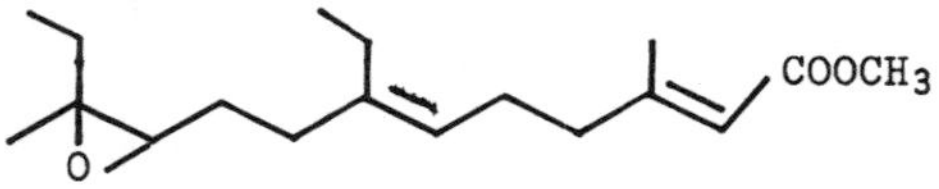

Juvenile Hormone (JH)

Experimental studies indicated that insects die or become sterile if one or other of these hormones is in excess at the wrong time. The possibility of using them, or, preferably, cheaper synthetic analogues, to control insect pests was apparent. A substantial number of compounds have, in fact, been discovered which have MH or JH activity. In the case of MH these are mainly steroid substances extracted from various plants, especially ecdysterone which is ecdysone with an OH group in position 20. Some steroids with MH activity have been synthesised from the readily available starting materials ergosterol and stigmasterol.

A considerable number of long chain unsaturated aliphatic compounds have been found to possess JH activity. The most easily accessible are those based on farnesol. The natural hormone is, in fact, essentially the epoxide of farnesol but with two methyl groups replaced by two ethyl groups.

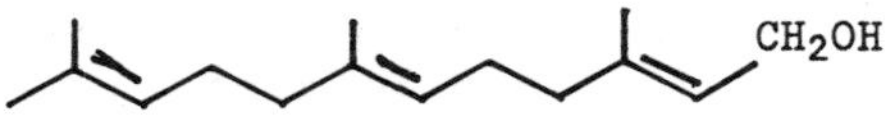

farnesol

One of the problems of using MH or JH or their "mimics" for insect control is that the compounds apparently cannot pass through the cuticle and are therefore not active when applied externally. This has, so far, proved a complete stumbling block in the case of MH mimics but some JH mimics have been discovered which are active by topical application and some of these are now being produced commercially, namely hydroprene, kinoprene, triprene and methoprene.

$(CH_3)_2CH(CH_2)_2CH_2CH(CH_3)CH_2CH{=}CHC(CH_3){=}CHCOOC_2H_5$ hydroprene

$(CH_3)_2CH(CH_2)_3CH(CH_3)CH_2CH{=}CHC(CH_3){=}CHCOOCH_2C{\equiv}CH$ kinoprene

$(CH_3)_2C(OCH_3)(CH_2)_3CH(CH_3)CH_2CH{=}CHC(CH_3){=}CHCOSC_2H_5$ triprene

$(CH_3)_2C(OCH_3)(CH_2)_3CH(CH_3)CH_2CH{=}CHC(CH_3){=}CHCOOCH(CH_3)_2$ methoprene

It has been found that synthetic mimics, in which part of the natural hormone is altered, are far more effective in disrupting insect growth and development than is an overdose of the natural hormone. One problem with their commercial development is that the compounds are difficult to synthesize and consequently expensive, although, if they were active enough, they might be cost-effective. Another disadvantage of JH is that its application retains the insect in the larval stage, which is often the one which does most damage to crops so, if it does not die rapidly, losses will continue. Hormones may, therefore, be most suitable for control of insects which are most damaging in their adult state and in which the larvae are exposed rather than hidden. Moreover, application of hormones has to be carefully timed and can be effective only when the development of the insect pest is synchronous across the whole of the area to be treated, i.e. that all insects in the area reach the same stage of development at the same time. The hormone mimics appear to be very non-toxic to men and animals and also have little potential for environmental damage but there is no assurance that they would be specific and not harm beneficial insects. It has been argued that resistance is unlikely to develop because the compounds used are analogues of natural hormones but there is no theoretical reason to believe that this is so and some cross-resistance to JH mimics has already been observed in some insects resistant to certain conventional insecticides.

Another point of difference between animals and insects is that insects have to form a new integument at each moult and, to do this, they have to synthesize chitin, a glucosamine polysaccharide, which is the main component of their horny shells. Some synthetic compounds have been discovered which inhibit chitin synthesis and deposition and leave the larvae with a fragile, malformed shell which easily ruptures and causes death.

The most promising compounds so far are N-(4-chlorophenylaminocarbonyl)-2,6-dichlorobenzamide and the corresponding 2,6-difluoro compound. Complete control of gypsy moth larvae has been obtained at the low application rate of 30 g/ha but, in order to be effective by ingestion, the particle size in the formulation must be less than 2 microns.

Recently a large number of synthetic compounds with widely varying and unrelated chemical structures have been found to disrupt insect growth to varying degrees. An example of a compound which inhibits pupal growth is shown below, but the precise mode of action of most of these compounds is as yet unknown.

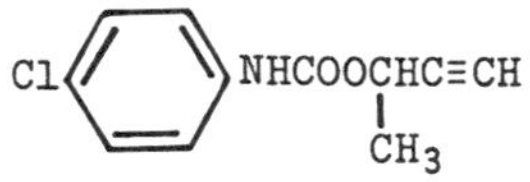

A further difference between the metabolism of animals and that of insects which might be exploited for pest control is that insects are incapable of synthesising steroids, which they require for larval growth and pupation. They have to obtain these from extraneous sources and herbivorous insects can transform plant sterols such as sitosterol and stigmasterol into the cholesterol they need. If compounds could be discovered which would prevent assimilation of these essential steroids from the insect's food they might provide very specific means of insect control. So far no such compounds have been discovered but there is some indication that this is not beyond the bounds of possibility in the observation that some antibiotics, such as filipin, are known to be insecticidal and are also known to form stable complexes with steroids, which may destroy their biological availability.

One difficulty in the way of recognizing and exploiting differences between animals and insects is that, although much is known about the physiology of most insect species, relatively little is known about their biochemical processes. Whereas most of the hormone and enzyme systems responsible for the major metabolic processes in man have been identified and characterised there is much less detailed knowledge of insect hormone and enzyme systems. Without this knowledge it is difficult to select targets for attack by synthetic chemicals.

The whole subject of the nature and function of amino acids in insect biochemistry and of the enzyme systems associated with them, such as monoamine oxidase, is currently being intensive studied and these studies may suggest methods of approach to new insecticides. Some advances have already been made in this direction and an example of a compound which kills insects by interfering with metabolism of 5-hydroxytryptamine and which may be suitable for insect control in the field, is shown below.

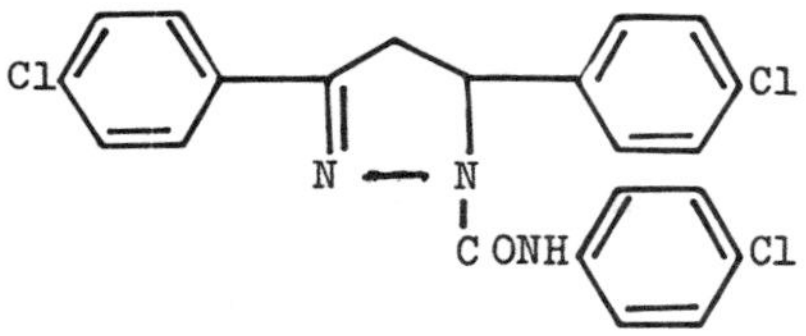

An attractive target for interference by synthetic chemicals is 4-ketoglutamic acid which appears to be an essential mediator in insect neurotransmission and is specific to insects, unlike acetylcholine which is common to insects and animals. A related insect mediator is γ-aminobutyric acid.

With respect to hormones, the insect clearly utilizes a great many more than the moulting hormone and juvenile hormone which have already been identified. Excretion in many insects is controlled by release into the blood of a powerful diuretic hormone. Water conservation is critical for insects because they have a large surface area to volume ratio. Their impermeable cuticles and excretory and respiratory systems are designed to keep water losses to a minimum. If compounds could be discovered which stimulated release of the diuretic hormone or which had diuretic hormone activity their application might lead to death by desiccation. Many insects take enormous amounts of food at one time and these appear to secrete a hormone which plasticises the abdominal cuticle and allows it to expand to accommodate these giant meals. Application of compounds which stimulated production of this hormone or which had plasticising hormone activity might lead to death literally by bursting. Such hormone-stimulators or hormone mimics might well be highly specific for the target species and thus environmentally safe.

There is no doubt that compounds which interfere with insect growth or development or which affect biochemical processes peculiar to insects - the so-called "third generation" insecticides - are a potentially very fruitful area for future research and will certainly lead to development of new insecticides with commercial utility, but their development will require a much greater knowledge of insect biochemistry than we have at present, and their effective use a greatly increased understanding of insect ecology and behaviour.

Microbial Control of Insects

Insects, like animals, suffer from a variety of diseases caused by microorganisms such as fungi, protozoa, rickettsia, viruses and bacteria. It is an attractive idea to control insects by artificially spreading diseases to which they are prone. Any microorganism which is to be used for insect control should be highly virulent and show no tendency to become attenuated in this respect; it should be specific to the target pest and harmless to other living organisms; it should be economical to produce and stable to storage; it should act rapidly to minimize crop damage. In particular, there must be no possibility that the organism might develop mutant forms pathogenic to men and animals because the adverse effects of a self-reproducing microorganism released into an environment if it took a "wrong turn" could be far more horrifying than the effects of any chemical substance.

There are over 400 species of fungi which infect insects but they rarely reduce insect populations below the levels at which they cause economic damage to crops, and it is not thought that they offer much prospect for practical pest control. Protozoa are, in general, not highly virulent nor do they cause rapid death so are also not regarded as candidates for control methods. The rickettsia likewise do not kill rapidly and it is suspected that they may be, or could easily become, dangerous to men and animals.

The bacteria offer much more possibilities and one in particular, Bacillus thuringiensis, has found widespread commercial use. It is easily produced on artificial media and the dried spores store well and remain potent for many years. It appears to be completely non-toxic to men and animals, and specific to lepidopterous larvae. The spore case (sporangium) of the bacteria contains, as well as the spore, a protein which rapidly paralyzes the gut of the insect and causes cessation of feeding a few minutes after injection, so it is very fast-acting.

For the future, it is probably the viruses which have the greatest potential for insect control, particularly the nuclear polyhedral viruses. They are very specific and highly virulent and can survive for years because of an inert protective coating. The ruptured body of one infected insect can release vast numbers of virus particles into the environment to infect other insects. Viruses have already been used on a number of occasions for highly successful pest control, for example, on the spruce sawfly in Canada and on the pine moth in France. The major obstacle to practical use of viruses has been the difficulty of their commercial production because they can be reared only on living insects and this is costly to do as the required insects may be available for only short periods during the year and have to be fed on artificial foliage which is also available only in limited amounts for a limited time. In recent years great developments have been made in mass rearing of insects all the year round on artificial media and, as a result, viruses may become practical and economical means of pest control. Experimental products are already available for use against cotton bollworm, tobacco budworm, cabbage looper and tussock moth.

The culture of a virus or bacterium on sufficient scale to be distributed as a pest-controlling agent is a process demanding chemical engineering development. The application of the product needs equipment similar to that for the application of orthodox pesticides. It is easy to foresee that, with the extension of these methods, the production and distribution will tend to be undertaken by the pesticide chemical industry, just as other micro-biological processes have been adopted in other branches of the industry.

More remotely in the future one can foresee the boundaries between industrial chemistry and applied biology becoming yet more indistinct - and not only for reasons of convenience. One of the most revolutionary developments in biochemistry in recent years has been the elucidation of part of the chemical mechanism of genetics. The molecular biologist, as he is now called, has gone a long way towards understanding how living organisms can repeat the most intricate of structures by fabricating exact copies of complex protein molecules. Very specific methods are now available to the organic chemist to synthesize polypeptides with pre-determined amino-acid sequences.

Knowledge is a prelude to interference and control. The possibilities of interference with genetic mechanisms are, from one aspect, horrifying. From another aspect they can be seen to promise the very high degree of specificity which is so much needed, because genetic chemistry is the most specific of all chemistry. The sterile male technique has exploited the extreme specificity of the mating process. The geneticist already considers that sterility is too limited a word. The male may still be in all respects sexually functional except that his genetic make-up contains some "lethal gene" so that he can give rise only to off-spring doomed to early death. It would be foolish to attempt prophesy at the present time. Non-selective application of mutagenic chemicals could lead to their condemnation or to a lost opportunity of exploiting their real potential. Just because the genetic mechanism is so specific, it is perhaps the place which the biochemists of the future should attack by highly specific exogenous chemicals. If man can get safely beyond the dangers attendant on a little learning, he may find himself in possession of the most specific of all weapons against his enemies.

Formulation of Pesticides

For many years formulation was an art rather than a science and most commercial formulations were developed empirically to give an acceptable combination of storage stability, rapid mixing and satisfactory performance in the field. There is still no well-developed theoretical background to enable accurate prediction of these properties to be made nor to describe the interactions of pesticides and formulation additives. There is no doubt that additives, particularly surfactants, can significantly alter the effects of a pesticide. The toxicity of a formulation to men and animals may be different from that of the pure active ingredient. Additives may change the amounts of residue left in the harvested crop and the amounts which persist in the environment, for example, by influencing the susceptibility of the pesticide to photodecomposition. They can affect penetration of herbicides into the plant and have, therefore, a potential for increasing efficiency of utilisation.

Future research will be directed to understanding exactly how surfactants work and to utilising this knowledge to produce improved formulations of pesticides. The outcome might well be that there is an optimum formulation for each crop/pest situation so that custom formulation would be the most desirable objective for most economical application and maximum safety. The increased cost of custom formulation would be a barrier to its acceptance by the farmer but it is possible that, at least in some cases, savings in material and application costs consequent on improved performance might tend to offset the extra costs of formulation.

The future may see development of some alternatives to the currently standard self-emulsifying concentrate, wettable powder and granular formulations. Foams which apply a pesticide as a frothy mass rather than as a spray might be capable of more accurate and precise direction and placement. True foams, which are stable and have an expansion ratio of up to 40 times, have not yet been used for application of pesticides, but may eventually be. Some use has already been made of air emulsions, which are unstable and have an expansion ratio of only about 2 times, but results with them have not been very promising.

Reference has already been made to the possibilities of microgranular formulations. The potential of modern techniques of microencapsulation is also being investigated.

The current tendency towards the use by farmers of combinations of pesticides in order to economize on application costs complicates the problems which face the formulation chemist in the future since it will not be easy to produce formulations which ensure that each of the active ingredients is applied at its optimal rate and that each gives its most effective performance. Even to attempt to approach this goal will require a great increase in basic understanding of formulation chemistry.

Application of Pesticides

As pesticides become more expensive there will be an increasing financial incentive for the farmer to use them as economically as possible. If the quantities of pesticide which need to be applied to give an acceptable degree of pest control could be decreased this would not only reduce the costs of materials and of application but would lessen any risk there might be of persistence of toxic residues and of unwanted environmental effects. Desirable objectives are to ensure that the pesticide does not drift away from the target area and that the target is covered evenly.

It is unfortunately true that many users of pesticides are casual or careless about calibration and maintenance of their application equipment and that this results in use of amounts often considerably in excess of those actually needed. This problem can be solved only by extensive education programmes to improve the ways in which application machines are used. Whatever new and more efficient machinery may be developed, existing equipment will continue to be used for many years by farmers because of the high capital costs of replacement. It is, therefore, pertinent to help them to use the equipment they have in the best way possible, and much more will be done in this respect in the future, particularly by agriculture advisory and extension services.

Large amounts of pesticides are applied from the air. The major consideration in design of agricultural aircraft has been to minimize operational costs, and insufficient attention has been given to the integration of aircraft design and distribution equipment to give optimum coverage with minimum drift. Most nozzles at present used on aircraft spraying equipment have been designed for ground use where there is little air shear. Future research will certainly be aimed at determining the relationships between drop size spectrum, vortex strength and distribution patterns for various combinations of crops, pests and pesticides and using this knowledge to design total aerial application systems which take all these factors into account.

Various types of equipment are used for ground application, including airblast machines, row crop machines and broadcast sprayers. Airblast spraying is a complicated process the results of which depend on a number of factors. Air velocity decreases rapidly with distance from the source and, as air is a poor carrier of liquid droplets, the large droplets fall quickly. The deposition rate is a function of the mass of individual droplets and the velocity at which they are travelling. Interfacial tension between the droplets and the target surface also influence rate of deposition. Evaporation of water from and, therefore, change in size of the droplets is dependent on humidity. Future research will be directed to gaining an understanding of the interplay of these various factors and to designing equipment in which they can be controlled to the best advantage. Users will then have to be taught how to operate this equipment in the most effective way.

A variety of row crop sprayers have been developed for various crop requirements and, in general, do not present problems of distribution, coverage or drift as great as those of aerial application and airblast spraying because they operate close to the target. Nevertheless, the modern tendency towards use of fine atomising sprays to reduce application volumes can lead to drift problems because of production of small droplets with a negligible gravitational settling rate. There is room for further improvements in design of equipment and in training of operators, if the economies of low-volume and ultra-low volume spraying are to be achieved with safety.

Broadcast sprayers are becoming larger and some modern machines can carry 5000 litres and operate a 20 metre boom. The main need for future research and development is to produce equipment in which the rate of dispersion of the spray is linked to the speed of movement of the machine, and in which automatic safeguards are provided to ensure that the pesticide concentrate is fully and uniformly mixed with the diluent before being discharged.

Drift

It will be appreciated that prevention of drift is one of the most significant problems to be solved in future application of pesticides. To place pesticide on the leaves of a crop some form of dispersion is necessary, particularly if uniform cover of the leaf surface is desired. The dispersed material cannot be aimed perfectly: some drifts away on the wind, often assisted by local draught created by the machine.

Direct economic loss because not all the chemical gets to the right gross target is rarely serious. Most pesticide molecules are wasted because they suffer some other fate than that incurred in initiating lethal biochemical damage in the target pest. The economic desirability, for the target crop, is to reduce this wastage to a minimum. At one extreme, pesticide dumped as a massive solid in the middle of the field would all stay there but it would do no useful job against the pest. At the other extreme, aerial spraying of very fine droplets would give the most uniform cover within the gross target (the field) but most of it would fall elsewhere. For most efficient kill of the real target within the gross target there must be an optimum degree of dispersal. Its position will depend on the pesticide (systemics need less dispersal than superficial pesticides) and on the pest (white fly must be attacked with smaller droplets than tobacco hornworm). Methods of application and formulation also influence the position of this optimum.

Drift may also cause damage outside the target area: this is of two kinds which must be kept distinct in planning its reduction. A high area density deposited just outside the target area may cause evident crop damage, a risk, obviously greatest with herbicides, which has led to the owner of the target crop being liable for compensation. Insecticides also can cause off-target damage as when mulberry (the food of the silkworm) is downwind of a cotton crop. Even sprayed motor cars have caused complaint.

Remote damage and general environmental contamination are much more complex and difficult to assess. Causatively they differ from near crop damage in important respects. Drops in the 50-100 microns range, if they escape from the delivered spray curtain, drift far enough on a light wind in inversion and high humidity to carry damaging amounts of pesticide on to an adjacent crop. Droplets in the 20-50 microns range, particularly if water-based and therefore rapidly decreasing in size in low humidity, will carry so far in turbulent wind (associated with low humidity) and impact so inefficiently that they contribute little to local damage but may contaminate a vast area at extremely low density. Much of their content will be lost by evaporation: much will suffer photochemical decomposition.

There used to be a rather general official recommendation, based only on unsound theory, to spray potentially damaging chemicals only at the evening and early morning inversion. It is now recognised that, while reducing the general effect (if there is one), this invites local damage. Accepted recommendations to reduce drift are to avoid unnecessarily fine spray, to deliver spray as closely as possible above the crop and to avoid strong cross-wind (because it releases more spray from the curtain). The most important improvement will come from cutting down the proportion of small droplets in the spray without unduly increasing mean dropsize. If, for a particular direct insecticide operation, 200 microns droplets are optional, 500 microns droplets will be largely wasted (they each carry 15 times as much and have much less change of scoring a hit). Most hydraulic nozzles currently in use were developed twenty or more years ago and produce drops in a wide size range. To have only a small proportion of droplets below 100 microns the mean size would have to be much above 200 microns. If a much closer range could be obtained a lower rate of application and less drift could be achieved together. For most spray applications a range of 100-150 microns droplet size is optimal. This is becoming possible now that practical commercial spinning-disc machines are being developed. These will make ultra-low volume spraying (ULV) a much more generally practicable technique.

Adjuvants which increase viscosity or retard evaporation can decrease the drift potential of hydrodynamic nozzles. Granules in the normal size range of 500-2000 microns if they are free from dust, can eliminate drift, but they are useful only for application to soil or water. Recent development of "microgranules" in the 50-300 microns range makes possible their use for foliar application. They provide the solid equivalent of the "Controlled Droplet Application" (CDA) made possible by the spinning disc. The physics of adhesion becomes of major importance. A flattened shape and some slow development of stickiness are desirable. By contrast, the spherical microgranules left by evaporation in flight of droplets of a wettable power suspension, are extremely non-adhesive and this property can help in decreasing the risk of damage from such spray to adjacent crops. Devices to impart a high electrostatic charge to dispersed particles are under investigation with the object of increasing attraction to the primary target, but they have not yet been shown to be practicable.

A problem not usually considered under "drift" but very directly related is the safety of the operator. He is for long periods close to the source of drift and, in a following wind, can be heavily contaminated, making protective clothing essential. In this respect, granules, including microgranules, if free from dust, are safer than spray. It is, incidentally, necessary that microgranules be free from dust in order to keep them free-flowing. Not only the operator, but also the machine, can become contaminated with spray liquid and, where a combination with seeder or fertilizer drill is used, this can cause mechanical trouble.

Pest Management

The proposition that the strategy for dealing with any crop protection and pest control problem should be selected from all the methods available in such a way as to produce maximum benefit at minimum cost would appear to be self-evident. (The term cost in this context and in the expression cost: benefit includes not only monetary expenditure but also all social and environmental effects). The major difficulty has been, and still is, a wide variance of opinions on the relative importance of the various benefits and the various costs and a tendency to emphasize some and underestimate others which has made it impossible to achieve a general concensus on what is the optimum strategy in any particular situation

Nevertheless, there is a general trend towards the view that control of pest and disease populations by a multiplicity of methods may be preferable to reliance on any one method and that maximum use should be made of natural regulatory forces inherent in the ecosystem. This would reduce the chances of evolutionary adaptation to the control method by the target pest and would avoid waste of human resources.

There are, however, considerable additional difficulties in the way of optimizing crop prevention and pest control strategies. Some of these have already been referred to in the previous chapter. The first is that we do not have nearly enough basic knowledge and understanding to do it. If it is to be done in the future, a vast amount of information will have to be obtained on population dynamics of pests and of their natural predators, of the ecology and economics of crop systems and of the complex inter-relationships of all life-forms in the general environment. The questions of where sufficient scientific resources are to be found to do this work, and who is to pay for it, have yet to be answered. Yet, without adequate understanding of the behaviour of the crop/pest/environment system which you are attempting to affect, cost:benefit comparisons of alternative strategies are impossible.

The second difficulty is that, even if sufficient knowledge were available to optimize crop protection or pest control strategies, it would be necessary to impart sufficient of this knowledge to farmers to enable them to make the right decisions or to make available to them instant expert advice. This is one aspect of the general problem of what is nowadays called "technology transfer". It is inconceivable that farmers could, or would be willing to, acquire all the relevant knowledge. Propagation of the concepts of pest management calls for a great increase in specialist advisors. Agricultural advisory and extension services would have to be greatly expanded with consequent greatly increased costs to public revenue to pay not only for their employment but also for their training. A subsidiary problem is how and where they are to be trained in sufficient numbers. Advice might be provided by private consultants or private advisory firms or by the pesticide industry extending its already large technical service to supply not just a product, but a crop protection service, to the farmer. In either case, this would involve extra costs to the farmer which would have to be met either by higher food prices or by direct Government subsidy and, in either case, would have to be borne ultimately by the public at large.

This brings us to what is probably the key issue, namely, that farmers are not motivated by an altruistic wish to produce more food but by a desire to maximize their personal profits and increase their standards of living. If they are asked to alter their current methods of crop protection they will have to be convinced that it is financially advantageous for them to do so. Also, if they are offered apparently cheaper alternatives they will need to be convinced that these will really be effective. Farmers will always attempt to optimize their strategies of pest management on the basis of monetary costs and returns. If this optimization differs from the optimum for society as a whole determined on a total cost:benefit basis, then society will have to meet the extra costs to the farmer in one way or another such as by higher food prices or increased taxes in order to give him the required incentive to change. The only alternative is to force the farmer to comply by law, which would either necessitate the public meeting the same increased costs or, if it did not, but left the farmer to bear them, might provoke a political reaction which would be unwelcome in a democratic country.

A further difficulty in the way of optimization of pest management strategies is that some methods, particularly biological control methods, cannot be applied to an individual farm but must be instituted over a whole area or region. This is particularly true of a project to eradicate a pest rather than just to control it. Such a project is being considered in the U.S.A. to eradicate the boll-weevil. The cost has been estimated at \$ 871 million. The question of who is to meet this cost and of how it can be compared with the costs to individual farmers of conventional control by pesticides has yet to be resolved. The technical feasibility of complete eradication on a pest is also in doubt. The idea that biological control methods, once the initial cost has been met, require no expenditure of maintenance is open to doubt. The so-called "eradication" of the screw-worm from the U.S.A. by biological means, which was initially apparently highly successful is currently costing \$ 10 million per year to prevent re-infestation. Apart from the technical problems there are social and political implications to taking crop protection and pest control out of the hands of the farmer, as there are to such possible solutions to a crop protection problem as recommending that a particular crop be no longer grown in a particular region and that some other crop be substituted.

Effects on the Chemical Industry

A major future problem for the pesticides industry is that the costs of discovery, development and registration of new active compounds are becoming rapidly greater, as described in Chapter 2. At the same time, large market opportunities for a new product are becoming more rare because comparatively cheap, large tonnage, commodity pesticides can provide a degree of control of most pest and disease infestations which is acceptable to many farmers. Future trends in crop protection and pest control are, as outlined earlier in this chapter, to evolve strategies which minimize the amount of pesticide which has to be used.

The consequence is that the number of companies in the world actually engaged in discovery and development of new pesticides is likely to decrease, and these will be the large international chemical companies with a diversity of interests and liquidity of assets sufficient to allow them to tie up large amounts of money for long periods without return and to buffer them against the costs of abortive development projects and sales which do not achieve expectations. Even such companies may be led to consider whether discovery and development of new pesticides is a justifiable investment compared with other ways in which they could use the money. The conclusion is that, in future years, there is likely to be less, rather than more, industrial research to produce new pesticides and what there is will most likely be targeted, for reasons of economy, towards products for use in major world crops such as small-grain cereals, corn, cotton, soya and sugar-beet.

There will be little incentive for a manufacturer to search specifically for a compound to deal with a pest problem in a minor crop or a minor pest problem in a major crop. It is true that discovery of new pesticides still depends almost entirely on empirical screening and that this is as likely to reveal a compound which is useful in artichokes as one which is useful in soya but, even if it does so, the company might not develop such a compound commercially because the costs of toxicological, residue and environmental studies are largely independent of the size of the ultimate market, and because production of small tonnages is relatively costly. There are already many compounds on manufacturers' shelves with demonstrated potential to deal with various minor crop or minor pest problems, but there they will stay. Even in the case of established pesticides, the small size of the market may make it uneconomical to meet the costs of obtaining a "label" for a minor crop use. This is an undesirable state of affairs, because the concept of "pest management" requires a wide range of active compounds so as to have as many options as possible, but it is not likely to be resolved unless it can be made adequately profitable for a manufacturer to develop and market the requisite pesticides. This might be done by allowing the price of minor crops to rise so that farmers could afford to pay more for crop protection, but the success of this tactic would depend on the willingness of the consumer to pay to maintain a wide variety of choice of foodstuffs. It might be done by national departments of agriculture, or possible associations of growers, financing the work needed for registration, although this could create difficulties with respect to the originating company's exclusive rights.

The dilemma is that registration authorities want pesticides which are highly specific for one pest or disease whereas industry needs pesticides with a broad enough spectrum of activity to provide sufficient markets to give an adequate return on their investment. There is already considerable evidence that growers of minor crops are not getting pesticides which are adequately tailored to their particular needs and it must always be remembered that what is a minor crop on a national scale may be the main economic mainstay of a particular region and the main source of livelihood to its inhabitants.

Another future problem for the pesticides industry is at the research level. Any search for highly selective compounds poses problems in screening. Screening of candidate pesticides is normally carried out on a representative selection of pest or diseases in a selection of major crops. To detect high specificity the range of screening organisms and crops will have to be greatly increased. The more selectivity depends on biochemical discrimination the more difficult will the screening problem become, for example, in searching for compounds which interfere specifically with an aspect of insect behaviour. The increase in screening and evaluation costs will exacerbate the financial difficulties already discussed.

It is likely that the main burden of screening for highly specific activities will need to be passed to crop-based organisations. Collaboration of these organisations with the pesticide industry has always been close but it will be necessary to arrange a different basis for it in the future. The crucial problem is that patent protection, of vital importance to a research-intensive industry, will be more difficult to arrange since many unpatented compounds will go for test and the evidence of value of the few found to be active will not have been provided within the organisation of the originating compound and, therefore, not patentable by them.

A requirement of "pest management" is that not only should there be available a wide range of pesticides but also of other techniques such as cultivational practices, biological control and plant breeding. If this is to be made a reality the means of applying such techniques must be made available. New types of agricultural machinery will need to be developed and commercially produced at reasonable prices. Factory rearing of predatory species and of microorganisms for control purposes will have to be developed on a vast scale using industrial mass-production methods. Since the customers for such products are the present customers of the pesticide industry it is logical that the pesticides industry should widen its scope to provide them by extending its activities on the one hand into biological production and on the other hand into machinery development. Many of the leading pesticide companies already have extensive interests in these areas and these will be extended in the future. For example, production of antibiotics for the pharmaceutical industry and, more recently, of protein from bacteria or yeasts grown on hydrocarbon feedstocks have given the chemical industry experience of handling microorganisms on an industrial scale, and biochemical engineering is now a recognized offshoot of chemical engineering. The media for mass production of insects, either as predators or as hosts for viruses, is likely to be provided from industrial sources. It is unlikely that the chemical industry would wish to enter into conventional plant-breeding but research on genetic engineering is being carried out within the industry and may lead to methods for factory production of new varieties.

A likely future trend, which has already been mentioned briefly in the previous chapter, is that pesticide producers will move towards providing a complete crop protection and pest control service for their customers. Most large companies which manufacture pesticides already operate an extensive technical advice service which would form the basis for such an extension of their operations. This would bring into the companies the full "added value" of the pesticide right up to its actual use, and this might make development and manufacture financially viable for some pesticides, such as those for minor crops or minor uses, which cannot, under present circumstances, justify the money which would have to be spent on them. The amounts of money at stake are not inconsiderable as the "mark up" in export markets is about $2\frac{1}{2}$ times, that is, the price to the farmer is about $2\frac{1}{2}$ times the price at the factory gate.

Since companies would be involved in the total "pest management" operation they would have the incentive to extend their activities to manufacture or production of the biological materials required and would have money available to finance the basic biological and ecological research required to make the ideas of "pest management" a practicable proposition. From the farmers' point of view it would be an advantage to be able to obtain all he required for crop protection and pest control from one source, conveniently from the commercial sources with whom he already deals. The possible objection that any one company would "push" its own products exclusively is not valid because existing companies already sell each others products under licence to a considerable extent and it would be in the interests of all of them to give the farmer the best advice possible.

In addition to meeting the challenge of current crop protection and pest control problems the chemical industry may be faced with new problems as a hungry world explores all potential new sources of food. Irrigation of hitherto uncultivated land by schemes such as those to direct the rivers in the U.S.S.R. which flow into the Arctic Sea to water the arid regions of central Asia or to tow polar icebergs to arid tropical areas, as well as by schemes to desalinate sea-water on a massive scale by use of solar energy, could present new pest and disease problems in the land thus opened up for crop production. About one-quarter of the total world area of land which is sufficiently flat and has enough soil for agriculture is at present infertile because of lack of water, but most of these are areas which receive the greatest amounts of sunshine and have, therefore, the greater potential for photosynthesis and crop growth.

At the other extreme, there are vast areas in the world which cannot be farmed during their very wet seasons because of difficulties of access. The possibility can be visualized of cultivating these areas entirely from the air by clearing the ground with herbicides, sowing seed and applying fertilizers and pesticides all from specially designed flying machines, leaving the only ground operation as harvesting of the final crop.

If any of these schemes come to fruition it will be necessary to obtain the highest possible yields of high value crops in order to justify the very large capital investment which will be needed so that pesticides will have a vital part to play.

Cultivation of food in the ocean as distinct from merely taking out what is there may assume greater importance as available terrestrial land becomes fully utilised. Chemical assistance will be required but will pose very difficult problems of selectivity and will demand extensive biological and ecological understanding of the marine environment.

It is certain that, whatever way food production develops, other species will continue to compete with man for sustenance. Whatever problems of crop protection and pest control arise, the ingenuity, adaptability and resources of the chemical industry will find solutions, always provided that the financial rewards for doing so are adequate.

Further Reading

Burges, W.D. & Hussey, N.W., Microbial Control of Insects and Mites, (Academic Press, London & New York, 1971)

Green, M.B., Pesticides: Boon or Bane? (Elek Books, London, 1976)

National Academy of Sciences, Contemporary Pest Control Practices and Prospects, (Washington, D.C., 1975)

National Academy of Sciences, Pest Control Strategies for the Future, (Washington,D.C., 1972)

Rabb, R.L. & Guthrie, F.E., (Ed)., Concepts of Pest Management, (North Carolina State University, 1970)

Summers, M., Eyles, R., Falcon, L.A. & Vail, P., (Eds)., Baculoviruses for Insect Pest Control, (American Society for Microbiology, Washington, D.C., 1975)

GLOSSARY

ACARICIDE. A chemical used to kill mites. From acaridae, the family of arachnidae (spiders) which includes the phytophagous mites and the ticks parasitic on animals.

ALGA (pl ALGAE). Primitive green aquatic plants ranging from unicellular species to giant seaweeds.

APHICIDE. An insecticide specially effective against sap-sucking plant-lice or aphids.

APOPLAST. More or less gelatinous aqueous matter outside the cell walls of plants, forming a tortuous continuum in which diffusion of water-soluble substances can occur.

ARTHROPODS. Members of the broad class of jointed-foot animals (arthropoda) to which insects, spiders, crustaceans and millipedes belong.

AXIL. The upper (usually acute) angle between a leaf-stalk and the stem from which it branches. Buds of new vegetative growth or flowers form in the axils. Adj. axillary.

BIOASSAY. Estimation of amount of toxicant in a sample by measurement of its effect on test organisms, usually estimation of amount necessary in a standard type of application to produce 50% kill.

BOOM. A bar, held parallel to the ground during operation, on which are mounted, at regular spacing, the nozzles of a crop sprayer or duster. The boom is often hollow, serving also as a supply tube for the spray liquid.

CARNIVORE. An animal (or, rarely, a plant) which eats animal flesh. Adj. carnivorous.

CHLOROPLASTS. The organelles (q.v.) inside plant leaf cells to which the chlorophyll is confined and in which the reactions of photosynthesis are carried out.

CHROMOSOME. The small rod-like stainable elements in the nucleus of a cell responsible for transmission of genetic information.

COMMODITY PRODUCT. An expression used in the trade to distinguish a patent-free substance from one over which a company exercises exclusive rights.

CUTICLE. The outer protective envelope of any living organism. In the case of mammals or arthropods the alternative names "skin" or "integument" are often used, but "cuticle" invariably in respect of green plants. Plant cuticle is a macromolecular structure of predominantly paraffinic composition over-lying cellulose and pectin.

CYST. Used here for a group of eggs contained within a tough protective envelope, the dormant stage of some species of nematode or eelworm. This cyst is actually a dead, swollen, female. Adj. encysted.

CYTOPLASM. The fluid or gelatinous content of living cells as distinct from the reproductive nucleus.

ECOLOGY. The study of an organism's mode of life and its relations to its surroundings.

ECTO-. A Greek-derived prefix meaning external. Used to qualify parasite, to distinguish, for example, fleas and ticks from endo- (internal) parasites such as lungworms and disease-producing micro-organisms.

ENDO-. Opposite of ecto- (q.v.). An endogenous chemical is one produced naturally by a plant itself as distinct from one applied externally, which

is called exogenous.
GENE. An individual unit of a chromosome which is responsible for transmission of some inherited characteristic.
GERRIS. A genus of bugs (hemiptera) with long legs, adapted to resting and moving upon the surface of water of normal surface tension. Water skaters.
HELMINTH. A member of the large group of primitive worms including nematodes (q.v.) and the worms parasitic in animal gut and tissues.
HERBICIDE. A chemical used to kill unwanted plants. A weed-killer.
HERBIVORE. An animal which eats plants. Adj. herbivorous.
HOST. An animal or plant supporting a parasite (q.v.).
HYDROPHILIC. Describes a molecule or group of atoms having, usually because of hydrogen bonding, a strong affinity for water.
HYDROPHOBIC. Used as opposite of hydrophilic (q.v.).
INOCULUM. Incoming fungus spores or other agents of disease which initiate the development of a parasite (q.v.) within the host.
INSTAR. A stage of development of an insect between two moults of the hard exoskeleton, but excluding the egg. An adult butterfly is the sixth instar. The first four are caterpillars, of increasing size, followed by a pupa.
INTEGUMENT. See cuticle.
INVERSION. Used in meteorology to describe the state of the lower atmosphere when the temperature is lowest near the ground. Frequent at night. The lower atmosphere is non-turbulent in inversion in contrast to its turbulent state under the opposite (lapse) condition when the ground is warmed by solar radiation.
IN VITRO. Strictly "in glass": refers to an experiment on living organisms, usually micro-organisms, carried out in an artificial laboratory environment - for example on parasitic fungi growing on an artificial culture medium.
IN VIVO. Refers to an experiment on living organisms growing in their natural environment - for example of parasitic fungi growing on their host. Contrast in vitro (q.v.).
LABEL (permissive). In the U.S.A. any agricultural chemical has to receive approval of the Environmental Protection Agency (EPA) for the purpose for which it is intended. Such purpose must be defined and limited by the wording of an agreed label. Hence the "granting of a label" has come to be of much more than verbal importance.
LABEL (radioactive). When a substance is made incorporating a radioactive isotope of one of its constituent atoms, in order to help trace the movement and chemical reactions of the substance, it is said to carry a (radioactive) label.
LARVA. A non-flying but mobile immature stage (instar, q.v.) of most insects, wholly different, structurally, from the adult form. Compare nymph (q.v.).
LEY. The name given to an annual, or short-period perennial, grass crop, usually introduced into an otherwise arable rotation (q.v.).
LIPOPHILIC. "Fat loving", used to describe a molecule or group of atoms tending to promote oil - rather than water-solubility. Contrast hydrophilic (q.v.).
LIPOPHOBIC. Used as opposite of lipophilic and therefore also of hydrophobic (q.v.).
MESOPHYLL. Refers to the tissue or cells in the interior of a plant leaf.
METABOLITE. A product of some action of biological chemistry on a compound introduced into an organism. More generally the introduced compound is a food-constituent, but, in the context of this book, it is a toxicant.
MUTANT. An individual having an abnormality of structure, properties of behaviour in which it differs distinctly from the type and which has internal (genetic) rather than environmental origin, so that there is probability of transmission to offspring.

MYCELIUM. The nutrient-seeking fibres of a fungus, equivalent to the roots of a green plant.
MYONEURAL JUNCTION. The site of transmission of an impulse from a nerve to a muscle.
NEMATODES. Small unsegmented worms, many of which are parasites on plant roots.
NYMPH. In insects, an immature stage (instar, q.v.) which is similar to the adult except for absence of functional wings and sexual organs. Distinct from larva (q.v.) of other insects, which is entirely different from the adult.
OBLIGATE. Adj. used to qualify parasite, to distinguish those which cannot live outside their hosts. Most parasites are obligate in nature, and the word is sometimes restricted to those which cannot (yet) be cultured in the laboratory on special media.
ORGANELLES. Organized microstructures inside cells to which certain biochemical processes are confined - e.g. mitochondria, chloroplasts (q.v.).
OVICIDE. A chemical which kills eggs before they hatch or one so used. Almost restricted to action on eggs of phytophagous mites (red spider). The word is almost universally accepted in this sense, but the classical scholar insists that it should mean "sheep (Latin, ovis)-killer" and that an egg-killer should be "ovacide".
OVIPOSITOR. The extensible tube on the abdomen of a female insect which she uses to place her eggs in suitable location. Often capable of remarkable drilling power. Hence oviposition - the act of egg-placing.
PARAMETER. Some varying factor which influences the nature of an observed effect.
PARASITE. A plant or animal living on or in another plant or animal, which is called its host (q.v.). The host continues, for some time at least, to live and feed its parasite (contrast predator q.v.), but may eventually be killed by it. A parasite of one host may itself be a host to another parasite, so the terms are relative. A crop-eating insect could be called a parasite or a herbivore, but is not usually so-called. It is usually simply called a pest, which may be subject to parasites, themselves perhaps subject to "hyper-parasites". A fungus living on the crop is always called a parasite. Adj. parasitic.
PARTHENOCARPY. The development of the fleshy part of a fruit without fertilization of the contained seed.
PHLOEM. A system of interconnected elongated living cells in plants adapted to convey the products of photosynthesis, particularly sucrose, from the leaves to growing tissues.
PHYTOPHAGOUS. Greek form of Latin-based adj. herbivorous (q.v.).
PHYTOPHARMACY. The treatment of diseases of plants.
PHYTOTOXIC. Toxic (q.v.) to plants, usually restricted to higher (green) plants. Hence phytotoxicity, which is used generally to include the action of intentional herbicides but by some authors restricted to adverse side-effects on the crop of fungicides, insecticides or formulating agents.
PREDATOR. An animal (or, rarely, a plant) which feeds upon other animals, called its prey. The prey is killed and consumed by its predator which generally devours many individuals during its life (contrast parasite, q.v.). There are borderline cases between predation and parasitism, e.g. species of wasp which inject a permanent paralysing drug into a caterpillar before laying eggs in it.
PRE-EMERGENT. Adj. describing a (chemical) treatment of a fully cultivated field after sowing of the crop seed but before the crop has emerged. The weed seeds may have emerged, so that the treatment, if herbicidal, may be

acting via leaves or roots. Sometimes loosely used in the sense of presowing (q.v.).

PRESOWING. Adj. describing a (chemical) treatment of a field before sowing the crop seed. A presowing treatment may include a cultivation operation to mix in the chemical, which, when the treatment has a herbicidal effect, must enter the plant from the soil. See also pre-emergent (q.v.).

PREY. An animal when fed upon by a predator (q.v.). A predator of one species can itself be prey to another.

PROPHYLACTIC. Adj. used to qualify a chemical or treatment used to prevent a disease-producing organism from invading the individual or crop so treated, as distinct from one used "curatively" on the diseased individual or crop. "Eradicant" is often used in the sense of "curative" when describing the treatment of fungus diseases of crops. "Protectant" is sometimes used in the sense of prophylactic.

POST-EMERGENT. Adj. describing a (chemical) treatment of a crop after it has emerged above the soil. Contrast pre-emergent (q.v.).

RESURGENCE. An increase of a (pest) population, after a period of decrease, to a level higher than its original one, especially when the decrease is caused by an intended control operation.

ROTATION. Applied to cropping, the practice of growing different crops on the same land in a regular, recurring, sequence. Rotation is adopted because of complementary effects, or demands, on the soil or for convenience of spreading times of peak labour demand. The most important reason, however, is to hinder the development of weeds, pests or fungi to damaging population levels. Other means of pest control make the last reason for rotation less important.

ROTAVATOR. A power-driven agricultural machine having an axle on which are mounted L-shaped blades and which rotates faster than the land wheels, so that the blades chop and stir the soil to a depth of a few inches. A rotary hoe. Hence rotavation, the treatment of soil with such implement, which provides the best means of mixing an applied substance into the surface soil.

SAPROPHYTE. A plant (usually fungus) living on dead tissue. Adj. saprophytic.

SARCOPHAGOUS. Greek form of Latin-based adj. carnivorous (q.v.).

SPIRACLE. Small openings in the abdominal segments of insects connected to a system of internal tubes (trachea) through which oxygen and carbon dioxide are exchanged with the atmosphere.

STOMA. (pl STOMATA). Small apertures in the surface of a leaf, opening or closing in response to light intensity, time of day and other factors, adapted to control exchange of carbon dioxide, oxygen and water vapour between active internal leaf cells and the atmosphere.

STYLET. A hollow tube-like organ, especially that possessed by aphids with which they probe the tissues of plants and through which the sugary contents of the phloem (q.v.) vessels are transmitted to the insect.

SYNERGISM. An activity of two or more agents which is greater than would be expected from summation of their single actions. Hence synergist, which has come to mean a compound itself of low toxicity, which increases the action of a toxicant with economic advantage.

SYSTEMIC. Used of a chemical which enters a plant either by roots or leaves and is translocated (q.v.) within it. Thus, a systemic insecticide renders a plant toxic to insects feeding on it and a systemic fungicide attacks the fungal organism actually whin the plant tissues.

TAXONOMY. The science of classification and naming of species.

TOXIC. Adj., having a poisonous action. Used here in general sense, requiring qualification if restricted to effect on mammals, insects, plants, etc. By some authors used in a sense restricted to higher animals. Hence, toxicant, a poison.

TRANSLOCATION. The movement of chemical substances, either endogenous (q.v.) or applied, within a plant.
VECTOR. A carrier. In the present context it refers to small animals, usually insects, which can carry pathogenic (disease-producing) micro-organisms from a diseased to a healthy plant or animal.
XYLEM. A system of more or less continuous channels in plants, formed of fused dead cells, which transport water and soluble minerals from roots to leaves.

GENERAL INDEX

This book is not a practical handbook of crop protection and pest control. Individual species of crop plants, fungi, insects and weeds are not indexed. For these, either the Weed Control Handbook (Fryer, J. & Makepeace, R., Blackwell, Oxford) or the Insecticide and Fungicide Handbook (Martin, H., Blackwell, Oxford) should be consulted.

INDEX OF PESTICIDES

All the pesticides mentioned in this book are indexed below under common names approved by or proposed to the British Standards Institution and, only if there is no such common name, under the chemical name. Where more than one reference is given, the main reference is underlined.

The type of activity is indicated by the following code: A, acaracide; At, attractant; F, Fungicide; FS, systemic fungicide; Fm, fumigant; GR, plant growth regulator; H, Herbicide; I, insecticide; IH, insect hormone mimic; IN, naturally-occurring insecticide; M, Molluscicide; N, nematicide; R, rodenticide; Rp, repellent.

abscisic acid	GR	173
acephate	I	55
acethion	I	61
acetophos	I	59, 60
acrolein	M	93
alachlor	H	15, 16, 156
aldicarb	I	66, 91, 92, 94
aldrin	I	16, 49, 222
allethrin	I	74
allethrolone	I	74
allidochlor	H	156
allyxycarb	I	66
alorac	H	152
ametryne	H	160
amidithion	I	60
aminocarb	I	66
aminopyridine	Rp	82
aminotriazole	H	164, 165
amiphos	I	60
amiprophos	H	166
amiton	I	33, 59
amitraz	A	89
ammonium fluosilicate	I	44
ammonium reineckate	I	44
ancymidol	GR	172
anthraquinone	Rp	82, 99
antu	R	96
asulam	H	154
atraton	H	159
atrazine	H	15, 16, 23, 27, 159, 160, 161
azauracil	FS	136
azidirachtin	IN	70
azinophos-ethyl	I	61
azinophos-methyl	I	16, 60

aziprotryne	H	160
azobenzene	A	88
azothoate	I	58
bacillus thuringiensis	IN	257
barban	H	153
basic copper carbonate	F	107
basic copper sulphate	F	107
basic cupric zinc sulphate complex	F	107
benazolin	H	164, 165
bendiocarb	I	67
benfluralin	H	144
benodanil	FS	131
benomyl	FS	138, 221
benquinox	F	123
bensulide	H	166
bentazone	H	164, 165
benzipram	H	157
benzoximate	A	90
benzoylprop-ethyl	H	157
benzthiazuron	H	159
BHC	see lindane	
binapacryl	A	89, 120
bioallethrin	I	74
bioresmethrin	I	75
blasticidin	FS	129
borax	I	44
Bordeaux mixture	F	104, 106
bromacil	H	162
bromadiolone	R	97
bromfenvinphos	I	56, 57
bromophos	I	58
bromoxynil	H	158
brompyrazone	H	162
bupirimate	FS	135
Burgundy mixture	F	106
butachlor	H	157
butam	H	157
buthiuron	H	89
butifos	H	166
butopyronoxyl	Rp	80
butoxypolypropylene glycol	Rp	80
butralin	H	144
butrizol	FS	136
buturon	H	159
butyl methylbenzoate	At	83
cacodylic acid	H	167
cadmium succinate	F	107
calcium cyanamide	GR	174
captafol	F	124
captan	F	15, 16, 23, 124
carbanolate	I	66
carbaryl	I	16, 23, 65, 92
carbasulam	H	154

carbendazol	F	126
carbofuran	I	16, 23, 67, 68
carbon disulphide	Fm	91, 208
carbon tetrachloride	Fm	208
carbophenothion	I	61
carboxin	FS	137
cellocidin	FS	129
chloralose	R	97, 99
chloramben	H	14, 16, 23, 150
chloraniformethan	FS	132
chloranil	F	69, 123
chloranocryl	H	155
chlorate	H	140
chlorazine	H	159
chlorazon	H	162
chlorbenside	A	88
chlorbenzilate	A	88
chlorbicyclen	I	49
chlorbromuron	H	159
chlorbufam	H	153
chlorcamphene	I	51
chlordane	I	49
chlordene	I	49
chlordimeform	A	89
chlorethalin	H	144
chloreturon	H	159
chlorfenethol	A	88
chlorfenson	A	88
chlorfenvinphos	I	56, 57
chlorflurecol-methyl	GR	172
chlormequat chloride	GR	171, 176
chloromebuform	A	89
chloroneb	FS	131
chloropicrin	F	91, 98, 208
chloroxuron	H	159
chlorphonium chloride	GR	172
chlorpropham	H	153, 172
chlorpyriphos-ethyl	I	58
chlorquinox	F	126, 136
chlorothalonil	F	122
chlorothiamid	H	158
chlortoluron	H	159
cisanilide	H	159
citronella oil	Rp	79, 80
clenpyrin	A	89
clobenprop-methyl	H	149
cloprop	Gr	173
copper naphthenate	F	40, 107
copper oxychloride	F	105, 106
copper sulphate	F	15, 19
coumachlor	R	97
coumaphos	I	58
coumatetralyl	R	97
coumithioate	I	58
4-CPA	GR	173

creosote	F	40, 140
cresyl acetate	F	119
crimidine	R	96
crotoxyphos	I	56, 57
cryolite	I	43
cuprous oxide	F	107
cyanatryn	H	160
cyanazine	H	160
cyanolate	I	55
cyanophos	I	58, 59
cycloheximide	FS	129
cycluron	H	159
cyhexatin	A	89, 112
cypermethrin	I	77
cyperquat	H	164, 165
cyprazine	H	159
cypromide	H	155
2,4-D	H	14, 16, 19, 23, 36, 145, 146, 246
dalapon	H	14, 152
daminozide	GR	171, 176
dazomet	N	91
2,4-DB	H	148
DCPM	A	88
DDDS	A	88
DDT	I	16, 19, 45, 94, 214, 222
decafentin	F	112
decamethrin	I	77
decarbofuran	I	67
DEET	Rp	80
demephion	I	58, 59
demeton-S-methyl	I	33, 58, 59, 63
derris	IN	70, 71
desmetryne	H	160, 161
dialifos	I	61
diallate	H	154
diamidofos	N	91
diazinon	I	16, 58, 219
dibromochloropropane	N	91, 208
dibutyl phthalate	Rp	80
dibutyl succinate	Rp	80
dicamba	H	23, 150
dichlobenil	H	32, 158
dichlofenthion	I	58
dichlofluanid	F	124
dichlone	F	69, 123
dichlormate	H	154
dichlorophen	F	119
dichloropropene	N	91
dichlorprop	H	148
dichlorvos	I	16, 56
diclofenthion	N	92
diclofop-methyl	H	149
dicloran	F	122, 220
dicofol	A	88

dicrotophos	I	56, 57
dieldrin	I	49, 222
diethacine-ethyl	H	157
difenacoum	R	97
difenoxuron	H	159
difenzoquat	H	164
dimefox	I	33, 54, 63
dimethametryn	H	160
dimethirimol	FS	134, 220
dimethoate	I	33, 60, 64
dimethyl phthalate	R	80
dimetilan	I	67
dimidazon	H	162
dinex	A	89
dinitramine	H	144
dinobuton	A	89, 120
dinocap	A	89, 120
dinocton	A	89, 128
dinonyl succinic acid	GR	177
dinoseb	H	23, 142
dinoterb	H	142
dioxacarb	I	67
dioxathion	I	61
diphacinone	R	98
diphenylamine	GR	173
dipropalin	H	144
dipropyl isocinchomeronate	Rp	80
diquat	H	23, 163, 174
disugran	GR	176
disulfoton	I	16, 61
ditalimfos	FS	130
dithianon	F	123
dithiochrophos	I	61
diuron	H	15, 23, 26, 159, 161
DMPA	H	166
DNOC	H	89, 120, 141, 174
dodemorph	FS	137
dodine	F	125, 220
drazoxolon	F	126
DSMA	H	167
DTA	I	86
eau celeste	F	106
ecdysone	IH	253
endosulfan	I	50
endothal	H	164, 165
endothion	I	59
endrin	I	16, 49, 222
epigriseofulvin	FS	129
epronaz	H	159
EPTC	H	154
erbon	H	152
etacelasil	H	165
ethalfluralin	H	144
ethephon	GR	176

ethidimuron	H	159
ethiolate	H	154
ethion	I	16, 61
ethiophencarb	I	66
ethirimol	FS	134, 135
ethoate-methyl	I	60
ethofumesate	H	164, 165
ethohexadiol	Rp	80
ethoxypyrithion	I	58
ethoxyquinol	GR	173
ethylene dibromide	Fm	91, 207, 208
ethylene dichloride	Fm	208
ethylene oxide	Fm	208
ethyl hydrogen 1-propylphosphate	GR	172, 176
ethyl mercury chloride (EMC)	F	110
ethyl mercury 4-toluenesulphonanilide	F	110
etrimfos	I	58
farnesol	IH	254
fenarimol	FS	134
fenazaflor	A	90
fenbutatin oxide	F	112
fenchlorphos	I	58
fenitrothion	I	58
fenobenzuron	H	159
fenoprop	H	148
fensulfothion	I	58
fenthion	I	58
fentin acetate	F	112
fentin hydroxide	F	112
fenuron	H	159
fenvalerate	I	77
ferbam	F	15, 23, 117
flamprop-isopropyl	H	157
fluchloralin	H	144
fluometuron	H	15, 159, 161
fluoroacetic acid	R	97
fluorodifen	H	15, 142
fluthiuron	H	159
folpet	F	124
fonophos	I	55
formaldehyde	Fm	91, 208
formothion	I	60
fospirate	I	56, 58
fuberidazole	FS	138
fumarin	R	97
furcarbanil	FS	132, 133
giberellic acid	GR	175
glyodin	F	125
glyphosate	H	23, 166
glyphosine	GR	176
griseofulvin	FS	129
guazatine	F	126
gyplure	At	84

heptachlor	I	49
hydrogen cyanide	F	96, 98, 206, 211
hydroprene	IH	254
hydroxymercury 2-chlorophenol	F	110
IAA	GR	145
IBA	GR	173
ioxynil	H	158
ipazine	H	159
isazophos	I	58
isoaxathion	I	58
isobenzan	I	50
isocil	H	162
isodrin	I	49
isoprocarb	I	65
isopropalin	H	144
isoproturon	H	159
juglone	F	70, 123
kadethrin	I	75
karbutilate	H	154, 159
kasugamycin	FS	129
kinetin	GR	172
kinoprene	IH	254
kitazin	FS	130
kitazin-P	FS	130
K-orthrin	I	75
lead arsenate	I	19, 43
lead chromate	I	44
leptophos	I	55
lethane	I	44
lime-sulphur	F	114, 115
lindane	I	19, 31, 48, 94, 214
linuron	H	15, 159, 161
lirimphate	I	58
lochnocarpus	IN	71
malathion	I	16, 60, 61, 64
maleic hydrazide	GR	172
mancozeb	F	118
maneb	F	15, 23, 117, 118
MCPA	H	23, 26, 36, 145, 146
MCPB	H	148
mebenil	FS	131
mecarbam	I	61
mecarphon	I	55
mecoprop	H	148
menazon	I	60
mephosfolan	I	55
mercurous chloride	F	108
merquinophos	I	58
metacrephos	H	166
metaldehyde	M	82, 92

metalkamate	I	16, 65
metflurazone	H	162
methabenzthiazuron	H	159
methacrifos	I	58
methamidophos	I	55
metham-sodium	F	117
methiobencarb	H	154
methiocarb	I	66, 92, 94
methidathion	I	60
methometon	H	159
methomyl	I	16, 66
methoprene	IH	254
methoprotryne	H	160
methoxychlor	I	16, 47
methoxyethyl mercury acetate (MEMA)	F	109, 110
methoxyethyl mercury silicate (MEMS)	F	109, 111
methyl bromide	F	32, 91, 96, 207
methyl eugenol	At	83, 86
methyl formate	Fm	208
methyl isothiocyanate	N	91, 117, 208
methyl nonyl ketone	Rp	82
methyl parathion	I	16, 23, 58
metiram	F	118
metobromuron	H	159
metoxuron	H	159
metribuzin	H	164, 165
mevinphos	I	56, 57
mexacarbate	I	66
mipafox	I	54
mirex	I	50
monalide	H	155
monocrotophos	I	56, 57
monolinuron	H	159
monuron	H	159
morphothion	I	60
MTMC	I	65
NAA	GR	173
nabam	F	117
naled	I	56
naphthylacetamide	GR	173
napropamide	H	157
naptalam	H	15, 151
neburon	H	159
neostanox	A	89
nickel salts	F	108
niclosamid	M	93, 94
nicotine	IN	70, 71
nisulam	H	154
nitralin	H	15, 144
nitrofen	H	142
nitrofluorfen	H	143
norflurazon	H	162
noruron	H	15, 159
nuarimol	FS	134

omethoate	I	59
oryzalin	H	144
oxadiazon	H	164, 165
oxapyrazone	H	162
oxine	F	119
oxycarboxin	FS	137
oxydemeton-S-methyl	I	59
oxydisulfoton	I	61
oxyfluorfen	H	143
paraoxon	I	56, 63
paraquat	H	23, 28, 163, 176
parathion	I	16, 19, 58, 63, 92
parinol	FS	133
Paris green	I	43
PCP	M	93, 118
penoxalin	H	144
pentanochlor	H	155
pepper	Rp	79
permethrin	I	76
phenmedipham	H	153
phenothrin	I	76
phenthoate	I	60
phenylmercury acetate (PMA)	F	109, 110
phenylmercury triethanolamine lactate	F	111
phenylphenol	F	119
phorate	I	16, 60, 61, 92
phosalone	I	61
phosazetim	R	98
phosfolan	I	55
phosmet	I	60
phosnichlor	I	58
phosphamidon	I	56, 57
phosphine	Fm	208
phoxim	I	58
picloram	H	151
pindone	R	98
piperalin	F	126
piperonyl butoxide	I	73
piracetaphos	I	58
pirimicarb	I	67
procyazine	H	159
prodiamine	H	144
profluralin	H	144
promecarb	I	65
prometon	H	159
prometryne	H	160, 161
propachlor	H	15, 16, 23, 156
propanil	H	15, 23, 155
propaphos	I	57
propathrin	I	75
propazine	H	159
propham	H	153
propineb	F	117
propoxur	I	66

propylphosphonic acid	GR	172, 176
prosulfalin	H	144
prothoate	I	61
prothrin	I	75
pyracarbolid	FS	133
pyrazone	H	162
pyrazophos	FS	130
pyrethrum	IN	70, 72, 81
pyridinitril	F	126
pyrimiphos-methyl	I	58
pyrimitate	I	58
quintozene	F	122
red squill	R	69, 96
resmethrin	I	74
rotenone	IN	71
ryania	IN	70, 71
ryanodine	IN	71
salicylanilide	F	119
sarin	I	53
schradan	I	33, 54, 63
sebuthylazine	H	159
secbumeton	H	159
siduron	H	159
silatrane	R	98
silicofluorides	I	43
simazine	H	159
simeton	H	159
simetryne	H	160
siprazine	H	159
sodium cacodylate	H	167
sodium fluoride	I	43
streptomycin	FS	128
strychnine	R	69, 98
sulfotep	I	55
sulphur	F	114
sulphur dioxide	Fm	96, 208
sulphuric acid	H	140
swep	H	153
2,4,5-T	H	19, 23, 27, 36, 147
tabun	I	53
tar oil	I	41
tazimcarb	I	67
TBA	H	36, 149
TCA	H	152
TCDD	H	147
TDE	I	16, 47
tecnazene	F	122, 172
TEPP	I	19, 55, 63
terbacil	H	162
terbucarb	H	154
terbuchlor	H	157

terbufos	I	60
terbumeton	H	159
terbuthylazine	H	159
terbutryne	H	160
tetraalkylthiourea	F	116
tetrachlorvinphos	I	56, 57
tetracopper calcium oxychloride	F	107
tetradifon	A	88
tetrafluron	H	159
tetramethrin	I	74
tetramethyl thiuram disulphide	Rp	82
TFP	H	152
thanite	I	44
thiabendazole	F	126, 138
thiazafluron	H	159
thiobencarb	H	154
thiocarbenil	H	154
thiocarboxime	I	<u>66</u>, 92
thiochrophos	I	59
thiocyanates	I	44
thiometon	I	60, 61
thionazin	N	92
thiophanate	FS	131, 132
thiophanate-methyl	FS	131, 132
thiophanox	I	67
thiophosvin	I	58
thioxamyl	I	66
thiram	F	117
tolylfluanid	F	124
toxaphene	I	16, 23, <u>51</u>
triadimefon	FS	136
triallate	H	154
triamiphos	FS	130
triarimol	FS	133, 134
tributyl phosphorothiolate	GR	174
tributyl tin oxide	F	111
tricamba	H	150
trichloronate	I	55
trichlorphon	I	54
triclopyr	H	151
tricyclotin	A	113
tridemorph	FS	137
trietazine	H	159
trifluralin	H	15, 16, 23, 27, 41, <u>143</u>
trifop-methyl	H	149
triforine	FS	135
triiodobenzoic acid	GR	172
triprene	IH	254
vamidothion	I	59
vernolate	H	154
zinc chromate	F	108
zinc phosphide	R	96
zineb	F	15, <u>117</u>, 118
ziram	F	117